D1333398

THE STRUCTURE AND PROPERTIES OF POLYMERIC MATERIALS

THE STRUCTURE AND PROPERTIES OF

Polymeric Materials

D.W. CLEGG
School of Engineering
and
A.A. COLLYER
School of Science
Sheffield Hallam University

THE INSTITUTE OF MATERIALS

Book 337

Published in 1993 by
The Institute of Materials
1 Carlton House Terrace
London SW1Y 5DB

British Library Cataloguing-in-Publication Data
Clegg, D.W.
Structure and Properties of Polymeric
Materials
I. Title II. Collyer, A.A.
620.1

ISBN 0-901716-39-1

Typeset by Inforum, Rowlands Castle, Hants

Printed and made in Great Britain by
The Bourne Press, Bournemouth

Contents

Preface

This work presents the basic principles underlying the properties and processing of polymeric materials, which comprise thermoplastics, thermosets and rubbers. The approach taken is to explain the mechanical and flow behaviour and how they may be manipulated in terms of the molecular characteristics of the polymer molecules.

Designing with plastics materials has always been regarded as more difficult than designing with metals because of the lower moduli and tensile strengths involved and the viscoelastic nature of the polymer molecules. Despite these problems, there is a great reward in the use of plastics and plastics composites due to the development of lighter products and the ease and cheapness by which they can be processed.

Theoretical analyses of polymer behaviour are highly mathematical and inappropriate as routine tools, or for readers unfamiliar with the subject area. As such, the treatments in the book are more approximate and at times graphical as these approaches are more suitable for product design and the prediction of processing behaviour.

In the first two chapters the various molecular and microstructural parameters are discussed as are the methods by which small molecules are polymerised to form polymer molecules. Chapter 3 examines the ways in which polymer architecture may be modified to improve engineering properties and the ways in which polymers are used in composites for improved performance. The viscoelastic behaviour of polymers and their mechanical properties are examined in Chapters 4 and 5. The rheological behaviour and methods of processing are covered in Chapters 6, 7, 8 and 9, with Chapter 10 describing joining techniques and Chapter 11 examining the ways in which mechanical property data available from suppliers are used in product design. This takes the study of polymeric materials from their structures through the various processes to the final product design.

The Structure and Properties of Polymeric Materials

The book is intended for engineering and science students with a knowledge of 'A' Level science and mathematics but no prior knowledge of plastics materials, and is set at first year HND and degree level. Scientists and engineers in industry and in educational establishments as well as students being introduced to the subject for the first time should find this book useful, especially where worked examples are given to illustrate various principles of design techniques. The subject matter in the book has been used for many years in the HND and degree courses run both in the Engineering and Science Departments.

D.W. Clegg
School of Engineering

A.A. Collyer
School of Science
Sheffield Hallam University

1 *The Nature of Polymeric Materials*

INTRODUCTION

The word polymer originates from the Greek word 'polymeros' and means many-membered. In polymer science it refers to molecules held together by covalent bonds and composed of small units which are repeated many times to form very large molecules. For example a common synthetic polymer is polypropylene or more correctly polypropene. This has the following structure:

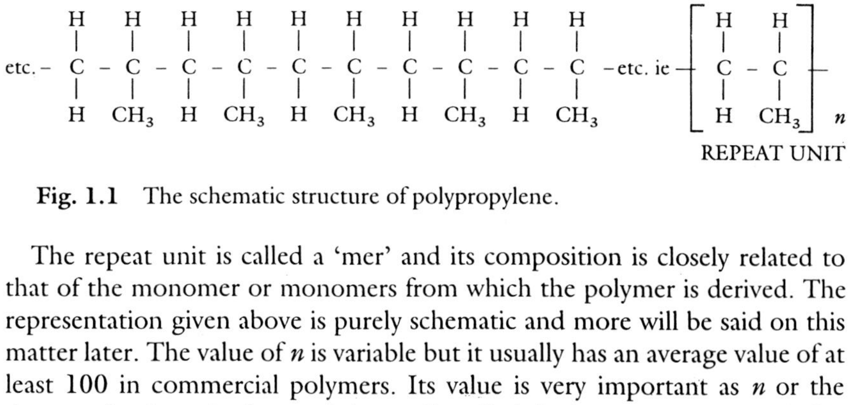

$$\begin{array}{ccccccccccccccccccccccc}
 & H & & H & & H & & H & & H & & H & & H & & H & & H & & H & & & \\
 & | & & | & & | & & | & & | & & | & & | & & | & & | & & | & & & \\
\text{etc.}- & C & - & C & - & C & - & C & - & C & - & C & - & C & - & C & - & C & - & C & -\text{etc. ie} & & \\
 & | & & | & & | & & | & & | & & | & & | & & | & & | & & | & & & \\
 & H & & CH_3 & & H & & CH_3 & & H & & CH_3 & & H & & CH_3 & & H & & CH_3 & & &
\end{array}
\quad
-\!\!\left[\begin{array}{ccc} H & & H \\ | & & | \\ C & - & C \\ | & & | \\ H & & CH_3 \end{array}\right]_n\!\!-$$

REPEAT UNIT

Fig. 1.1 The schematic structure of polypropylene.

The repeat unit is called a 'mer' and its composition is closely related to that of the monomer or monomers from which the polymer is derived. The representation given above is purely schematic and more will be said on this matter later. The value of n is variable but it usually has an average value of at least 100 in commercial polymers. Its value is very important as n or the degree of polymerisation, as it is usually called, is the most important influence on polymer properties.

Polymer structures vary considerably from one type to another and these variations lie behind one of the most basic distinctions between types of polymeric materials. This distinction can be made from observations of thermal properties and has resulted in the terms 'thermoplastic' and 'thermosetting plastic', or 'thermoset' for short. Thermoplastics soften and flow when heated but will reharden on cooling. This process can be repeated a large number of times. Thermosets, on the other hand, may soften on the first heating cycle and then cure or set. Heating is not always necessary, but in all cases, thermosets cannot subsequently be resoftened to the point of flowing.

Thermosets are amorphous, i.e. non-crystalline, but thermoplastics are in many cases partially crystalline. All polymeric materials can be subdivided as shown below and it should be noted that the thermosets include materials as diverse as rubbers and cross-linked epoxy resins.

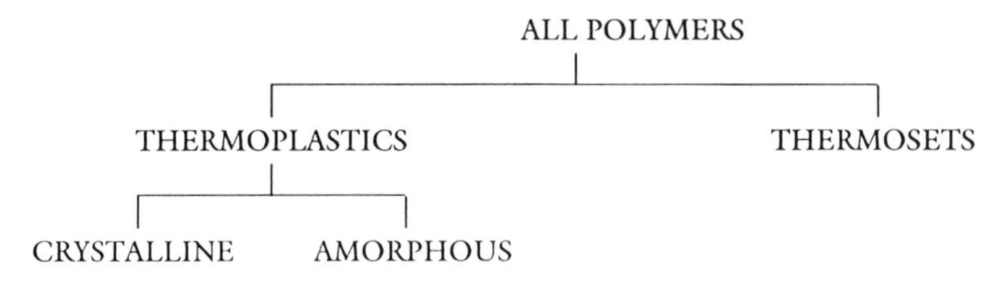

In all cases the molecular structure of a polymer affects its mechanical, physical, chemical and processing properties. Processing operations may

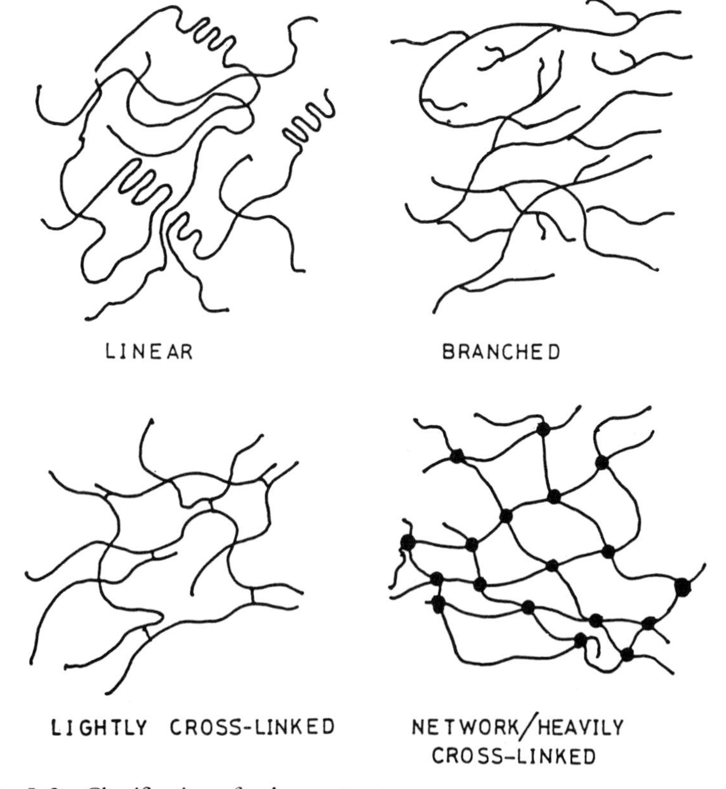

Fig. 1.2 Classification of polymer structures.

themselves affect the molecular structure and microstructure of a polymeric material to a considerable extent. Thus there is a complex inter-relationship between molecular structure, processing and properties.

THE CLASSIFICATION OF POLYMER MOLECULAR STRUCTURES

Polymers can be classified into four distinctly different structural variations. These are linear, amorphous, partially crystalline, lightly cross-linked and network (and heavily crosslinked) types of polymer structures as shown in Fig. 1.2.

Linear Polymers

These are produced by polymerising difunctional monomers, i.e. monomers which react at only two positions. Propylene is such a monomer and yields polypropylene when polymerised as shown in Fig. 1.3.

Polymerisation is made possible by the reactive double bond in the monomer. Such monomers are said to be unsaturated.

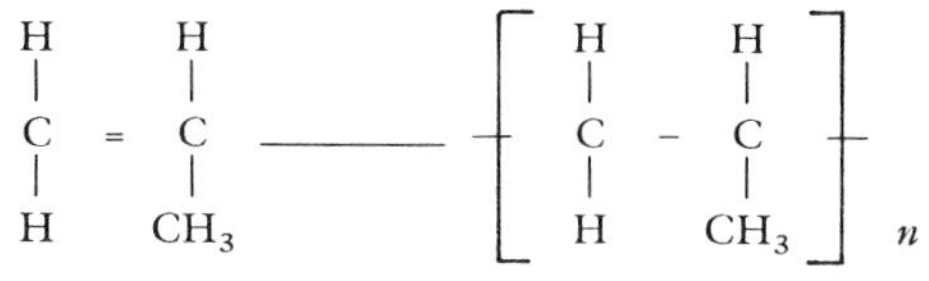

$$\underset{\substack{|\\ H}}{\overset{\substack{H\\ |}}{C}} = \underset{\substack{|\\ CH_3}}{\overset{\substack{H\\ |}}{C}} \longrightarrow \left[\underset{\substack{|\\ H}}{\overset{\substack{H\\ |}}{C}} - \underset{\substack{|\\ CH_3}}{\overset{\substack{H\\ |}}{C}} \right]_n$$

Fig. 1.3

Other examples are linear polyethene (i.e. polyethylene or polythene), polymethylmethacrylate (i.e. PMMA, Perspex or Plexiglass), polystyrene etc. Some of these polymers may crystallise partially, e.g. polyethylene.

Branched Polymers

Many apparently linear polymers are in fact branched as a result of subsidiary reactions occurring during polymerisation. Polyethylene is the best known example and is available with a wide range of structures and hence properties. The low density (LDPE) types are extensively branched whilst high density variations (HDPE) are essentially linear. Medium density (MDPE) and linear low density (LLDPE) types fall in between these two extremes. All these grades have their uses ranging from packaging in the case of the flexible lower density types to semi-structural applications for the stiffer high density polyethylenes.

Branching can be a problem in many polymers as it may occur during polymerisation leading to unacceptably massive chains and unprocessable material, e.g. rubbers which must be extensively compounded and worked during subsequent processing.

Lightly Crosslinked Polymers

These polymers include most rubbers in their fully fabricated condition. Rubbers are crosslinked or cured during the final stages of their processing and forming operations. Sulphur is the usual crosslinking chemical employed. At this point rubbers pass from being thermoplastic gums to elastic thermosets. Normally the crosslinking is limited to only one crosslink per chain on average. As a result rubbers are flexible and elastic. Plastics may also be crosslinked to modify their properties. For example polyethylene and PVC are sometimes crosslinked using electron beams or gamma radiation. Heat shrinkable cable insulation sleeving is made in this way.

Network and Heavily Crosslinked Polymers

This category includes the truly network polymers such as phenol formaldehyde (the basis of Bakelite), where the network can be formed during the polymerisation process. However, in practice an intermediate stage resin is usually prepared and the network structure is completed during fabrication, at which stage the material thermosets fully.

Other examples are crosslinked epoxy resins, polyester resins and heavily crosslinked rubber, e.g. ebonite. Some of these materials can be formulated to cure at ambient temperatures and are very useful as matrix materials in the manufacture of composites and laminates.

This group of polymers are hard and rigid and retain their properties to fairly high temperatures. They are used extensively in the manufacture of electrical devices.

NATURAL POLYMERS

The major growth of synthetic plastics has occurred since about 1930 but natural rubber was established as a significant commercial material from the middle of the nineteenth century. In terms of a general picture of materials utilisation these developments seem to be very recent, with ceramics and metals being in use for many thousands of years. However, some natural polymeric materials have been known and used by man for almost as long. For example, the use of bitumen, a complex mixture of the heavier petroleum fractions, is mentioned in the Bible as a mortar for bonding bricks. Amber was known in ancient Rome. It is the fossilised resin of pine trees and is really only of interest as a decorative semi-precious gem, being either gum-like or brittle. Another material, shellac, was in use about 1000 BC. It is derived from the lac insect and has been used as recently as 1950 in the manufacture of gramophone records but is now replaced by PVC/PVA copolymers for this purpose. Shellac was originally used as a coating material and later, in Europe, as a moulding material and as sealing wax.

About 1600, a plant derivative called 'gutta percha' was brought to Europe. It is a high molar mass trans isomer of polyisoprene with low elasticity. However, it found uses as a moulding material during the industrial revolution and particularly as a cable insulation for underwater applications as recently as 1940. The cis isomer of polyisoprene, natural rubber, has been known to the natives of Central and South America for hundreds of years and used by them for waterproofing cloth. It was introduced to Europe in the early part of the nineteenth century. Developments by Thomas Hancock in England and particularly by Charles Goodyear in the USA led to its widespread use. Hancock discovered that intensive shearing made it mouldable by reducing its molar mass, and Goodyear is usually credited with the invention of cross-linking, curing or vulcanisation of natural rubber with sulphur. This process 'cured' the rubber of its poor elasticity, stickiness and other undesirable properties. Natural rubber cured with small percentages (2–3%) of sulphur yielded highly elastic materials whilst large amounts of sulphur were used to produce hard rubber or ebonite, the first true thermosetting plastic.

Both the highly elastic rubbers and ebonite involve chemical modifications in their manufacture and thus mark the beginnings of the modern synthetic plastics and rubber industries. As a result, since the first patents for vulcanisation (1843 in England and 1844 in the USA) steady progress has been made in the development of plastics and rubbers leading to the introduction of celluloid about 1900. This was derived from natural cellulose by producing cellulose nitrate which was then plasticised with camphor. The exploitation of natural polymers fueled expansion of new industries. The 1871 Elsevier's Report on England, for instance, spoke of the rapid increase in india rubber and gutta percha manufacturers.

The only natural product that has achieved commercial significance as a plastic, as opposed to rubber, is casein which is a protein obtained from cow's milk. When cross-linked with formaldehyde, casein has a pleasant horn-like texture making it very suitable for button making. However, buttons are now made from polyesters. Natural polymers have rapidly diminished in importance, particularly from 1930 onwards. Today the only natural polymer of real commercial importance is natural rubber, and synthetic plastics and rubbers dominate in all areas.

CHARACTERISATION OF MOLECULAR SIZE

Introduction

As previously mentioned the properties of polymers are closely related to the sizes of the molecules present in the samples studied. Clearly a method of describing the dimensions of polymer molecules is necessary. This can be done by stating the number of repeating units in an individual polymer chain, i.e. the degree of polymerisation, or by giving the molar mass of the molecule. Unfortunately this apparently simple procedure is complicated by the fact that most polymers consist of a range of molecular sizes, instead of molecules with just a single size. Consequently any characterisation of molecular size must be capable of incorporating the observation that there may well be a broad distribution of molecular sizes in a sample being described as shown in Fig. 1.4.

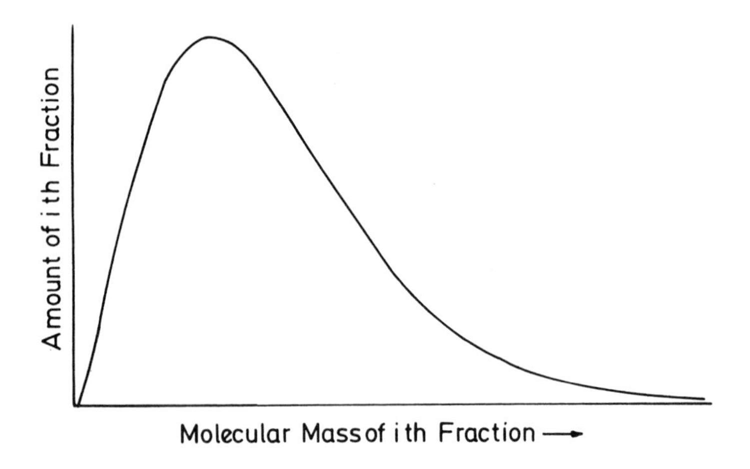

Fig. 1.4 The distribution of molecular size in a typical polymer.

In many instances the nature of the distribution will not be known in detail and so it is convenient to make use of appropriate averaging procedures. In the case of compounds containing small molecules the molar mass is given by:

$$M = \frac{m}{N} \tag{1.1}$$

where m = total mass of sample, N = number of moles in the sample.

A mole of a substance is the amount of the substance which contains as many molecules as there are atoms in 0.012 kg of carbon 12. The units of molar mass are thus kg mol^{-1}. The term molecular weight is still in common usage but its use should be discouraged. However, it is used interchangeably with molar mass in this monograph.

The degree of polymerisation is defined as:

$$x = \frac{M}{M_r} \tag{1.2}$$

where M_r = molar mass of the repeat unit.

In the case of polymers where a distribution of molecular sizes almost always exists it is usual to describe the distribution in terms of a number average molar mass, $\overline{M}_n$, or a weight average molar mass $\overline{M}_w$, both of which

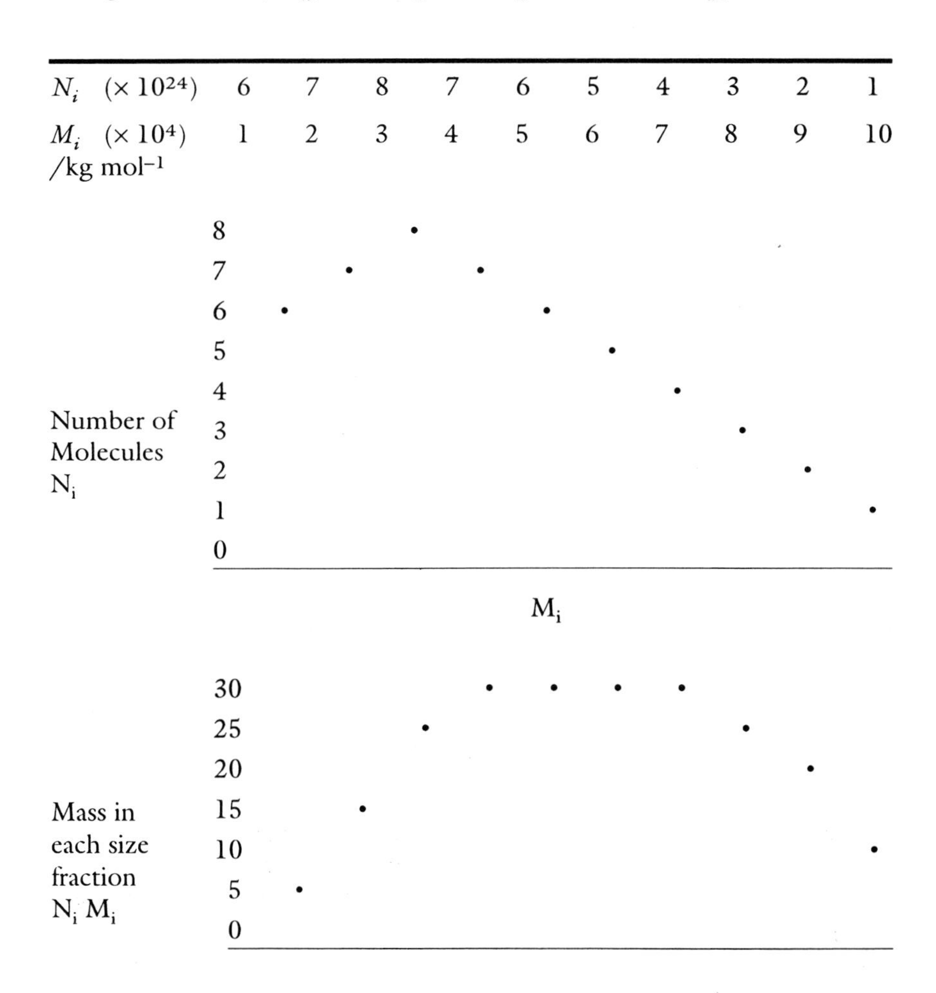

N_i ($\times 10^{24}$)	6	7	8	7	6	5	4	3	2	1
M_i ($\times 10^4$) /kg mol^{-1}	1	2	3	4	5	6	7	8	9	10

Fig. 1.5 Illustrative example of a very simple size distribution.

can be determined by a variety of analytical techniques. To simplify the calculation of $\overline{M}_n$ and $\overline{M}_w$ the molar mass distribution is assumed to be separated into discrete molar mass fractions. Following on from equation (1.1) $\overline{M}_n$ may be obtained by summing the products of the molar mass of each fraction and its mole fraction:

$$\overline{M}_n = \Sigma M_i \frac{N_i}{N} = \frac{\Sigma M_i N_i}{\Sigma N_i} \quad (1.3)$$

where M_i = molar mass of the ith fraction
N_i = number of molecules in i^{th} fraction
N = total number of molecules.

Numbers of molecules are rather difficult to deal with and it is more convenient to express $\overline{M}_n$ in terms of mass fractions, W_i. Hence:

$$W_i = \frac{N_i M_i}{\Sigma N_i M_i} \quad (1.4)$$

$$\Sigma \frac{W_i}{M_i} = \frac{\Sigma N_i}{\Sigma N_i M_i} \quad (1.5)$$

thus $\overline{M}_n = \dfrac{1}{\Sigma \frac{W_i}{M_i}}$ by combining equations (1.3) and (1.5) (1.6)

As an alternative it is possible to define an average as the sum of the products of the molar mass of each size fraction and its weight fraction, W_i. This is the weight average molar mass, $\overline{M}_w$.

$$\overline{M}_w = \Sigma W_i M_i \quad (1.7)$$

Combining equations (1.4) and (1.6) yields an expression for $\overline{M}_w$ in terms of the numbers of molecules present:

$$\overline{M}_w = \frac{\Sigma N_i M_i^2}{\Sigma N_i M_i} \quad (1.8)$$

The $\overline{M}_n$ value is the first moment of the molar mass distribution which is analogous to the centre of gravity in mechanics. The $\overline{M}_w$ value is the second moment of the distribution (radius of gyration) and a so-called third moment of the distribution is defined as $\overline{M}_z$ and is used occasionally. A viscosity average, $\overline{M}_v$, is also encountered.

In certain cases the degree of polymerisation is used, as defined in equation (1.2). Averaging is necessary and a number average degree of polymerisation, $\overline{x}_n$, and a weight average value, $\overline{x}_w$, may be defined simply as:

$$\bar{x}_n = \frac{\overline{M}_n}{M_r} \tag{1.9}$$

$$\text{and } \bar{x}_w = \frac{\overline{M}_w}{M_r} \tag{1.10}$$

In all cases $\overline{M}_z > \overline{M}_w > \overline{M}_v > \overline{M}_n$. The breadth of a molar mass distribution may be described by the ratio:

$$\frac{\overline{M}_w}{\overline{M}_n}$$

This would be unity for a monodisperse polymer where $\overline{M}_w = \overline{M}_n$. However, real values range from around 1.02 for specially prepared samples to over 20 for many commercial polymers where the distribution of molecular sizes is very broad.

In real polymers the molecules do not occupy discrete size fractions but have a continuous range of sizes. Consequently the summation signs in the preceding equations should be replaced by integral signs. However, it is far simpler to think in terms of discrete size fractions and the following example will help to illustrate the difference between weight average and number average molar masses.

Consider a sample of a polymer in which the molecules are present in only ten size fractions. This is of course very unrealistic. The situation is illustrated in Fig. 1.5. The numbers and masses of molecules in the ith fraction are given, i.e. N_i and M_i. The number of molecules N_i in each fraction and the mass in each fraction are then plotted in turn against M_i.

It is clear that the distributions obtained are quite different and are likely to yield different molar mass averages. The calculations given below use equations (1.6) and (1.7) to derive $\overline{M}_n$ and $\overline{M}_w$ respectively.

Example

Consider the polymer shown in Table 1.1.

The range of molecular sizes or polydispersity for this example can be indicated by the ratio:

$$\frac{\overline{M}_w}{\overline{M}_n} = \frac{5.8 \times 10^4}{4.33 \times 10^4}$$

This yields a value of 1.34. This is a low value and a typical commercial polymer would contain a much wider size range of polymer molecules with longer tails at both ends of the distributions. A value of 20 is typical.

$N_i M_i$ (× 10^{28})	N_i (× 10^{24})
6	6
14	7
24	8
28	7
30	6
30	5
28	4
24	3
18	2
10	1
$N_i M_i = 212 \times 10^{28}$	$N_i = 49 \times 10^{24}$

$$\overline{M}_n = \frac{\Sigma N_i M_i}{\Sigma N_i} = \frac{212 \times 10^{28}}{49 \times 10^{24}}$$

$$\overline{M}_n = 4.33 \times 10^{4} \text{ kg mol}^{-1}$$

$N_i M_i^2$ (× 10^{32})
6
28
72
112
150
150
196
192
162
100
1222×10^{32}

$$\overline{M}_w = \frac{\Sigma N_i M_i^2}{\Sigma N_i M_i} = \frac{1222 \times 10^{32}}{212 \times 10^{28}}$$

$$\overline{M}_w = 5.8 \times 10^{4} \text{ kg mol}^{-1}$$

Table 1.1

Effects of Molar Mass on Properties

Chain size is one of the most important influences on the properties of bulk polymers as illustrated in Table 1.2. It can be seen that to produce a useful polymer for load bearing applications molecules containing at least about 500 repeat units must be synthesised.

Table 1.2 The influence of chain length on character at 25 °.

Number of $-CH_2-CH_2$ repeat units per chain (Degree of Polymerisation)	Molar Mass kg mol^{-1}	Softening Temperature °C	Character at 25 °C and 1 at
1	28	−169	Gas
6	170	− 12	Liquid
35	1 000	37	Grease
140	4 000	93	Wax
430	12 000	104	Resin
1 350	38 000	112	Hard Resin

Properties such as density, transparency and refractive index vary very little with molar mass. However, properties such as softening temperature, melting temperature, melt viscosity, tensile strength, elastic modulus and toughness vary considerably with molar mass. This can be seen for cross-linked (cured or vulcanised) rubber in Fig. 1.6.

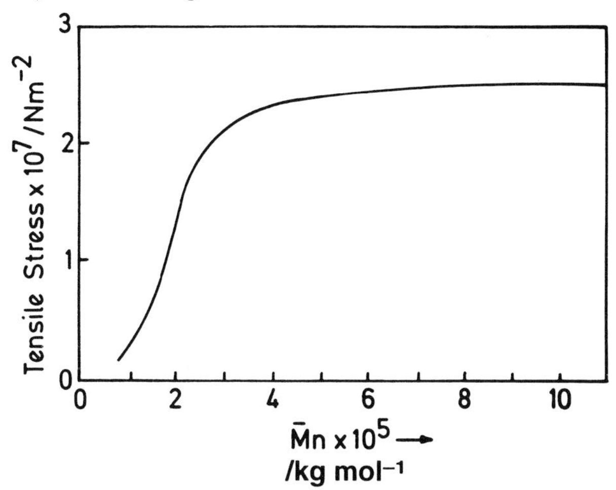

Fig. 1.6 The tensile strength of a lightly cross-linked rubber as a function of $\overline{M}_n$.

Most of these properties change rapidly with molar mass over the lower molar mass range but stabilise at higher values. It is important to realise that for polydisperse systems it is important to plot the property of interest against the appropriate molar mass average, usually $\overline{M}_n$ or $\overline{M}_w$. If the inappropriate average is used then a systematic variation of property with molar mass will not be obtained.

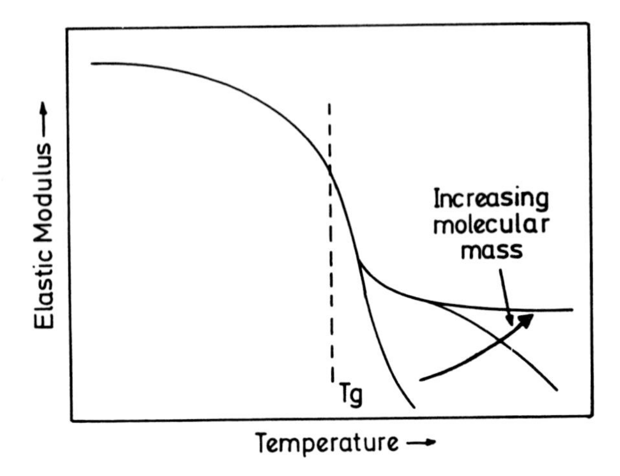

Fig. 1.7 The effects of molar mass on the elastic modulus of a linear amorphous polymer above its T_g.

Fig. 1.7 shows the effects of molar mass on the modulus of linear amorphous polymers. It can be seen that above the glass transition temperature the modulus drops to a very small value if the molar mass is low. However, as the molar mass is increased, a plateau is seen where the polymer behaves in a rubbery way. For very large molar masses this plateau extends up to temperatures where the polymer degrades. The reason for this effect is that the higher the molar mass the more the chains become entangled. Higher stresses are then required to disentangle and extend them. The entanglements behave rather like cross-links between the molecules.

STEREOISOMERISM

Introduction

The ways in which polymer molecules are arranged internally can have a significant effect on polymer properties. Most polymers have backbones consisting of chains of carbon atoms linked by covalent bonds. Carbon atoms are tetravalent and in methane, for example, a carbon atom is covalently linked to four hydrogen atoms. The bonding arrangement is symmetrical in space producing a tetrahedral arrangement as shown in Fig. 1.8(i). The angle between adjacent bonds is approximately 108.5 degrees. This arrangement is present in polymer chains and a polyethylene chain would be a planar zig-zag, if fully extended, with hydrogen atoms lying on either side (Fig. 1.8(ii)).

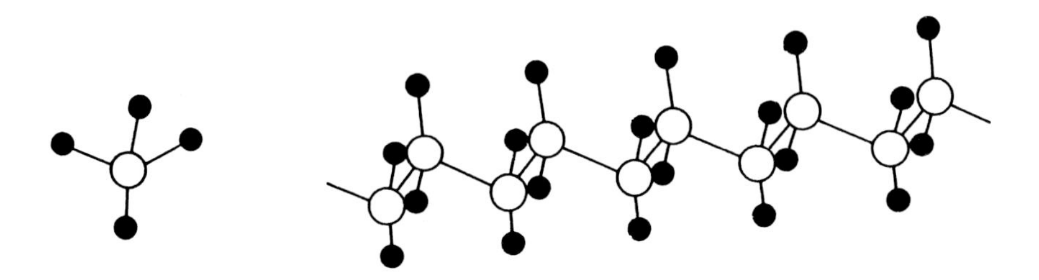

Fig. 1.8 (i) Methane (ii) Polyethylene – extended chain

Vinyl Polymers

Polyethylene has a symmetrical repeat unit but in general vinyl polymers do not. Vinyl polymers have repeat units with general structure

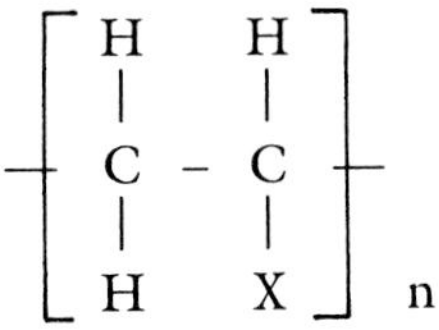

$$\left[- \overset{\displaystyle H}{\underset{\displaystyle H}{\overset{|}{\underset{|}{C}}}} - \overset{\displaystyle H}{\underset{\displaystyle X}{\overset{|}{\underset{|}{C}}}} - \right]_n$$

where X can be one of a number of atoms or groups of atoms.

Some common vinyl monomers are shown in Table 1.2.

Isomers

The substituent groups on the carbon atoms linked by double bonds can in reality be on the same side of the chain. This is the cis isomer. Alternatively they may be on opposite sides. This is the trans isomer. The isomers of 1.4 polyisoprene are shown in Fig. 1.9.

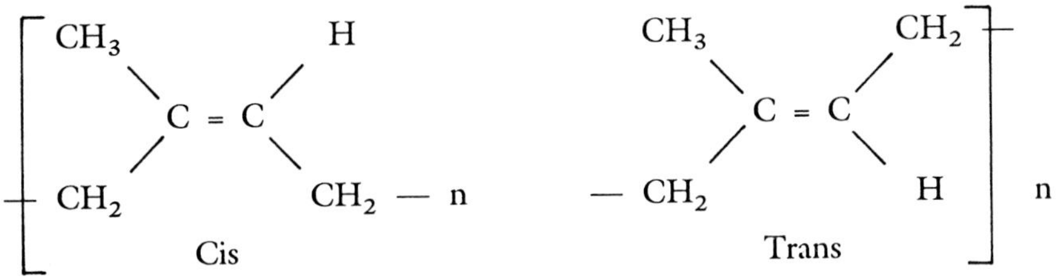

$$\left[\begin{array}{ccc} CH_3 & & H \\ & C = C & \\ -CH_2 & & CH_2- \end{array} \right]_n \quad \text{Cis} \qquad \left[\begin{array}{ccc} CH_3 & & CH_2- \\ & C = C & \\ -CH_2 & & H \end{array} \right]_n \quad \text{Trans}$$

Fig. 1.9 The Cis and Trans 1.4 isomers of polyisoprene.

The trans isomer has a regular chain structure and occurs naturally as gutta percha. It is substantially crystalline and as a result it is hard and tough.

Table 1.3 Structures of common vinyl polymers.

Monomer	X	Structure
Ethylene	H	$H_2C{=}CH_2$ (H–C(H)=C(H)–H)
Propylene	CH_3	H–C(H)=C(H)–CH_3X
Vinyl Chloride	Cl	H–C(H)=C(H)–Cl
Styrene	(benzene ring)	H–C(H)=C(H)–(benzene ring)

The cis isomer does not crystallise as easily and occurs naturally as natural rubber which is highly elastic.

Tacticity

Because vinyl polymers are not symmetrical there are three ways in which the *X* group can be arranged with respect to the carbon backbone plane. However, it can be assumed that vinyl monomers polymerise in a head to tail manner. The arrangement of the *X* group is built in during the polymerisation and is a permanent feature. It cannot be altered by rotation of the chain backbone. The three possible arrangements yield three so-called stereoisomers as follows:

i) ATACTIC – the *X* groups are randomly arranged along the chain. This irregularity interferes with crystallisation in atactic polymers.
ii) ISOTACTIC – all of the *X* groups are arranged on the same side of the backbone. This is a regular structure and helps crystallisation.
iii) SYNDIOTACTIC – The *X* groups are arranged in alternative positions along the chain.

These structures are illustrated in Fig. 1.10.

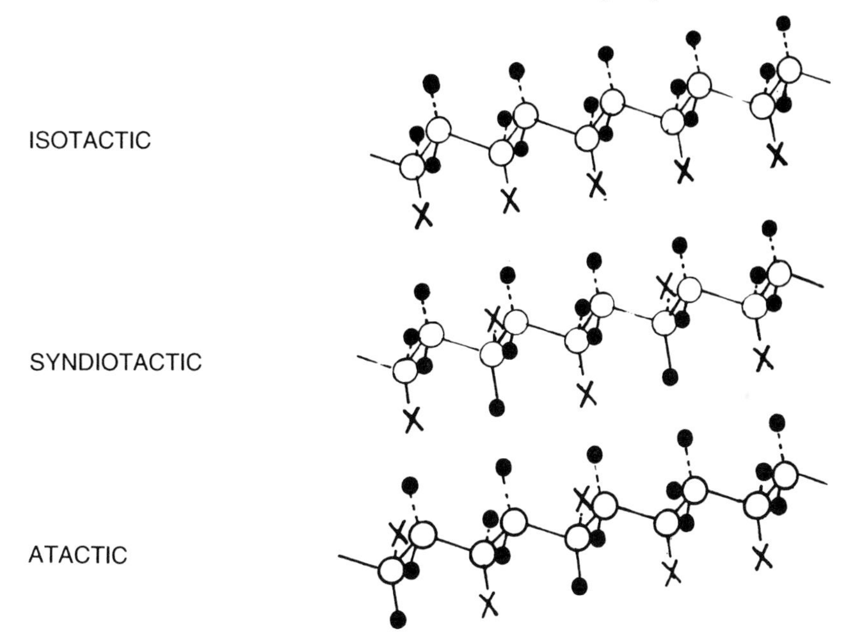

Fig. 1.10 The stereoisomers of a vinyl polymer.

Diene Polymers

Diene polymers have the general structure

$$\begin{array}{ccccccccc} & W & & X & & Y & & Z & \\ & | & & | & & | & & | & \\ - & C & = & C & - & C & = & C & - \\ & 1 & & 2 & & 3 & & 4 & \end{array}$$

The carbon atoms are numbered as shown. If polymerisation takes place at the 1 and 4 positions the resulting polymers can exhibit isomerism. The repeat unit in the polymer chain contains a double bond, i.e. the polymer is unsaturated as shown schematically in Fig. 1.11. Polybutadiene is cited as an example.

CRYSTALLINITY

Introduction

Some polymers will crystallise under favourable conditions. However, in no case is the crystallinity greater than 98% and so these polymers should really be referred to as partially crystalline. Metals are normally almost 100% crystalline because they contain a lower proportion of defects than the crystalline polymers. Non crystalline or amorphous polymers are transparent in the pure state but crystalline polymers are opaque or translucent. This is because they

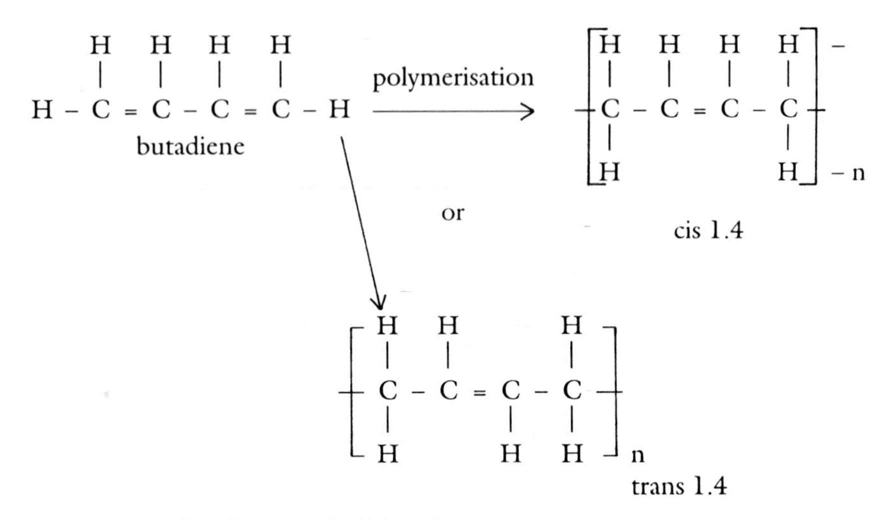

Fig. 1.11 Two isomers of polybutadiene.

contain crystallites surrounded by amorphous material. The denser crystallites have a higher refractive index and so light is scattered instead of passing through the polymer. Crystalline polymers are resistant to solvents and, because the crystalline regions are relatively rigid, the stiffness of the material increases with the amount of crystallinity.

Factors Affecting Crystallisation

Thermal energy tends to lead to disorder and so strong secondary bonds are necessary to attract polymer molecules in order to produce crystalline arrangements. Therefore hydrogen bonding and strong dipole interactions are important. One example is nylon 6 where the molecules are linked by hydrogen bonds. The oxygen atoms of one molecule always occur next to the NH groups of a neighbouring molecule as shown in Fig. 1.12.

Molecules with ordered, regular structures with no bulky side groups and a minimum of chain branching will usually crystallise. The following sources of irregularity discourage crystallisation:

i) random copolymerisation
ii) bulky side groups

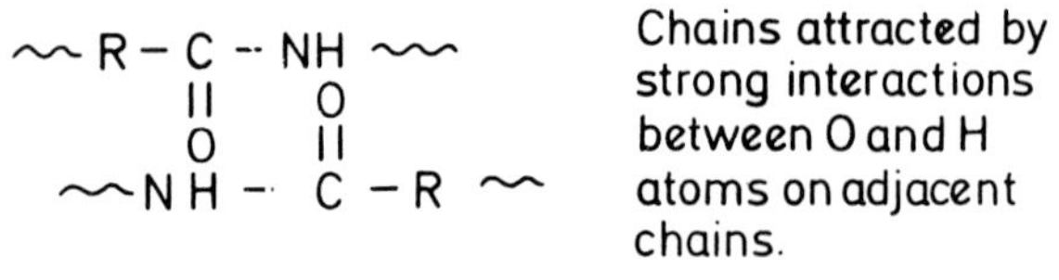

~ R – C -- NH ~
      ‖   O
      O   ‖
~ NH -- C – R ~

Chains attracted by strong interactions between O and H atoms on adjacent chains.

Fig. 1.12 Nylon 6, showing the role of hydrogen bonding on crystallisation.

Table 1.3 Some Partially Crystalline Polymers.

Polymer	T_g, °C	T_m, °C
Polyethylene	− 20	+120
Polypropylene	+ 5	+150
Polyamide 6	+ 50	+215
Polyetheretherketone	+144	+335

iii) chain branching
iv) lack of stereoregularity i.e., atacticity
v) certain geometrical isomers

Some examples of crystalline polymers are given in Table 1.3.

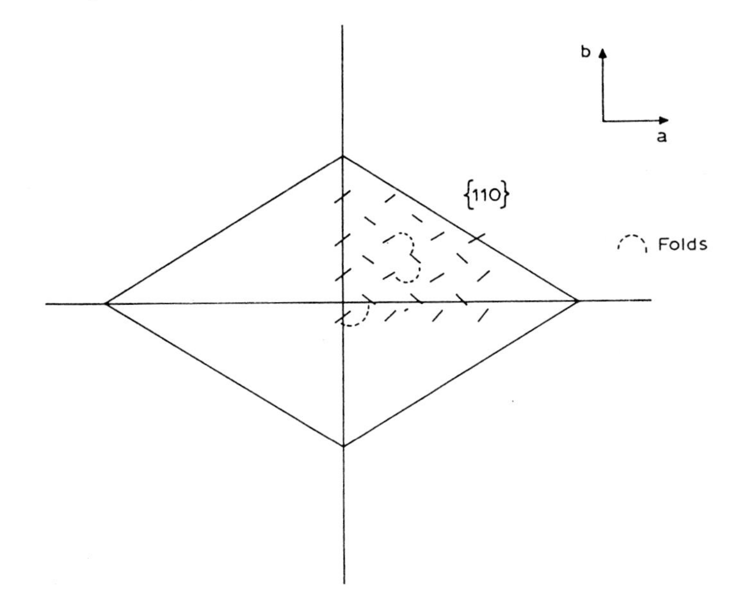

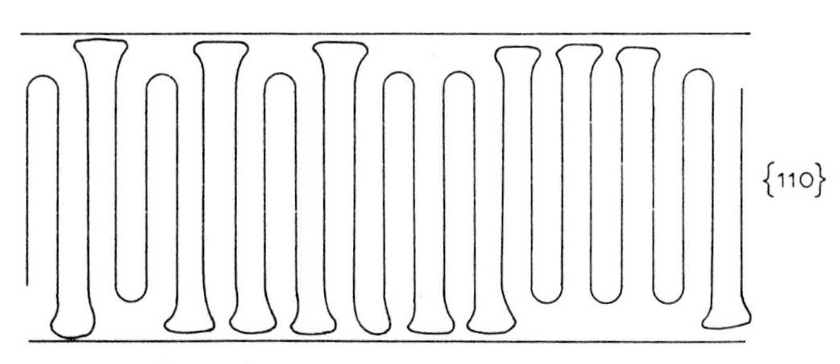

Fig. 1.13 Lamella showing folds and re-entry.

Polymer Single Crystals

In 1953 it was discovered that polymer single crystals could be grown from polymer solutions. Single crystals of polymers all seem to consist of flat or pyramidal lamellae with a thickness of about 100 nm and lateral dimensions of several μm. X-ray investigations showed that chains are arranged perpendicular to the flat faces of the crystal. As polymer molecules are in the region of 100 nm long it was concluded that they must be folded as shown in Fig. 1.13. Lamellae are about 50–60 carbon atoms thick and each fold contains about 5 carbon atoms. The folds are not part of the regular crystal structure and the re-entry pattern is probably irregular as shown.

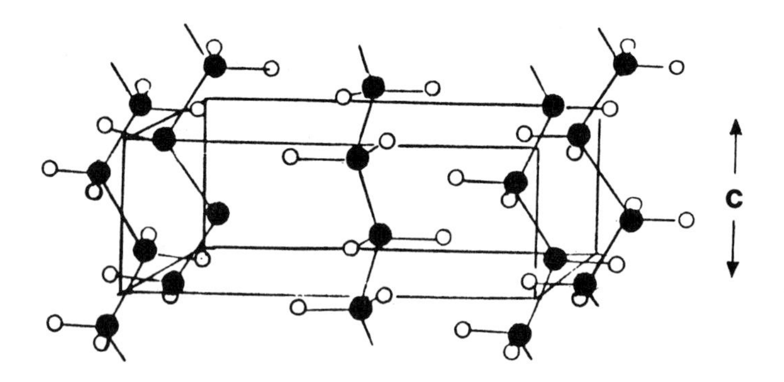

ORTHORHOMBIC POLYETHYLENE

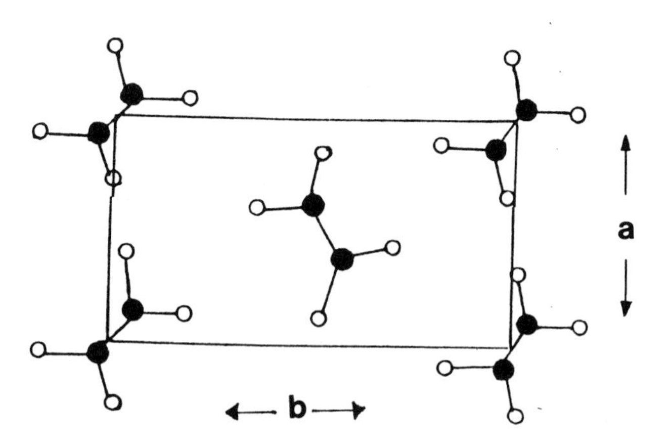

Fig. 1.14 The crystal structure of orthorhombic polyethylene.

The crystallography of polymer crystals has been studied and their unit cells characterised. For instance polythylene is orthorhombic with a = 0.741 nm, b = 4.94 nm and c = 2.55 nm (the chain repeat distance). The structure is shown in Fig. 1.14.

Melt Crystallised Polymers

Polymers crystallised from the melt contain similar structures to solution crystallised polymers. However, X-ray diffraction patterns show broad diffuse rings with backgrounds of diffuse scattering. This suggests that crystals are present together with some amorphous material. This is in fact the case. Randomly oriented crystals are embedded in and bonded to an amorphous matrix with polymer chains passing through crystallites into the amorphous region producing strong interphase bonding.

There is evidence to suggest that extended chain crystals also exist especially as a result of crystallisation from melts subjected to elongation flow.

Spherulites

In many cases crystals are arranged in larger aggregates known as spherulites. These grow radially from nucleation points until they meet other spherulites.

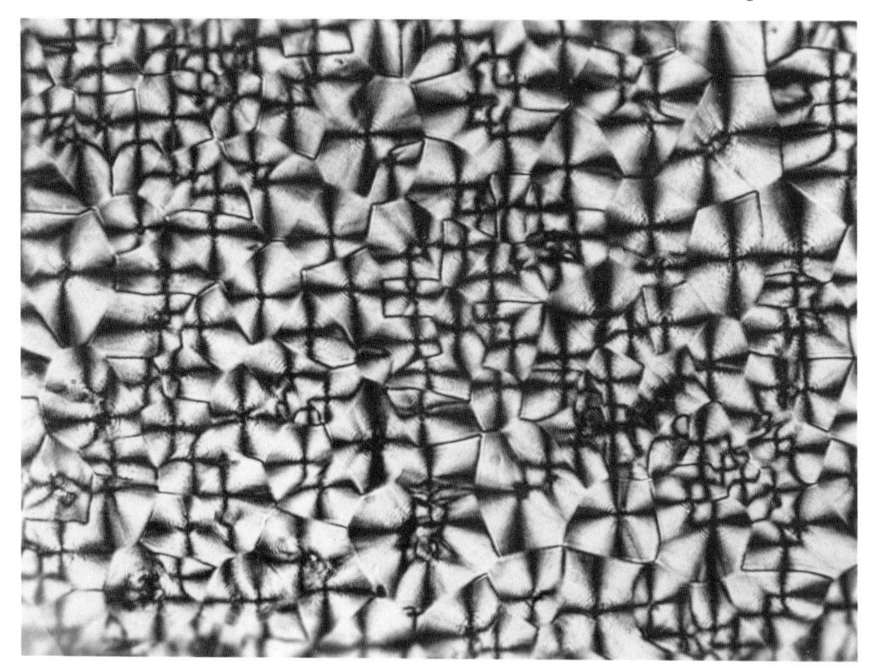

Fig. 1.15 Spherulites in low density polyethylene. (x 100)

New lamellae form between gaps in the lamellae as growth progresses. They are similar to grains in metals with diameters of the order of 0.01 mm. Spherulites have a maltese cross appearance when thin polymer films are viewed under crossed polarising filters as shown in Fig. 1.15.

Spherulites contain branching chain folded crystals which grow outwards from each nucleating centre in Fig. 1.16. Large spherulites are undesirable as they promote brittleness, as do large grains in metals.

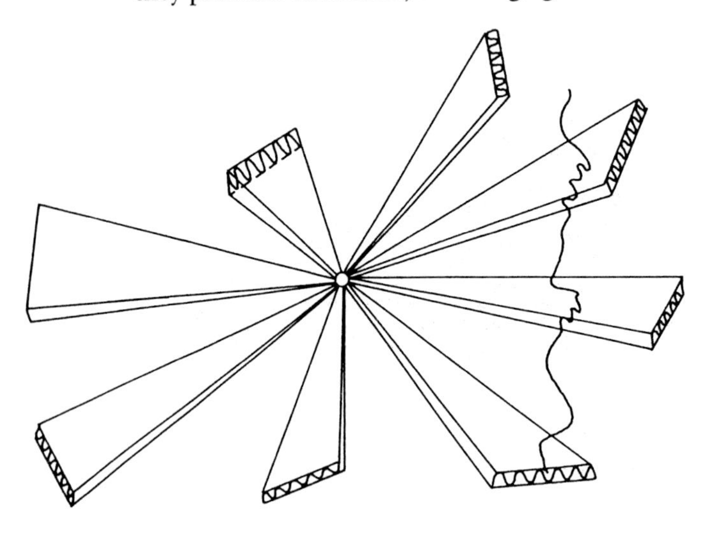

Note: Intercrystalline links lamellae may be twisted

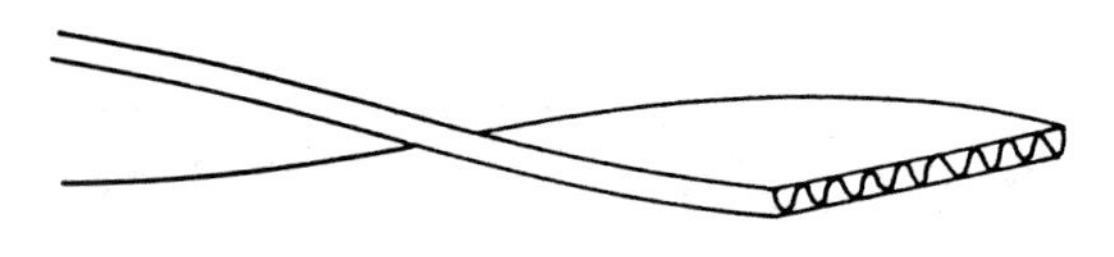

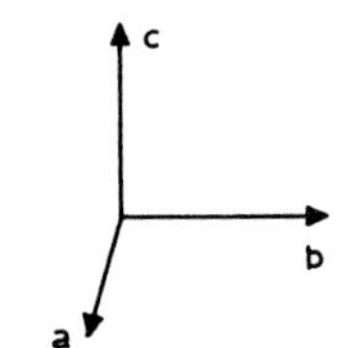

Fig. 1.16 Schematic diagram of a partially crystalline polymer.

Degree of Crystallinity

The degree of crystallinity may be determined from measurements of specific volume or density. The specific volume of the fully crystalline material may be calculated from a knowledge of the unit cell and that for the fully amorphous material may be obtained by extrapolating data obtained from polymer melts. Assuming that degree of crystallinity is directly proportional to specific volume then the degree of crystallinity W_c is given by:

$$W_c = \frac{(V_a - V)}{(V_a - V_c)} \tag{1.11}$$

where V_a, V_c and V are the fully amorphous, fully crystalline and partially crystalline specific volumes respectively.

Other methods of measuring crystallinity are also used. These include analysis of X-ray diffraction patterns and infrared absorption bands.

Kinetics of Crystallisation

For crystallisation to occur, considerable molecular mobility is necessary. Thus the temperature range for crystallisation is between the glass transition temperature T_g (*see* section on TRANSITIONS) and the crystalline melting temperature T_m. The overall formation rate of the crystals is equal to the product of the nucleation rate $\dot{n}$ and the growth rate $\dot{g}$. The growth rate is diffusion controlled and is a maximum just below T_m. The nucleation rate is a maximum just above T_g. Therefore the overall formation rate is a maximum about halfway between T_g and T_m as shown in Fig. 1.17. Increasing the molecular mass of a polymer decreases its rate of crystallisation because diffusion rates and hence crystal growth rates are reduced.

In many polymers the progress of crystallisation from the melt is described by the Avrami equation. This can accommodate different nucleation mechanisms and takes into account impingement of crystals during the later stages

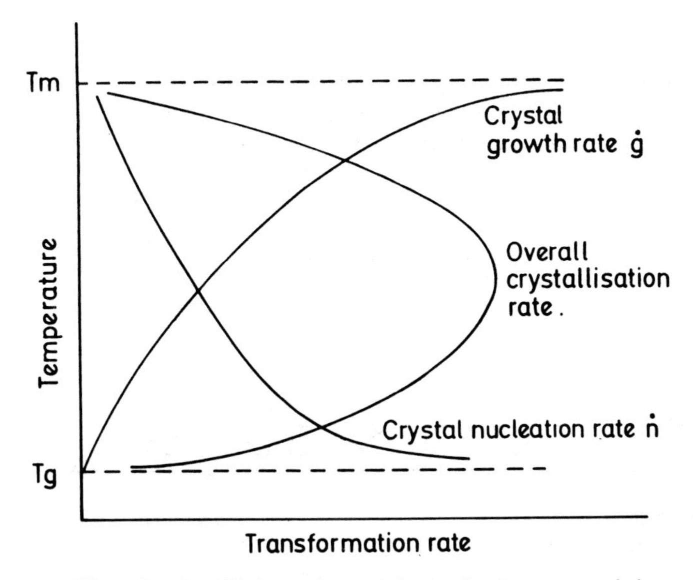

Fig. 1.17 Plots showing the growth rate, the nucleation rate and the overall rate of crystallisation versus Temperature for a polymer.

of crystallisation. It assumes homogeneous nucleation but in reality nucleation is more likely to be heterogeneous.

$$X(t) = 1 - \exp(-kt^n) \tag{1.12}$$

where: $X(t)$ = mass fraction of crystals at time t
t = time
k = constant describing the rate of crystallisation
and n = the Avrami exponent

If all the nuclei are formed instantaneously $n = 3$. If nucleation occurs progressively then $n = 4$, since both the spherulitic volume change and the nucleation rate have to be taken into account.

Effect of Crystallisation on Properties

The degree of crystallinity and the size of the spherulites affect properties. Spherulite size can be reduced by using nucleating agents, and tougher polymers are produced as a result.

The effect of the degree of crystallinity on properties is well illustrated in Figure 1.18 which gives property data for low, medium and high density polyethylenes. The higher the density the greater the crystallinity due to differences in the amounts and nature of side branches (higher density grades are more linear in structure).

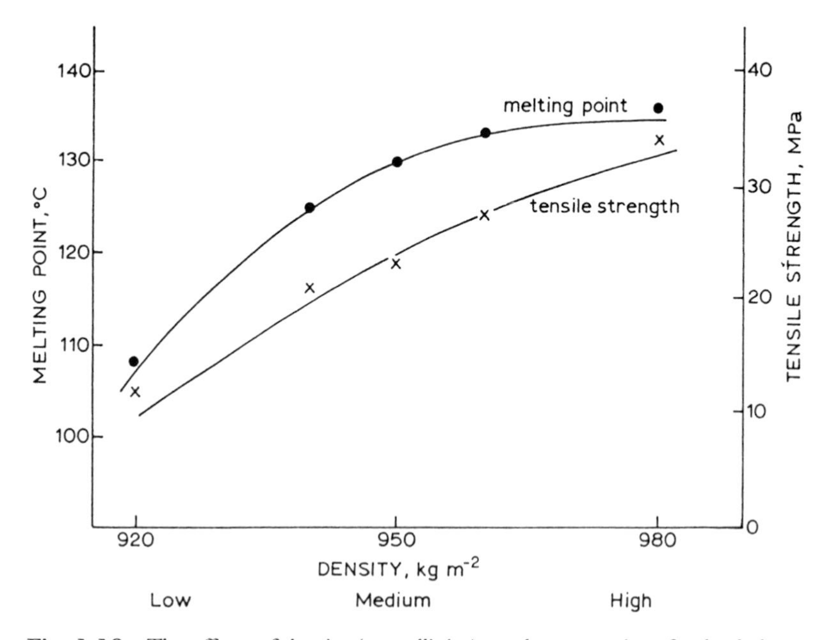

Fig. 1.18 The effects of density (crystallinity) on the properties of polyethylene.

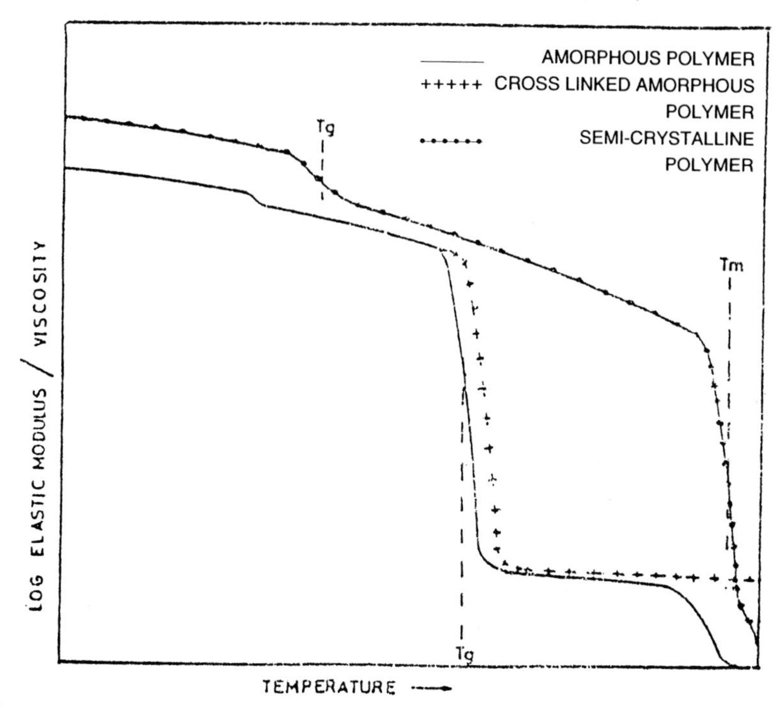

Fig. 1.19 Changes in properties with temperature for amorphous and partially crystalline polymers. (See also Fig. 1.7.)

TRANSITIONS

Introduction

Polymers exhibit various transitions in properties as the temperature is varied. The most important are the melting point T_m for crystalline polymers and the glass transition T_g for amorphous polymers.

The Glass Transition Temperature

As a polymer which does not crystallise is cooled from the melt it becomes progressively more viscous and then rubbery. If cooling is continued it will become hard and elastic as shown in Fig. 1.19. It has changed from a rubbery state to an elastic state at its glass transition temperature, T_g. Many properties change significantly at T_g including the elastic modulus, specific volume, heat capacity and thermal expansion coefficient. However, the exact value of T_g depends on the rate of cooling. The lower the cooling rate the lower the value of T_g. Partially crystalline polymers show limited property changes at T_g due to changes occurring in the amorphous phase.

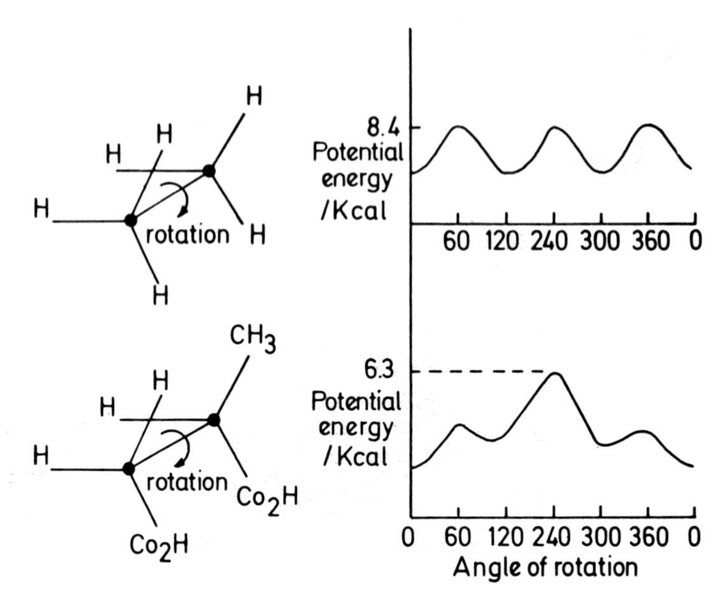

Fig. 1.20 The ethane and methyl succinic acid molecules and their rotational energies.

The glass transition temperature is the temperature for a particular polymer at which molecular rotation about single bonds in the backbone becomes possible. Rotation cannot occur about double bonds if they are present. Owing to the inherent zig-zag nature of the backbone even a few rotations along the backbone of a molecule can lead to significant changes in shape (*see* Fig. 4.1). This occurs if stresses are applied and the polymer is rubbery and capable of flow.

Definition: The temperature at which molecular rotation about single bonds becomes restricted as the temperature is reduced.

Rotation is thermally activated and the easier it is, then the lower the value of T_g for a polymer. The structural features which affect the ease of rotation fall into two categories, i.e. those which affect the intrinsic mobility of a single isolated chain and those which come into play when the molecules are in close proximity as they are in bulk polymers.

An example of the first category is apparent if the barriers to rotation in simple small molecules are calculated. Consider ethane and methyl succinic acid as shown in Fig. 1.20.

In ethane the barrier to rotation is just 0.5 kJ mol^{-1} but is increased to

3.5 kJ mol^{-1} in methyl succinic acid due to the steric interference between the bulky CH and CO_2H groups during rotation, which is maximised in the eclipsed positions. The potential energy as a function of rotation is also shown in Fig. 1.20. The lowest energy positions are the most stable. In the case of polymers the higher the energy of rotation the higher the T_g.

Summary of factors affecting T_g:

i) Bulky groups attached to the backbone which affect the energy of rotation and increase T_g.
ii) Rigid structures incorporated into the chain backbone, e.g. aromatic ring structures, increase T_g.
iii) Presence of divalent atoms such as oxygen or conjugated bonds, i.e. alternate double and single bonds in the backbone. The net result is to reduce the number of side groups and hence reduce the rotational energy and hence T_g.
iv) Size effects of side groups. If side atoms or groups are relatively large then the chain mobility may be reduced and the T_g increased as in polytetrafluoroethylene (PTFE). In this case the fluorine atoms are tightly packed around the backbone causing the chain to take up a rigid twisted zig-zag confirmation.
v) The stronger the secondary bonds between chains the higher is the T_g.
vi) Cross-linking between chains increases T_g.
vii) Side branches increase T_g.
viii) T_g increases with molar mass particularly at low molar mass values. The effect is slight at higher values of molar mass.
ix) Copolymerisation can be used to control T_g. In general if a random copolymer is produced from two monomers the chain stiffness hence T_g, will be intermediate between the values for homopolymers made from the same monomers.
x) Plasticisers reduce T_g as they increase the space or free volume between chains increasing chain mobility. They are used primarily with PVC and themselves have much lower T_g or crystallisation temperatures than pure PVC.

The concept of free volume is very useful in discussions of T_g values. The free volume is the space between the chains in a polymer. It increases rapidly above T_g. In general, if the free volume is increased, for instance by chain branching, reducing molar mass or plasticisation, the T_g is decreased as greater chain mobility is possible. For example the following empirical equation predicts the variation of T_g with molar mass:

$$T_g = T_g - K \frac{1}{\overline{M}_n} \qquad (1.13)$$

where T_g = the T_g of a sample containing molecules of infinite molar mass, $\overline{M}_n$ = the number average molar mass, and K = a constant.

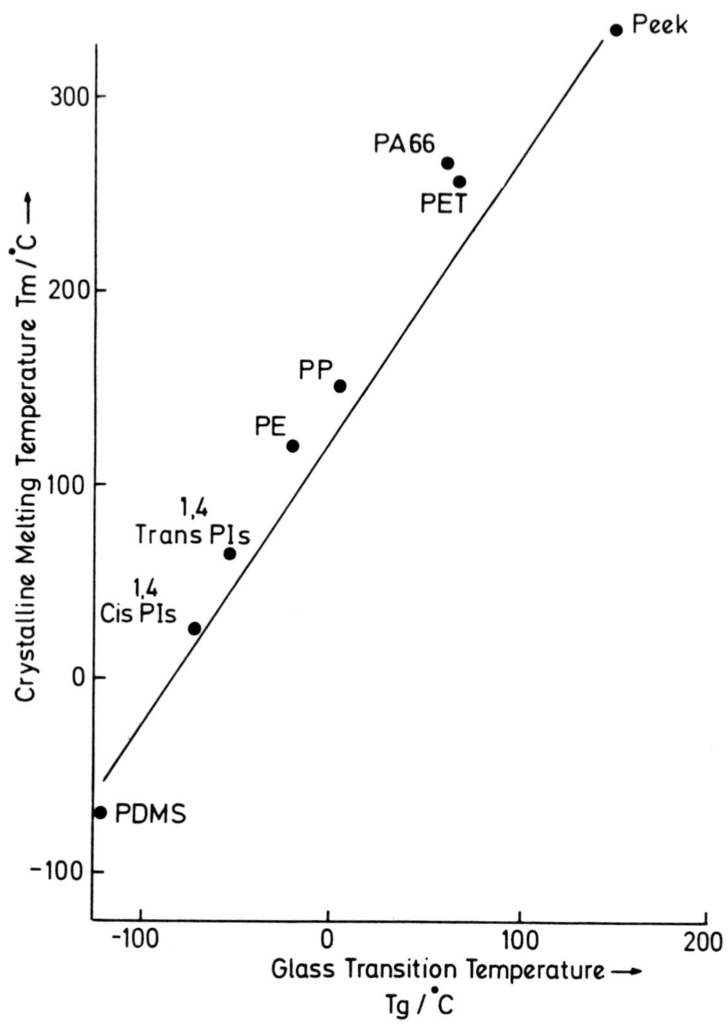

Fig. 1.21 The relationship between T_g and T_m.

The Crystalline Melting Temperature

Melting of polymers is not sharp as in metals but takes place over a temperature range. The melting temperature T_m is affected by the specimen history, the crystallisation temperature and the heating rate. T_m is always greater than the prior crystallisation temperature. However, extrapolation of experimental data suggests that if a polymer is crystallised at an infinitely slow rate then the crystallisation temperature and T_m would coincide. This value represents the equilibrium value of T_m. Melting behaviour depends on the thicknesses of the

Table 1.4 Values of T_g and T_m for typical polymers.

Repeat Unit	Name	T_g/°C	T_m/°C
$-CH_2-CH_2-$	Polyethylene (PE)	−20	+120
$-CH_2-CH(CH_3)-$	Polypropylene (isotactic) (PP)	+ 5	+150
$-CH_2-CH(C_6H_5)-$	Polystyrene (atactic) (PS)	+100	–
$-CH_2-C(CH_3)(COOCH_3)-$	Polymethyl methacrylate (PMMA, Perspex etc)	+ 99	–
$-CH_2-CH(Cl)-$	Polyvinyl chloride (PVC)	+ 80	–
$-CH_2-C(CH_3)=CH-CH_2-$ (trans)	trans polyisoprene Gutta Percha	− 53	+ 65
$-CH_2-C(CH_3)=CH-CH_2-$ (cis)	cis polyisoprene Natural Rubber	− 73	+ 25
$-Si(CH_3)_2-O-$	Polydimethyl siloxane (silicone rubber)	−123	− 70

$-\langle O \rangle-C(CH_3)_2-\langle O \rangle-O-C(=O)-O-$	Polycarbonate (PC)	+149	(+225)
$-O-CH_2-CH_2-OOC-$ O $-COO-CH_2-CH_2-O-$	Polyethyleneterephthate	+ 67	+256
$-\langle O \rangle-CO-\langle O \rangle-O-\langle O \rangle-O-$	Polyethereterketone	+144	+335
NH $(CH_2)_6$ NHOC $(CH_2)_4$ CO	Nylon 66	+ 60	+264

crystal lamellae. T_m increases with thickness and it is interesting to note that polymers which are heated slowly or annealed, particularly just below T_m, show a general increase in lamellae thickness. This process is enabled by increased chain mobility at higher temperatures and the accompanying reductions in lamellae surface energy.

In crystallisable polymers T_m is more important than T_g. T_m limits the use of such polymers and considerable efforts have been made to produce polymers with high melting temperatures. A knowledge of the controlling factors is required. In general the same factors affect T_m as affect T_g. Consequently it is found that for most polymers T_g is related to T_m, and $T_g = 2/3\ T_m$ approximately (*see* Fig. 1.21).

The stiffer the backbone of a polymer the higher is T_m. Bulky side groups and strong secondary bonds increase T_m as they do for T_g. Increases in molar mass and decreases in chain branching increase T_m. This is because chain ends and branches cause defects in the crystalline structure. The greater the severity of defects the lower the T_m. As a polymer contains a range of types and numbers of defects its melting temperature is distributed over a finite temperature range rather than being sharp. Examples of values of T_g and T_m for typical polymers are given in Table 1.4.

REFERENCES

D.C. BASSETT: *Principles of Polymer Morphology*, Cambridge Solid State Science Series, Cambridge University Press, 1981.

F. BILLMEYER: *Textbook of Polymer Science*, 3 edn, Wiley Interscience, New York, 1984.

A.W. BIRLEY, B. HAWORTH and J. BATCHELOR: *Physics of Plastics*, Hanser, Munich, 1991.

J.A. BRYDSON: *Plastics Materials*, 4 edn, Butterworth Scientific, London, 1982.

L. MASCIA: *Thermoplastics: Materials Engineering*, Elsevier Applied Science, London, 1982.

2 *Polymer Synthesis and Manufacture*

INTRODUCTION

The importance of polymer synthesis cannot be overemphasized as it is during this stage of a polymer's history that its most important characteristics are established. The molar mass and its distribution can be controlled along with other characteristics such as the degree of chain branching. Also copolymers of varying types can be produced, i.e. random alternating, block and graft copolymers. Even interpenetrating networks of polymer molecules can be obtained by careful control of the appropriate chemical reactions. In this chapter the most important features of polymerisation mechanisms will be examined and this will provide a worthwhile introduction and background for materials technologists.

There are two main types of polymer synthesis reactions. One is condensation or step growth polymerisation. The other is chain addition polymerisation which can be further sub-divided into free radical chain addition and non-radical addition polymerisation. The purpose of studying these reactions is to determine ways in which the molar mass and its distribution can be controlled and so to produce polymer of required specification.

FREE RADICAL CHAIN ADDITION POLYMERISATION

This process takes place in three distinguishable steps: **Initiation, Polymerisation** and **Termination**. The reaction is initiated by the generation of free radicals which, although electrically neutral, contain an unshared electron. Free radicals are required for the initiation stage and are normally produced by decomposing organic peroxides or azo compounds or by using redox systems. For example:

(i)

$$\langle O \rangle\text{-}\overset{\overset{\displaystyle O}{\|}}{C}\text{-}O\text{-}O\text{-}\overset{\overset{\displaystyle O}{\|}}{C}\text{-}\langle O \rangle \rightarrow 2\,\langle O \rangle\text{-}\overset{\overset{\displaystyle O}{\|}}{C}\text{-}O^{*} \rightarrow 2\,\langle O \rangle^{*} + 2\,CO_2$$

benzoyl peroxide free radicals

(ii)

$$(CH_3)_2 - \underset{\underset{\displaystyle CN}{|}}{C} - N = N - \underset{\underset{\displaystyle CN}{|}}{C} - C - (CH_3)_2 \rightarrow 2\,(CH_3)_2\,\underset{\underset{\displaystyle CN}{|}}{C^{*}} + N_2$$

Azobisisobutyronitrile free radicals

Suitable monomers contain double bonds which are unstable and are readily broken by free radicals to form active centres. This is the initiation stage and is followed by rapid addition of monomer molecules during propagation or polymerisation. The process is eventually halted by a termination reaction at an active chain end. Because the propagation stage is very rapid, long chain molecules are present at an early stage of the process of conversion of monomer into polymer. The process is statistical and so a wide distribution of chain lengths are usually produced. Examples of some monomers used in free radical addition polymerisation are shown in Table 2.1.

Mechanism and Kinetics

If the initiator is represented by I then the initiator decomposition reaction can be represented by:

$$I \xrightarrow{K_d} R^{*} \tag{2.1}$$

where K_d is the rate constant for decomposition.

This is a slow reaction and determines the rate of the whole polymerisation, as soon as an initiator radical forms it combines with a monomer molecule to form an active centre or monomer radical and polymerisation is initiated:

$$R + M \xrightarrow{K_a} RM^{*} \tag{2.2}$$

where K_a is the rate constant for activation.

Equation (2.1) is the rate controlling equation but not all radicals generated initiate chain formation. Only a certain fraction, f, are effective and the overall rate equation for initiation is that for simple first order decay;

$$r_i = \frac{d[R^{*}]}{dt} = 2\,k_d f[I] \tag{2.3}$$

where $[R^{*}]$ is radical concentration and $[I]$ is initiator concentration.

Polymerisation involves the addition of monomer as follows:

$$RM_1^{*} + M \xrightarrow{K_p} RM_2 + M \xrightarrow{K_p} RM_i^{*} \tag{2.4}$$

Table 2.1 Typical monomers, their structures and resultant polymers.

	Monomer		Polymer	
Ethylene	$HHC=CHH$	→	$-[CHH-CHH]_n-$	Polyethylene (PE)
Propylene	$HHC=CH(CH_3)$	→	$-[CHH-CH(CH_3)]_n-$	Polypropylene (PP)
Vinyl chloride	$HHC=CH(Cl)$	→	$-[CHH-CH(Cl)]_n-$	Polyvinylchloride (PVC)
Methyl methacrylate	$HHC=C(CH_3)(COOCH_3)$	→	$-[C-C(CH_3)(COOCH_3)-]_n$	Polymethylmethacrylate (PMMA)
Acrylonitrile	$HHC=CH(CN)$	→	$-[CHH-CH(CN)]_n-$	Polyacrylonitrile (PAN)

The polymerisation rate constant is assumed to be independent of the growing chain length. The polymerisation rate r_p is proportional to the monomer concentration $[M]$ and the total growing chain radical concentration $[RM_i{}^*]$:

$$r_p = K_p\,[M]\,[RM_i{}^*] \qquad (2.5)$$

where K_p is the rate constant for polymerisation.

Chain radicals disappear by termination reactions. Termination usually takes the form of combination or disproportionation as shown in Fig. 2.1.

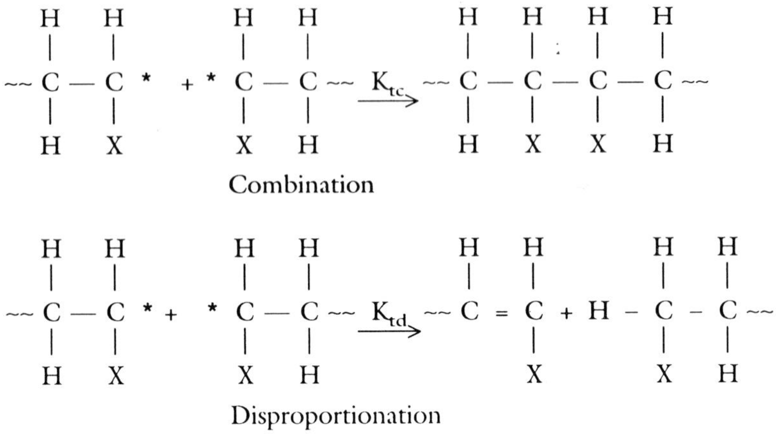

$$\sim\sim \underset{H}{\overset{H}{C}} - \underset{X}{\overset{H}{C}}{}^{*} \; + {}^{*}\underset{X}{\overset{H}{C}} - \underset{H}{\overset{H}{C}} \sim\sim \xrightarrow{K_{tc}} \sim\sim \underset{H}{\overset{H}{C}} - \underset{X}{\overset{H}{C}} - \underset{X}{\overset{H}{C}} - \underset{H}{\overset{H}{C}} \sim\sim$$

Combination

$$\sim\sim \underset{H}{\overset{H}{C}} - \underset{X}{\overset{H}{C}}{}^{*} \; + {}^{*}\underset{X}{\overset{H}{C}} - \underset{H}{\overset{H}{C}} \sim\sim \xrightarrow{K_{td}} \sim\sim \overset{H}{C} = \underset{X}{\overset{H}{C}} + H - \underset{X}{\overset{H}{C}} - \underset{H}{\overset{H}{C}} \sim\sim$$

Disproportionation

Fig. 2.1 Termination mechanisms.

In most cases one or other of the termination reactions predominates. In either case two growing chains are consumed in the reaction and the termination rate is:

$$r_t = 2\ K_t\,[RM_i^*]^2 \tag{2.6}$$

where K_t is the appropriate rate constant for termination.

The main problem in calculating the rates of polymerisation reactions is that $[RM_i^*]$ is not known. However, it can be assumed that under steady state conditions $[RM_i^*]$ remains constant and that chain radicals are generated as fast as they are removed, i.e. $r_i = r_t$ and:

$$2\,f K_d[I] = 2\ K_t[RM_i^*]^2 \tag{2.7}$$

Hence

$$[RM_i^*] = \frac{f K_d[I]}{K_t}^{1/2} \tag{2.8}$$

and from equation (2.5)

$$r_p = K_p\,\frac{f K_d[I]^{1/2}[M]}{K_t} \tag{2.9}$$

Equation (2.9) gives the rate of polymerisation and applies well to many polymerisations although it is misleading in certain commercially important cases such as emulsion polymerisation. Equation (2.9) can be integrated to yield an expression for the monomer conversion ratio

$$\frac{[Mo]-[M]}{[Mo]}$$

This is of considerable interest as the general aim in a polymerisation is to achieve as great a conversion of monomer to polymer as possible. It is inaccurate to assume that the initiator concentration, $[I]$, in equation (2.9) remains at its initial value $[I_o]$. It can be assumed that $[I_o]$ decays by a first order reaction with $[I] = [I_o]$ at zero time, i.e.

$$\frac{d[I]}{dt} = -K_d [I] \qquad (2.10)$$

and

$$[I] = [I_o]\, e^{-K_d t} \qquad (2.11)$$

If this is substituted into equation (2.9) and an integration carried out then:

$$\ln \frac{[M]}{[M_o]} = \frac{2\,K_p}{K_d} \frac{f K_d [I_o]}{K_t}^{1/2} [e^{\frac{-K_d t}{2}} - 1] \qquad (2.12)$$

where $[M_o]$ = monomer concentration at t, 0
and $[M]$ = monomer concentration at time, t.

At infinite time there is a maximum conversion given by:

$$\frac{[M_o] - [M]}{[M_o]} = 1 - \exp \frac{-2}{K_d} \left[\frac{K_p^2\ fK_d\,[I_o]}{K_t}\right]^{1/2} \qquad (2.13)$$

This is shown in Fig. 2.2. Figure 2.2a shows the effects of changing the rate constant for decomposition of the initiator, K_d, on the conversion. Figure 2.2b shows the effect of changing the initiator concentration, $[I_o]$.

The rate of polymerisation may be affected by the use of retarders or inhibitors. An inhibitor delays the start of a reaction but once started it proceeds at the normal rate. On the other hand retarders slow down the reaction. These effects are shown in Fig. 2.3. Inhibitors are used during storage of monomers to prevent premature polymerisation, e.g. benzoquinone for styrene. Oxygen inhibits vinyl monomer polymerisation and so atmospheres of nitrogen are used in the reactors.

Kinetic Chain Length

The degree of polymerisation of the polymer molecules produced during free radical chain addition polymerisation is given by the kinetic chain length, ν

$$\nu = \frac{r_p}{r_i} = \frac{K_p f K_d [I]}{K_t}^{1/2} \frac{[M]}{2\,K_d f\,[I]} = \frac{K_p [M]}{2\,fK_d K_t^{1/2}\,[I]^{1/2}} \qquad (2.14)$$

The number average degree of polymerisation, $\bar{x}_n$, is related to ν in a way which depends on the mechanism of termination. If combination occurs solely then $\bar{x}_n = 2\,\nu$; if disproportionation is the only mechanism them $\bar{x}_n = \nu$.

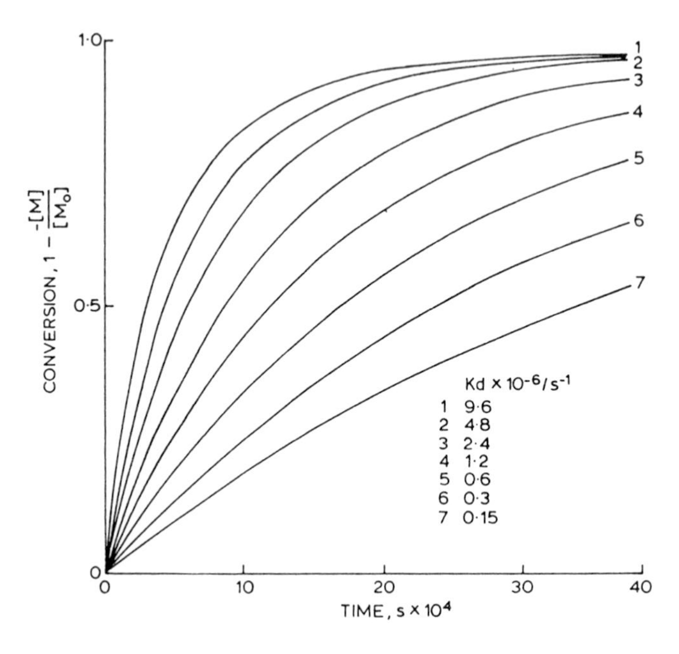

Fig. 2.2a The effect of K_d on conversion as a function of time.

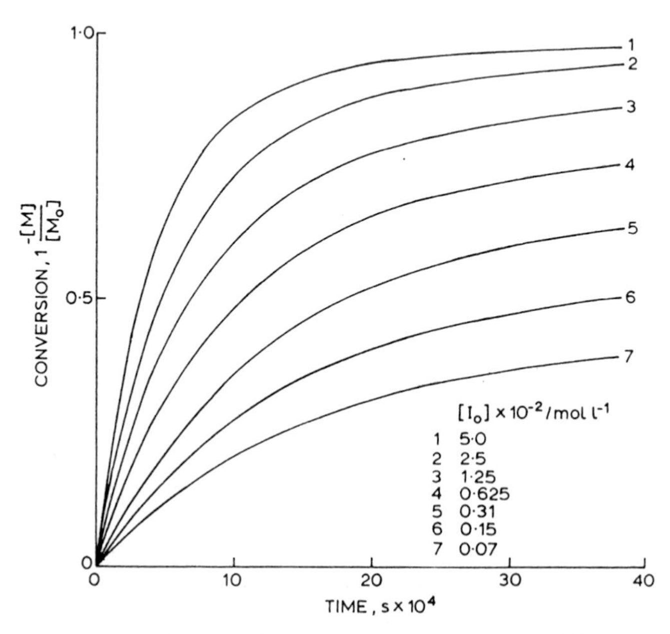

Fig. 2.2b The effect of $[I_o]$ on conversion as a function of time.

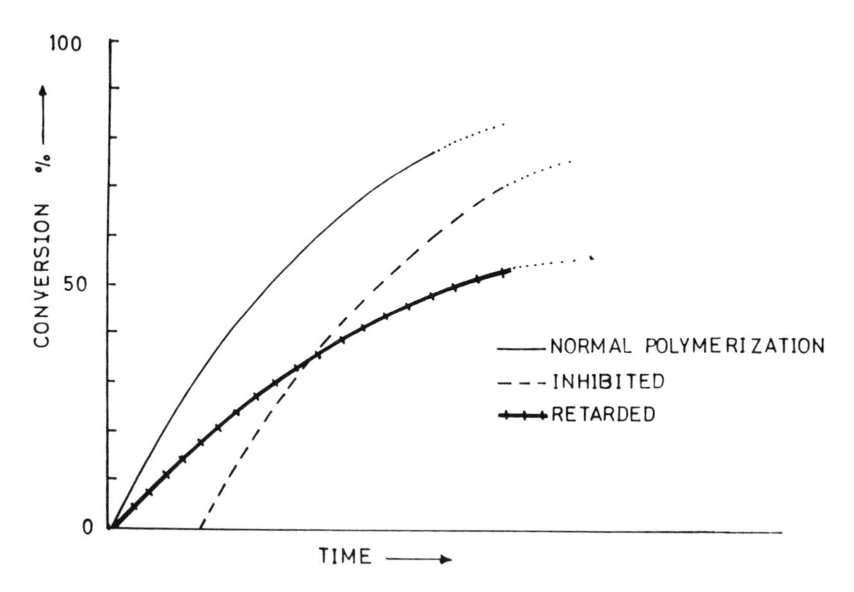

Fig. 2.3 The effects of inhibitors and retarders on polymerisation.

Generally a situation somewhere between these two extremes exists, i.e. $x_n = T_v$ where T is between 1 and 2.

Temperature Effects

The rate constants introduced above, in fact, vary with temperature according to an Arrhenius relationship:

$$k = A \exp(-E/RT) \tag{2.15}$$

The rate of polymerisation, kinetic chain length and degree of polymerisation must also vary with temperature:

$$r_p = K_p \frac{fK_d [I]}{K_t}^{1/2} [M] = C_1 \exp - \left[\frac{E_p + E_d/2 - E_t/2}{RT} \right] \tag{2.16}$$

where C_1, is a constant, E_p, E_d and E_t are the appropriate activation energies and f is assumed to be temperature independent. The exponent term is normally found to be negative and so r_p increases with temperature.

The degree of polymerisation $\bar{x}_n$ is given by:

$$\bar{x}_n = \frac{T K_p [M]}{2 (fK_d K_t)^{1/2} [I]^{1/2}} = C_2 \exp \left[\frac{E_i + E_t/2 - E_p}{RT} \right] \tag{2.17}$$

where C_2 is a constant.

In this case the exponent term is positive and so an increase in temperature leads to a reduction in $\bar{x}_n$ while r_p increases. Thus it can be concluded that increasing temperature results in an increase in reaction rate but a decrease in molar mass. The latter is certainly undesirable in most cases.

CHAIN TRANSFER

As well as termination reactions such as combination and disproportionation an alternative type of reaction may result in terminated polymer chains. This is called a chain transfer reaction as reactivity is transferred from a growing chain to a new active centre. The overall rate of polymer chain production is unlikely to be affected by chain transfer reactions but a useful outcome is that the molar mass of the resultant polymer can be controlled by appropriate chain transfer reactions. A typical chain transfer reaction is:

$$RM_i^* + XP \rightarrow R\,M_iX + P^*$$
$$\text{then } P^* + M \rightarrow PM^*, \text{etc} \qquad (2.18)$$

X_P is the chain transfer agent and may be molecules of initiator, monomer, solvent, terminated polymer or deliberately added chain transfer chemical (modifier or regulator). Chain transfer must be allowed for in polymerisation reactions. It can have rather annoying results such as chain branching. This is illustrated in Fig. 2.4 for polyethylene where a growing molecule may 'bite its own back' or reactivity may be transferred to a terminated chain. The outcome of both reactions is that active centres are formed on molecules and these can grow into branches of perhaps five or six carbon atoms long.

Where chain transfer reactions are present the degree of polymerisation is given by:

$$\bar{x}_n = \frac{\text{rate of growth of polymer chains}}{\Sigma \text{ rates of all reactions leading to terminated chains}} \qquad (2.19)$$

where the denominator is made up of the sum of all, or some of, the following rate terms:

i) termination – $2fK_d\,[I]$ (equation 2.7)
ii) transfer to monomer – K tr-m $[M]$ [RM*]
iii) transfer to solvent – K tr-s $[S]$ [RM*]
iv) transfer to initiator – K tr-i $[I]$ [RM*]
v) transfer to modifier – K tr-mod [MOD] [RM*]

Transfer constants are defined as follows:

$$C_m = \frac{K_{tr-m}}{K_p} \qquad Cs = \frac{K_{tr-s}}{K_p} \qquad C_i = \frac{K_{tr-i}}{K_p} \qquad C_{mod} = \frac{K_{tr-mod}}{K_p}$$

where K_p = rate constant for polymerisation

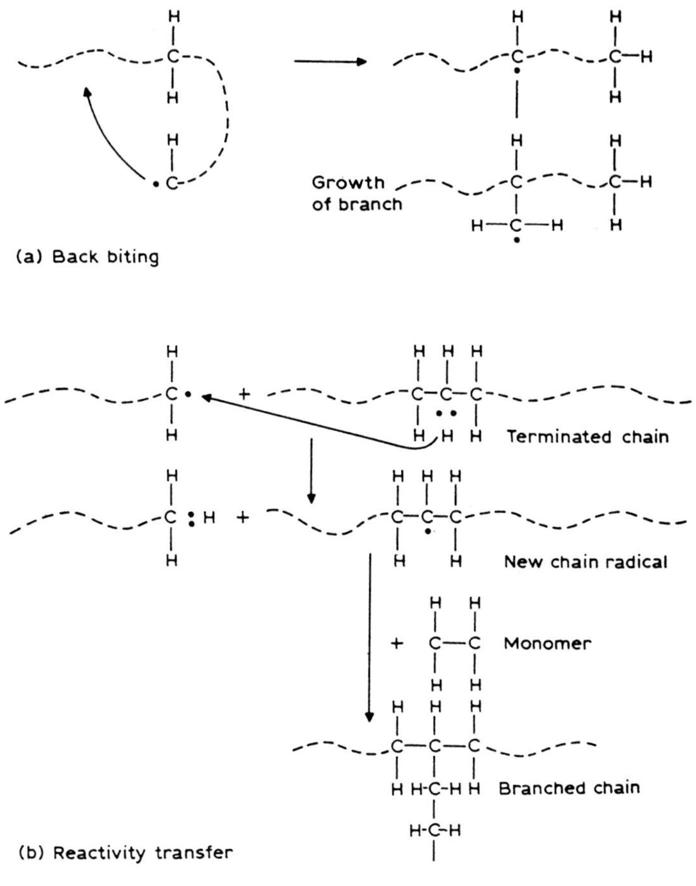

Growth of branch

(a) Back biting

Terminated chain

New chain radical

Monomer

Branched chain

(b) Reactivity transfer

Fig. 2.4 Chain branching in polyethelene.

If a modifier is used and all other transfer reactions are kept to a minimum it can be shown that:

$$\frac{1}{\bar{x}_n} = \left(\frac{1}{X_{no}} \right) + C_{mod} \frac{[MOD]}{[M]} \qquad (2.20)$$

where $\bar{x}_n$ is the number average degree of polymerisation in the absence of modifier. Equation (2.20) is shown graphically in Fig. 2.5. If the chain transfer constant C_{mod} is greater than unity then the modifier is used up too quickly in the reaction and is not available at the later stages of conversion to control the molar mass. Values less than unity mean that consumption is slow

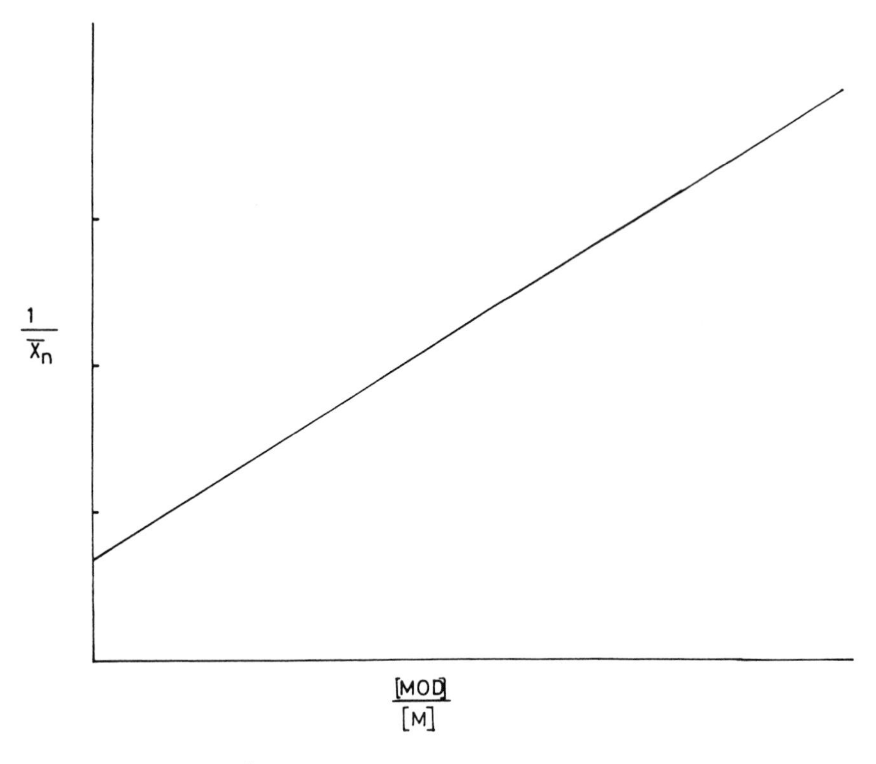

Fig. 2.5 Plot of $1/\bar{x}_n$ versus modifier concentration, [*MOD*].

and that residual modifiers will be left at the end of the reaction. If $C = 1$ the ratio of modifier to monomer remain constant throughout the reaction. This is the ideal situation but is difficult to achieve in practice. Some values of C for different monomers are shown in Table 2.2. Mercaptans are often added to polymerising systems to keep molar mass low, as in the polymerisation of diene rubbers.

Table 2.2 Some chain transfer constants C for vinyl monomers and transfer agents.

	CHAIN TRANSFER CONSTANT, $C \times 10^{-4}$		
TRANSFER AGENT	STYRENE	METHYL METHACRYLATE	VINYL ACETATE
Benzene	0.018	0.075	3.0
Toluene	0.125	0.525	21.0
Carbon tetrachloride	92	2.4	
1,1,1-tetrachlorethane		0.6	
n-butyl mercaptan	220000	6700	4800

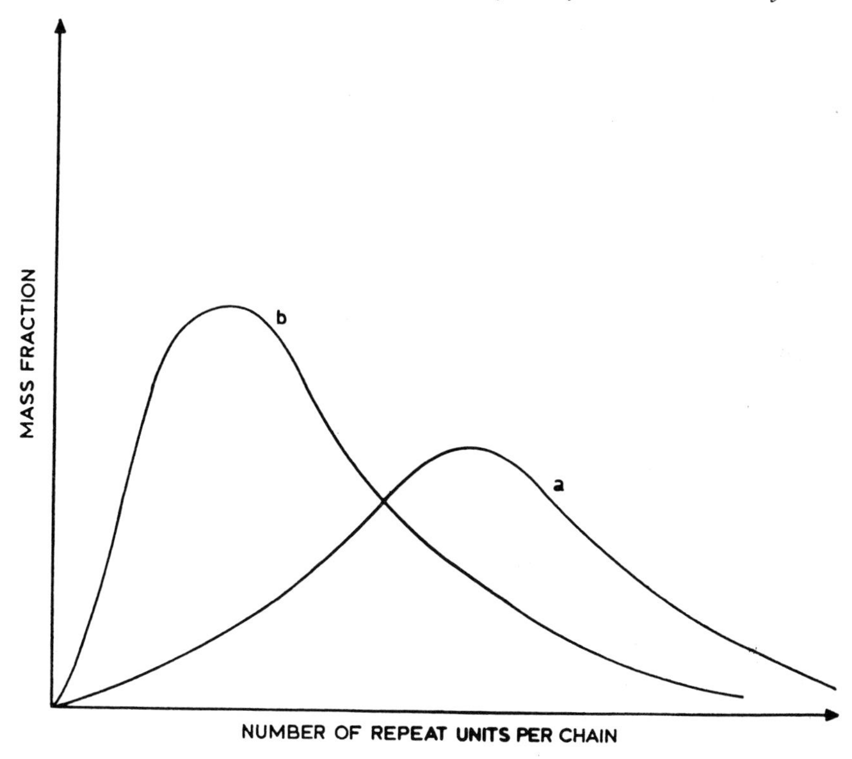

Fig. 2.6 Distribution of chain lengths for different termination mechanisms (a) combination and (b) disproportionation.

DISTRIBUTION OF CHAIN LENGTHS

Growing chains in a polymerisation reactor are terminated principally by combination, disproportionation or chain transfer reactions. It is possible to predict the distribution of chain lengths produced at any instant during the reaction. Termination by combination yields a different distribution from termination by disproportionation or chain transfer. Consideration of the probabilities that chains will grow or terminate under different conditions can lead to quantitative conclusions. Two different cases are illustrated in Fig. 2.6. It can be seen that if termination is by combination alone the distribution of chain lengths is relatively sharp as the probability of two small or two long chains combining is, in fact, quite low. Broader distributions are obtained if disproportionation and/or chain transfer mechanisms operate.

It is reasonably straightforward to calculate instantaneous quantities but during the course of a polymerisation conditions such as the concentration of modifier change, and so the instantaneously produced chain lengths vary as time passes. As a consequence it is the cumulative distribution of chain

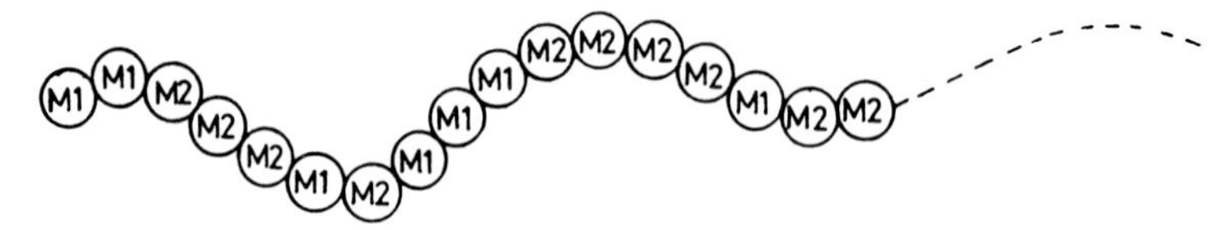

Fig. 2.7 Random copolymer consisting of M_1 and M_2 monomer units.

lengths that is important. This concept also applies to the rate of polymerisation because [M] and [I] usually change during the course of reactions. It is possible to predict cumulative quantities as a function of conversion, and knowledge of this type is vital for proper control of polymerisation reactors. The reader is referred to more advanced works on polymer synthesis for further information in this area.

COPOLYMERISATION

If two different monomers M_1 and M_2 are polymerised together in a free radical process the result may be a random mixture of repeat units in the resulting polymer chains, as shown in Fig. 2.7. However, this is not always the case.

In addition a particular ratio of monomers in the starting mixture does not necessarily result in the same ratio of monomers in the copolymer produced. These effects will be investigated here.

Copolymers produced in this way commercially are usually more or less random and are very important where it is necessary to tailor properties to engineering requirements. For instance, conventional discs for record players are made from essentially random copolymers of vinyl chloride and vinyl acetate in ratio of approximately 3:1. Another example is styrene butadiene rubber (SBR) used in tyre construction. In this case the styrene–butadiene ratio is about 1:6 and results in a rubber with the correct resilience and other properties.-

The Copolymer Equation

When two monomers are polymerised we can assume that the nature of the reactivity of a growing chain depends only on the active end unit on the chain. There are four possible ways in which monomer units can be added to a growing chain:

REACTION	*RATE*	
$RM_1\cdot + M_1 \rightarrow RM_1\cdot$	$K_{11}\,[RM_1\cdot]\,[M_1]$	(2.21)
$RM_1\cdot + M_2 \rightarrow RM_2\cdot$	$K_{12}\,[RM_1\cdot]\,[M_2]$	(2.22)
$RM_2\cdot + M_1 \rightarrow RM_1\cdot$	$K_{21}\,[RM_2\cdot]\,[M_1]$	(2.23)
$RM_2\cdot + M_2 \rightarrow RM_2\cdot$	$K_{22}\,[RM_2\cdot]\,[M_2]$	(2.24)

Assuming that the concentrations of $RM_1{\boldsymbol\cdot}$ and $RM_2{\boldsymbol\cdot}$ remain constant, the rate of conversion of $RM_1{\boldsymbol\cdot}$ to $RM_2{\boldsymbol\cdot}$ must be equal to the rate of conversion of $RM_2{\boldsymbol\cdot}$ to $RM_1{\boldsymbol\cdot}$, i.e.

$$K_{21}\,[RM_2{\boldsymbol\cdot}]\,[M_1] = K_{12}\,[RM_1{\boldsymbol\cdot}]\,[M_2] \qquad (2.25)$$

The rates of disappearance of the two monomer species are given by:

$$-\frac{d[M_1]}{dt} = K_{11}\,[RM_1{\boldsymbol\cdot}]\,[M_1] + K_{21}\,[RM_2{\boldsymbol\cdot}]\,[M_1] \qquad (2.26)$$

$$-\frac{d[M_2]}{dt} = K_{12}\,[RM_1{\boldsymbol\cdot}]\,[M_2] + K_{22}\,[RM_2{\boldsymbol\cdot}]\,[M_2] \qquad (2.27)$$

Reactivity ratios r_1 and r_2 can be defined in terms of the appropriate rate constants, K_{11}, K_{12}, K_{21} and K_{22}:

$$r_1 = \frac{K_{11}}{K_{12}} \quad \text{and} \quad r_2 = \frac{K_{22}}{K_{21}}$$

Combining equations (2.25), (2.26) and (2.27) the COPOLYMER EQUATION is obtained:

$$\frac{d[M_1]}{d[M_2]} = \frac{[M_1]}{[M_2]}\,\frac{r_1\,[M_1] + [M\ ]}{[M_1] + r_2\,[M_2]} \qquad (2.28)$$

This relates the ratio of the two monomer units in the increment of copolymer $\frac{d[M_1]}{d[M_2]}$ formed at an instant when the ratio of unreacted monomers is $\frac{[M_1]}{[M_2]}$.

The monomer reactivity ratios r_1 and r_2 express the preference of an active chain radical adding one type of monomer unit over the other. If $r>1$ then $RM_1{\boldsymbol\cdot}$ prefers to add M_1. If $r<_1$ then $RM_1{\boldsymbol\cdot}$ prefers to add M_2 and similarly for r_2. Reactivity ratios have been determined experimentally for a large number of cases. Typical values are given in Table 2.3.

Table 2.3 A selection of monomer reactivity ratios.

M_1	M_2	r_1	r_2
Styrene	Butadiene	0.78	1.39
Styrene	Vinyl Acetate	55	0.01
Styrene	Methyl Methacrylate	0.52	0.46
Methyl Methacrylate	Vinyl Chloride	12	~0
Methyl Methacrylate	Acrylonitrile	1.2	0.15
Methyl Methacrylate	Vinyl Acetate	20	0.015
Maleic Anhydride	Isopropenyl Acetate	0.002	0.032

From a practical point of view it is more convenient to re-express the copolymer equation in terms of f_1, the mole faction of M_1 in the monomer mixture and F_1, the mole faction of M_1 in the polymer formed from the monomer mixture at a particular instant.

$$F_1 = 1 - F_2 = \frac{d[M_1]}{d([M_1] + [M_2])} \tag{2.29}$$

$$f_1 = 1 - f_2 = \frac{[M_1]}{([M_1] + [M_2])} \tag{2.30}$$

If equations (2.28), (2.29) and (2.30) are combined a more useful form of the copolymer equation is produced:

$$F_1 = \frac{r_1 f_1 + f_1 f_2}{r_1 f_1^2 + 2 r_1 r_2 + r_2 f_2^2} \tag{2.31}$$

Consequently the instantaneous composition of copolymer formed at any

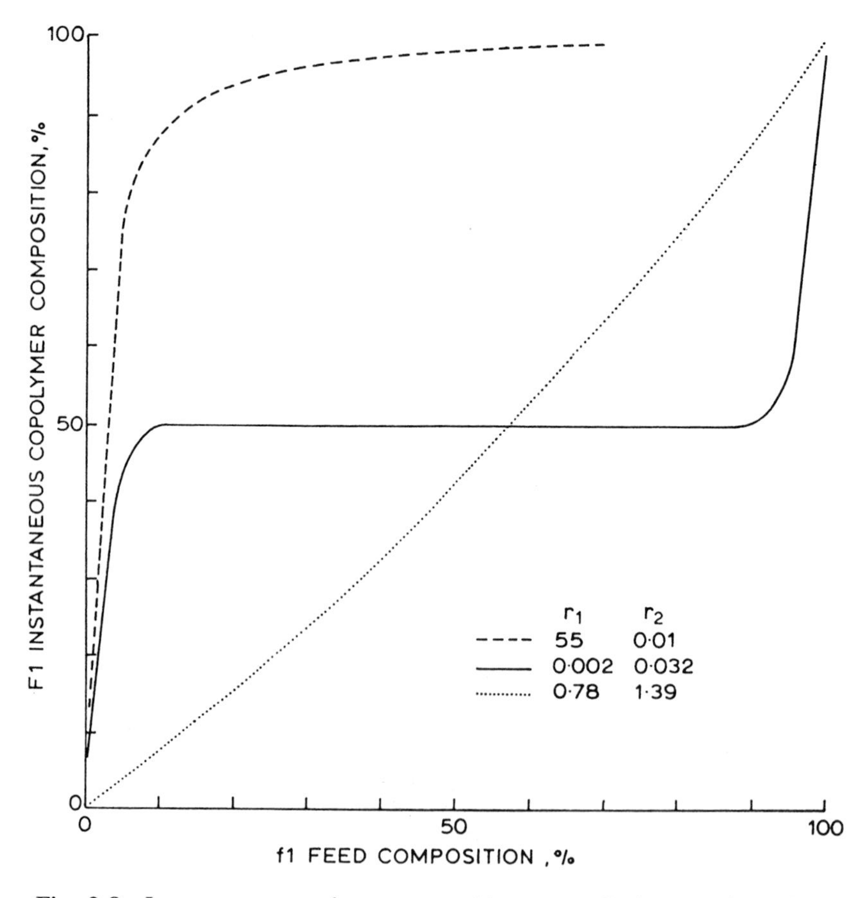

Fig. 2.8 Instantanous copolymer composition versus feed composition for different monomer reactivity ratios.

instant can be predicted from f_1, f_2, r_1 and r_2. This is illustrated graphically in Fig. 2.8 for some real cases of copolymerisation.

The reactivity ratios determine the type of copolymer formed, not just its composition, and hence the final properties. Some important cases and their practical results are shown in Tabe 2.4.

Table 2.4 Relationships between reactivity ratios and copolymer formed.

Reactivity Relationship	Polymer Formed
$r_1 = r_2 = 0$	Perfectly *alternating* copolymer, $f_1 = 0.5$
$r_1 = r_2 = \infty$	Mixture of homopolymers 1 and 2
$r_1 = r_2 = 1$	Completely random copolymer. Chain radicals cannot distinguish between the two monomers and so $f_1 = f_2$
$r_1 = \frac{1}{r_2}$	Both growing chains radicals show equal affinity for M_1 and M_2. IDEAL copolymerisation.

A final observation is of considerable importance. It will be obvious to the observant reader that, in many cases, as a polymerisation progesses the composition of the monomer mixture f_1, changes and so the copolymer composition, F_1, forming from it also changes. Thus a range of copolymer compositions are produced and at the end of the reaction the cumulative

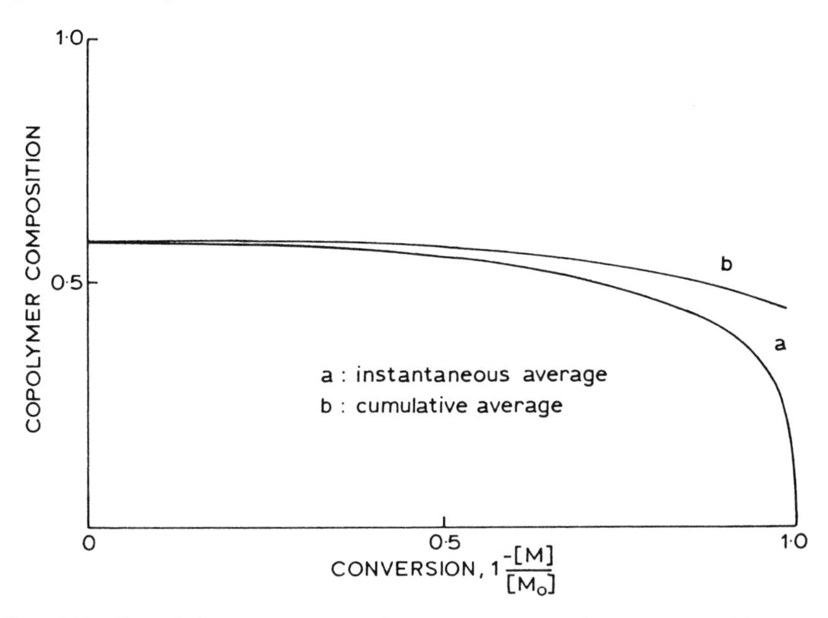

Fig. 2.9 Cumulative average and instantaneous copolymer composition versus conversion for styrene–butadiene. ($r_1 = 1.39$, $r_2 = 0.78$). Concentration of monomers at start is 1:1 (molar ratio).

composition is of considerable interest. Of course the final copolymer will consist of copolymer molecules with considerably different compositions. An exception for this situation is where r_1 and r_2 are close to zero (*see* Fig. 2.8) for maleic anhydride and isopropenyl acetate. However the cumulative copolymer composition can always be predicted and a general example of such a prediction is given in Fig. 2.9 for stryene and butadiene.

NON-RADICAL OR IONIC POLYMERISATION

Not only free radicals are able to initiate polymerisation. Ionic sites are also very useful active centres. The ionic sites can be cationic (positive) or anionic (negative) but, unlike free radical chain addition, it is difficult to generalise about reaction mechanisms. Different initiators, monomers and solvents can all affect the detailed reaction mechanisms. Ionic polymerisation has been developed for several reasons. For instance, some monomers containing double bonds can, somewhat surprisingly, only be polymerised by ionic mechanisms. Also, ionic polymerisations occur rapidly at low temperatures and very good control over stereoregularity and molar mass and its distribution can be obtained. Block copolymers and novel polymers can be produced easily and economically.

Cationic Polymerisation

This may be initiated by strong Lewis acids (electron acceptors) such as BF_3 or $AlCl_3$. A co-catalyst is required and this enables a proton (H+) to be donated to a monomer molecule. The H^+ is the initiator:

$$\begin{array}{ccccccccc}
\mathrm{H} & & \mathrm{Cl} & & \mathrm{Cl} & \mathrm{H} & & \mathrm{Cl} & \\
\ddot{\mathrm{O}}: & + & \ddot{\mathrm{Al}} : \mathrm{Cl} & \rightarrow & \mathrm{Cl} : \ddot{\mathrm{Al}} : & \ddot{\mathrm{O}} : & \rightarrow & \mathrm{H^+[Cl} : \ddot{\mathrm{Al}} \mathrm{:O:H).]^-} & \quad (2.32) \\
\mathrm{H} & & \mathrm{Cl} & & \mathrm{Cl} & \mathrm{H} & & \mathrm{Cl} &
\end{array}$$

CO-CATAYLST

$$\begin{array}{ccccc}
 & \mathrm{H} \quad \mathrm{CH_3} & & \mathrm{H} \quad \mathrm{CH_3} & \\
 & | \qquad | & & & \\
\mathrm{H^+[HO\!-\!AlCl_3]^-} + & \mathrm{C} :: \mathrm{C} & \rightarrow & [\,\mathrm{H} : \ddot{\mathrm{C}} : \mathrm{C}]^+ & + \mathrm{[HO\!-\!AlCl_3]^-} \quad (2.33) \\
 & | \qquad | & & & \\
 & \mathrm{H} \quad \mathrm{CH_3} & & \mathrm{H} \quad \mathrm{CH_3} & \\
 & \text{ISOBUTYLENE} & & \text{CARBONIUM ION} & \text{COUNTER OR GEGEN ION}
\end{array}$$

The above case shows the scheme for the polymerisation of isobutylene to butyl rubber which is the major industrial application of cationic polymerisation. It is interesting to note that the counter or gegen ion is held near the end of the growing chain and has a controlling effect on the way the successive monomer units are added. Propagation is very rapid and, in the case of isobutylene, occurs in a few seconds at −100°C to yield a molar mass of several million. Monomer units are added in a head to tail fashion.

Termination of growing chains may occur in two ways. A disproportionation type mechanism may occur which completely regenerates the catalyst. Alternatively chain transfer can occur to monomer, polymer, solvent or impurity molecules.

The kinetics of cationic polymerisation are not well understood but may be treated in a broadly similar way to free radical chain addition as previously described. The rate of polymerisation r_p for termination by disproportionation is given by:

$$r_p = \frac{K_i K_p \, [C]\,[M]^2}{K_t} \tag{2.34}$$

and for termination by chain transfer to monomer:

$$r_p = \frac{K_i K_p \, [C]\,[M]}{K_{tr}} \tag{2.35}$$

where $[C]$ is the catalyst (initiator) concentration. As termination by combination does not occur the degree of polymerisation $\bar{x}_n$ is the same as the kinetic chain length:

$$\bar{x}_n = \frac{\text{rate of polymerisation}}{\text{rate of initiation}} \tag{2.36}$$

$$\text{rate of initiation, } r_i = k_i\,[C]\,[M] \tag{2.37}$$

$$\bar{x}_n = \frac{K_p \, [M]}{K_t} \quad \text{for disproportionation} \tag{2.38}$$

$$\text{and } \bar{x}_n = \frac{K_p}{K_{tr}} \quad \text{for termination by chain transfer to monomer} \tag{2.39}$$

Comparing this with equation (2.14) and its implications it can be seen that, unlike for free radical polymerisation, $\bar{x}_n$ in the cationic case is independent of initiator concentration. Only in the case of termination by disproportionation to monomer is $\bar{x}_n$ dependent on $[M]$. One interesting observation is that in some cationic polymerisations $\bar{x}_n$ and r_p increase as the temperature is reduced. This may be explained by referring to the analogous situation for free radical polymerisation given by equations (2.16) and (2.17). The main difference is that the activation energy for cationic propagation, E_p, is much lower than that for the free radical case. This can make the overall activation energy term positive and the associated rate will then increase as the temperature is reduced. Similar reasoning can be applied to $\bar{x}_n$.

Anionic Polymerisation

In recent years anionic polymerisation has become extremely important because it enables the polymer chemist to control molecular structure in precise ways not before possible. Sometimes it is possible to anionically polymerise monomers containing double bonds which cannot be polymerised by the free radical route.

Early work in Germany before World War II led to the polymerisation of butadiene with sodium and lithium metals. Later this work resulted in the production of cis 1.4 polybutadiene and polyisoprene in the USA. Cis 1.4 polyisoprene is the synthetic counterpart of natural rubber and requires that close steric control is exercised during its polymerisation.

The anionic mechanism is illustrated by the polymerisation of styrene by sodium. As in the cationic case a counter or gegen ion is associated with the growing chains and brings about close steric control:

$$Na + H_2C = \underset{\underset{C_6H_5}{|}}{CH} \rightarrow H_2\overset{*}{C} - \underset{\underset{C_6H_5}{|}}{CH^-}\ Na^+ \quad \text{one electron transfer} \tag{2.40}$$

$$2\,H_2C - \underset{\underset{C_6H_5}{|}}{\bar{C}H}\ Na^+ \rightarrow Na^+ - \underset{\underset{C_6H_5}{|}}{\bar{C}H} - CH_2 - CH_2 - \underset{\underset{C_6H_5}{|}}{\bar{C}H}\ Na^+ \quad \text{DIANION} \tag{2.41}$$

Propagation occurs from both ends of the dianion. In certain cases and where impurities are absent no termination step occurs.

Living Polymers and Block Copolymers

In several important cases of anionic polymerisation the rate of chain initiation, r_i, is similar to the rate of propagation, r_t. Thus chains are initiated rapidly and grow until all the monomer is used up in those cases where there is no termination step. If more monomer is added then the polymerisation can be restarted. Hence these polymers are referred to as 'living' polymers. This property is important as it is possible to produce polymers with special

Fig. 2.10 AB and $(AB)_n$ copolymers.

end groups and, in particular it is possible to produce block copolymers. Block copolymers consist of polymer molecules built up from blocks of different repeat units. e.g. AB, ABA and ABABA, repeating block sequences may, for instance, be made up from two repeat units as shown below:

It is possible to incorporate more than two monomers and hence more than two types of blocks. A scheme for producting poly(styrene–butadiene–styrene), i.e. SBS block copolymer is shown below using butyllithium as initiator:

$$\overset{\delta-}{Bu} - \overset{\delta+}{Li} + n\,H_2C = \underset{C_6H_5}{\underset{|}{CH}} \rightarrow BU \left[CH_2.\underset{C_6H_5}{\underset{|}{CH}} \right]_{n-1} CH_2.\underset{C_6H_5}{\underset{|}{\bar{C}H}}\ Li^+$$

$$m\,H_2C = CH.CH = CH_2 \text{ butadiene}$$

$$Bu \left[CH_2.\underset{C_6H_5}{\underset{|}{CH}} \right]_n \left[CH_2.CH = CH.CH_2 \right]_{m-1} CH_2.CH = \bar{C}H_2\ Li^+$$

$$\downarrow n\,H_2C = \underset{C_6H_5}{\underset{|}{CH}} \text{ styrene}$$

$$Bu \left[CH_2\ \underset{C_6H_5}{\underset{|}{CH}} \right] \left[CH_2.CH = CH.CH_2 \right]_m \left[CH_2.\underset{|}{CH} \right]_{n-1} CH_2.\underset{C_6H_5}{\underset{|}{\bar{C}H}}\ Li^+ \qquad (2.42)$$

$$\downarrow CH_3OH \text{ termination}$$

This material is a thermoplastic rubber forming a two phase material of hard polystyrene regions in a flexible polyutadiene rubber.

Kinetics of Anionic Polymerisation

The kinetics of anionic polymerisation are complicated by the presence of some solvents but in general the rate of propagation, r_p, is given by:

$$r_p = K_p [M] [\text{Bu}\ \bar{M}_i\ \text{Li}^+] \qquad (2.43)$$

for polymerisations initiated by butyllithium as in equation (2.42).

If there is no termination then the number of living chains is equal to the number of initiator molecules present, therefore:

$$r_p = K_p [M] [I] \qquad (2.44)$$

The chain length increases with conversion. If $r_i = r_p$ then all chains start growing at the same time and to approximately the same length. The degree of polymerisation, $\bar{x}_n$, is given by:

$$\bar{x}_n = \frac{[M_p]}{[I]} \text{ (constant volume)} \qquad (2.45)$$

where M_p is the amount of monomer polymerised and I the initiator concentration.

Consequently anionic polymerisation provides the only way of producing monodisperse homopolymers and copolymers containing blocks of uniform size.

Ziegler Natta and Supported Catalysts

Although only brief mention will be made here of this topic the importance of these catalysts cannot be overemphasized. Their use enables, for instance, the production of isotactic (and therefore useful) polypropylene and substantially linear polyethylenes.

Ziegler and Natta found that, for example, combinations of metal alkyls with metal halides act as catalysts which can produce stereoregular polymers. A typical reaction is as follows:

i) A solid catalyst is formed:

$TiCl_3 + Al(C_2H_5)_3 \rightarrow$ Complex crystalline precipitate in deep violet or brown liquor

ii) The monomer is chemically absorbed on the surface of the precipitate and is oriented for stereoregular addition.

iii) The polymer chain forms progressively by a complex reaction sequence involving an anionic type mechanism. Stereospecific polymers such as isotactic polypropylene are produced.

There are considerable problems with preparation, handling and storage of Ziegler Natta catalysts. Polymers with a polydispersity of about 20 are produced although the reasons for this are obscure.

Another type of stereospecific catalyst is CrO, supported on alumina or silica. These were developed by Phillips Petroleum, and are often referred to as Phillips catalysts. These catalysts are particularly useful in producing linear polyethylene. They can also be used to produce high molar mass and stereoregular polymers.

STEP GROWTH (CONDENSATION) POLYMERISATION

This type of polymerisation takes place in clearly identifiable steps and hence should be referred to as a step growth reaction. However, it can also be viewed as a reaction in which the polymer product condenses out from the reactants. It is thus more usually referred to as condensation polymerisation. Condensation polymerisations can be represented as follows:

$$n_i \text{ mers} + n_j \text{ mers} \rightarrow (\ (n_i + n_j)\ ) \text{ mers} \qquad (2.46)$$

n_i and n_j may be any value from 1 upwards. The participating units must be end-capped with functional (reactive) groups which will inter-react. Unlike addition reactions, long chain molecules can react with one another. In the addition case monomer units are added one at a time. In condensation reactions molecules increase in size progressively, whereas in addition reactions large molecules are present almost from the start of the polymerisation. The starting materials in condensation polymerisations are monomer units and these monomers can be identical or dissimilar. The repeat unit in the resultant polymer may be identified as the starting monomer, where the starting monomers are identical, or a combination of two monomer units where the starting monomers are dissimilar.

Functionality

A typical condensation reaction is that between an acid chloride and an amine to produce a polyamide (nylon):

Difunctional acid chloride — Amine — Catalyst NaOH — Repeat Unit

$$\underset{\text{Monomer 1}}{\text{Cl}\overset{O}{\overset{||}{C}}(CH_2)_4\overset{O}{\overset{||}{C}}\text{Cl}} + \underset{\text{Monomer 2}}{H_2N(CH_2)_6NH_2} \rightarrow \underset{\text{Dimer}}{\text{Cl}\overset{O}{\overset{||}{C}}(CH_2)_4\overset{O}{\overset{||}{C}}NH(CH_2)_6NH_2}$$

(in the printed formulae the =O groups are drawn below each C)

$$\text{Dimer} + \text{Monomer (1 or 2)} \xrightarrow{\text{NaOH}} \text{Trimer}$$

$$\text{Dimer} + \text{Dimer} \xrightarrow{\text{NaOH}} \text{Tetramer etc}$$

In general:

$$\underset{\text{Monomer 1}}{n\,\text{Cl}\overset{O}{\overset{||}{C}}(CH_2)_4\overset{O}{\overset{||}{C}}\text{Cl}} + \underset{\text{Monomer 2}}{nH_2N(CH_2)_6NH_2} \xrightarrow{2\,\text{NaOH}} \underset{\text{Polymer}}{\left[\overset{O}{\overset{||}{C}}(CH_2)_4\overset{O}{\overset{||}{C}}NH((CH_2)_6NH\right]_n}_{\text{Repeat Unit}} + 2n\text{NaCl} + nH_2O \qquad (2.46)$$

Note the stepwise build up of chain length.

It is clear that in this case the reacting monomer units must be capable of reacting at two sites. Thus to produce a linear polymer a functionality of 2 is required. If the functionality of at least one of the monomers is greater than 3

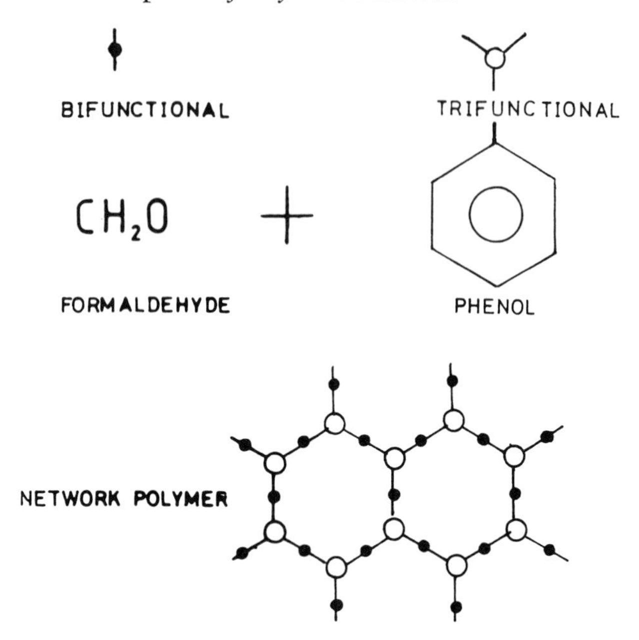

Fig. 2.11 Formation of a network polymer from a trifunctional monomer.

then it is possible under ideal conditions to produce a network polymer, e.g. phenol formaldehyde (Bakelite) (*see* Fig. 2.11).

In most condensation reactions a by-product (or products) is formed as a result of reactions between the functional groups. This product is a small molecule, often water.

Carothers Equation

Consider the production of a linear polymer. For instance a linear polyester can be produced by condensation of a hydroxy acid. Only one monomer is involved and the reaction takes place by reactions between the OH groups:

Hydroxy Acid

$$n\ \underset{\text{Monomer}}{\text{HO–R–}\overset{\overset{\displaystyle O}{||}}{\text{C}}\text{–OH}} \rightarrow \underset{\text{Dimer}}{\text{HO–R–}\overset{\overset{\displaystyle O}{||}}{\text{C}}\text{–O–R–}\overset{\overset{\displaystyle O}{||}}{\text{C}}\text{–OH}} + H_2O$$

$$\downarrow$$

$$\underset{\text{Polymer}}{\left[\text{R–}\overset{\overset{\displaystyle O}{||}}{\text{C}}\text{–O}\right]_n}$$

As the number of molecules decreases the average chain length increases. If N_o molecules were present at time = 0 and N remain at time = t then the number of functional OH groups reacted at time t is $N_o - N$. The extent of reaction p, is expressed as:

$$p = \frac{N_o - N}{N} \qquad (2.47)$$

Therefore:

$$\frac{N_o}{N} = \frac{1}{(1 - p)} \qquad (2.48)$$

The number average degree of polymerisation, $\bar{x}_n$, is given by:

$$\bar{x}_n = \frac{N_o}{N} \qquad (2.49)$$

because N_o monomer molecules are distributed between N molecules. Thus:

$$\bar{x}_n = \frac{1}{(1 - p)} \quad \text{[Carothers equation]} \qquad (2.50)$$

$\bar{x}_n$ values of at least 100 are required in order to obtain useful mechanical properties. Consequently, p must be at least 0.99 as an be seen in Table 2.5. This is very difficult to achieve and requires near perfect stochiometry of reactants, i.e. the ratio r of the numbers of the two reacting monomer molecules must be 1:1.

Table 2.5

Extent of Reaction, p	0.50	0.90	0.95	0.98	0.99	0.999	1.0
No Av. Deg. of Polm., $\bar{x}_n$	2	10	20	50	100	1000	∞

If r is not exactly 1:1 then a modified Carothers equation must be used as follows:

$$\bar{x}_n = \frac{(1 + r)}{1 + r - 2rp} \qquad (2.51)$$

when $p = 1$

$$\bar{x}_n = \frac{1 + r}{1 - r} \qquad (2.52)$$

In practice careful control of the ratio of reactants is required. Where it is necessary to limit the molar mass 'quenching' with a slight excess of one monomer or adding a small amount of a monofunctional reagent may be required.

Distribution of Chain Lengths

The probabilities of finding molecules with finite chain lengths can be calculated. The number fraction distribution of chain length is given by:

$$\frac{N_x}{N_o} = p^{(x-1)}\,(1-p)^2 \qquad (2.53)$$

where x is the chain length.

The mass fraction distribution of chain lengths is given by:

$$W_x = x\,p^{(x-1)}\,(1-p)^2 \qquad (2.54)$$

where $W_x = \dfrac{\text{mass of molecules with length } i}{\text{total mass of molecules}}$

The molar mass distributions are readily calculated. Useful relationships are given below:

$$\overline{M}_n = \frac{M_o}{(1-p)} \qquad (2.55)$$

$$\overline{M}_w = \frac{M_o\,(1+p)}{(1-p)} \qquad (2.56)$$

where M_o is the molar mass of the monomer unit.

The polydispersity is expressed as:

$$\frac{\overline{M}_w}{N_n} = (1+p) \qquad (2.57)$$

This tends to a value of 2 when p is 1.

Gel Point Prediction

A gel is a cross-linked network and a polymer in this state is unworkable and insoluble. Network polymers can be thought as being one infinite molecule. They are formed when one of the monomers has a functionality greater than 2. A network will form at a critical extent of reaction. Clearly it is often important to halt a condensation reaction before gelling occurs. The gelling and network formation can be completed during subsequent processing. Statistical theory has shown that the critical extent of reaction for gelling, P_{gc}, is given by:

$$P_{gc} = [1 + \lambda\,(f-2)]^{-\frac{1}{2}} \qquad (2.58)$$

where f is the functionality of the branching monomer units and λ is a statistically derived coefficient. The ultimate gelled state is shown in Fig. 2.12.

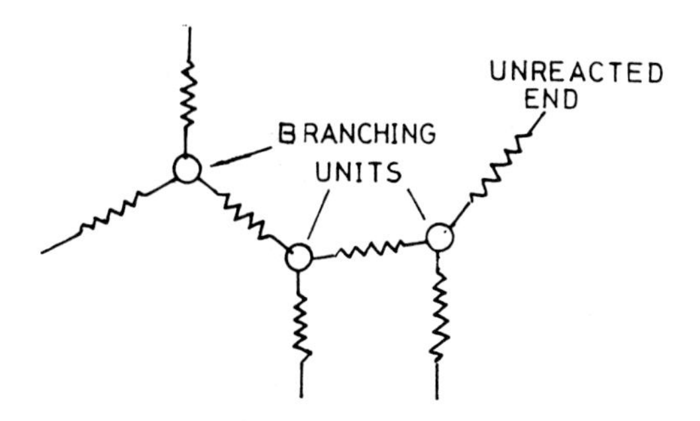

Fig. 2.12 The ultimate gelled state.

Kinetics of Condensation Polymerisations

As chain length is related to the extent of reaction which is time dependent it is possible to control $\bar{x}_n$ by stopping the reaction.

It may be assumed that the reactivity of functional units is independent of chain length. Catalysts are usually employed and the rate of reaction is, in many cases, proportional to the product of three concentration terms, i.e. functional group 1, functional group 2 and catalyst — C_1, C_2 and C_3.

This can be written as:

$$R = k\ C_1,\ C_2,\ C_3 \qquad (2.59)$$

where k is the fundamental rate constant

$$\text{OR}\quad R = k',\ C_1{}^2 \qquad (2.60)$$

where $C_1 = C_2$

and k' includes a term representing the catalyst concentration. Integration yields:

$$\text{Co}\ k'\ t = \frac{1}{(1-p)} - 1 \qquad (2.70)$$

where k' is assumed to remain constant.

This has been shown to be valid for many polycondensation reactions by plotting $\frac{1}{(1-p)}$ versus t when a linear plot is obtained.

PRACTICAL POLYMERISATION ROUTES

The purpose of a commercial polymerisation process is to produce polymer to an acceptable specification at as low a cost as possible. It is seldom necessary to produce polymer of a narrow molar mass range and so broad distributions of molar mass are normally produced. This makes the task considerably easier. However, good process control is necessary and, in particular, the product must be consistent in character. This is important as small differences in, for example rheological properties, are detectable by modern microprocessor controlled injection moulding machines.

Batch or Continuous Flow Processes

Polymerisation can be brought about by adding the necessary ingredients to a closed vessel and so bringing about the reaction. The polymer product is then removed after the necessary reaction time. This approach is useful for smaller

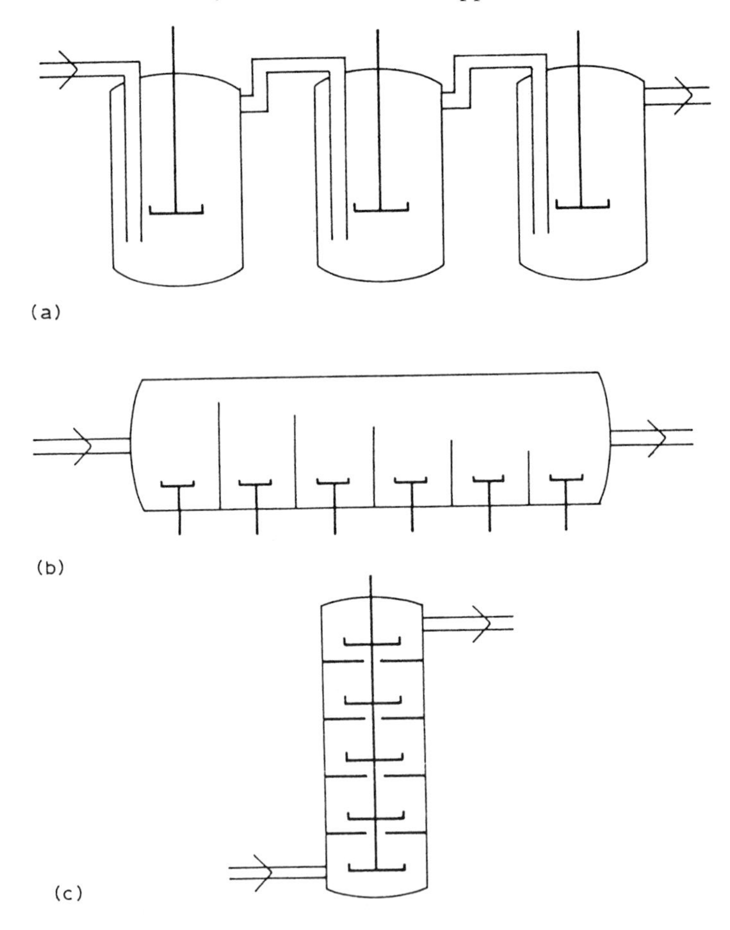

Fig. 2.13 Types of continuous flow reactor (a) continuous stirred tank reactors – CSTRs (b) overflow reactor (c) vertical mixer–settler reactor.

quantities of specialised polymers but is less suitable for bulk polymers. Variations from batch to batch may occur and time is lost in loading and unloading the reaction vessels. Thus the cost is relatively high. In addition, scale up is difficult as polymerisations are exothermic and large reactors become difficult to control. Stirring becomes difficult as the size of the reactor increases.

Continuous flow processes are preferred for bulk production. The reaction mixtures are prepared continuously and pass sequentially through a series of stirred reactors or other forms of continuous flow reactors. There are a very large number of continuous reactor types such as continuous stirred tank reactors (CSTRs), plug flow reactors and tubular reactors. Some examples are shown in Fig. 2.13. The reaction products are recovered continuously and unreacted monomer and solvents or water, where used, are extracted. Solvents and monomer are then recycled.

Classification of Polymerisation Processes

The monomer may be polymerised in bulk in the liquid, vapour or solid phase. However, in the case of liquid monomers there are a number of modifications which can be made to the way in which the process is carried out. The major problems of stirring and thermal control can be overcome by dissolving the monomer in a suitable solvent or by dispersing or emulsifying it in water. These strategies are widely used.

Some examples of commercial polymerisation processes are given in the following sections.

Bulk Polymerisation of Styrene

Styrene is polymerised in bulk but considerable problems have to be overcome to cope with the increasing viscosity during polymerisation. The 'tower process' uses stirred vessels to bring about a prepolymerisation, at 80°C, to approximately 35% conversion, see Fig. 2.14. This is the viscosity limit for practical stirring. Careful temperature control is necessary and a nitrogen atmosphere is used. Monomer is added continuously and the vessels discharge into the top of a tower. The partially polymerised mixture passes down the tower over a series of baffles. A temperature gradient is maintained by rigorous control pocedures. The temperature at the top is 100°C and at the bottom about 180°C. Typically a number average molar mass of 175 000 is achieved. Rubber toughened modifications can be made by adding SBR to the prepolymerisation reactors.

There are many examples of bulk polymerisations. For instance methyl methacrylate may be rapidly converted to solid polymer containing low levels of residual monomer using a peroxide initiator. The examples given here have been addition polymerisations but condensation polymerisations such as the production of nylons and polyurethanes can be carried out in bulk directly from the appropriate monomers or precursors.

Solution Polymerisation of Polypropylene

Dissolving of the monomer in a solvent enables good heat transfer and stirring at high conversions. As mentioned earlier complications arise in terms of transfer reactions between the solvent and polymer molecules but allowances can be made for these reactions in reactor control models. An interesting example is the polymerisation of propylene. The main problem is to produce the isotactic and hence crystallisable form and to do this a Natta catalyst must be used. The Natta catalyst is formed by the interaction of titanium trichloride with aluminium triethyl or aluminium diethyl monochloride under nitrogen. The catalyst is formed as a 10% slurry in naptha and this is further diluted with naptha to form a 0.5% dispersion for use in the CSTRs. These are stainless steel vessels which are electrically heated and stirred and connected in series. The mixture passes from one vessel to another as the polymerisation proceeds.

The propylene monomer and catalyst dispersion are metered and charged into the first reactor. The temperature is maintained at around 60°C and the pressure at 2–4 atm. Conversion to 80% polymer takes in the region of 8 hours after which the unreacted monomer is extracted in a flash tank. At this stage the polymer consists of approximately 92% isotactic polypropylene, the remainder being atactic polypropylene. The atactic form is in solution but the isotactic polypropylene is insoluble and so is precipitated out as beads as it forms. The isotactic polymer is separated from the 'gummy' atactic polymer by centrifuging. It is then washed in methanol and further centrifuged to remove the remaining catalyst which would discolour the polypropylene.

The polypropylene is then dried, compounded with stabilisers and other additives and then converted into pellets by extrusion and cutting. A weight average molar mass in the range 150,000–200,000 is typical.

Butyl rubbers and several other addition polymers, including ionically polymerised types, are polymerised in solution. Condensation polymerisation reactions may also be conveniently carried out in suitable solvent systems.

Suspension Polymerisation of PVC

Solvents are a problem in that they must be recovered and may represent a toxic and fire hazard. An alternative possibility is suspension polymerisation. The monomer is suspended as small droplets in water and the only significant problem is removing the water at the end of the process. There is no problem of separating unreacted monomer from the recovered solvent. A large amount of PVC is produced by suspension polymerisation in batch reactors. A typical charge composition is as follows:

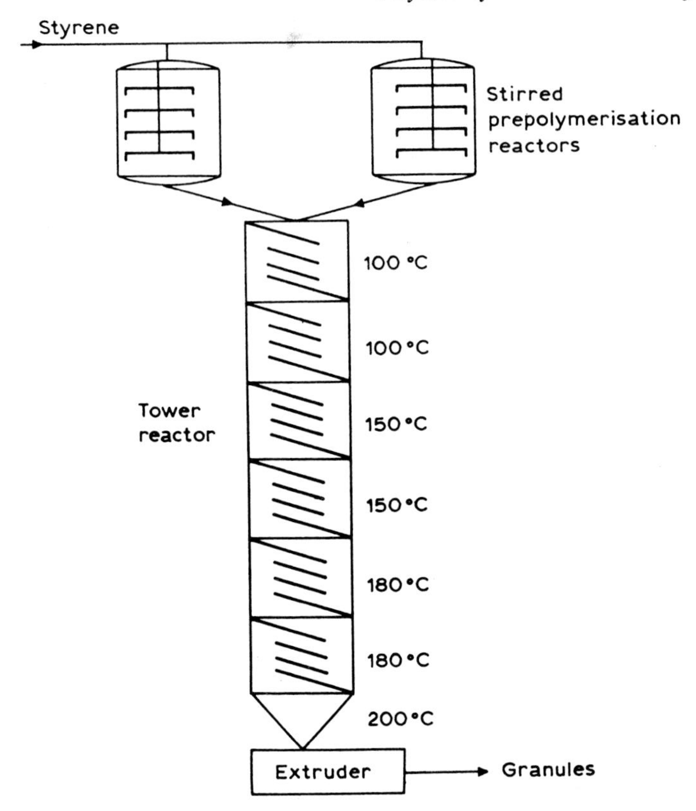

Fig. 2.14 Schematic diagram of the 'tower' process for the polymerisation of styrene.

	parts by mass
Monomer: vinyl chloride	30–50
Initiator: e.g. caprylol peroxide	0.001
Dispersing agent: e.g. gelatine	0.001
Modifier: trichloroethylene	0.1
Dispersion medium: demineralised water	90

The polymerisation is carried out under nitrogen at approximately 50°C in a batch reactor, see Fig. 2.15. This produces a vinyl chloride pressure of 7 atm. The reaction is allowed to run until the pressure drops to 2–3 atm corresponding to 85–90% conversion in about 12 hours. The polymer is insoluble in the monomer in this case and forms as irregularly shaped beads which are only slightly contaminated with dispersing agents. The ionic strength is an important variable in terms of maintaining the suspension. It can be controlled by adding calcium phosphate. Coalescence of the droplets is a problem in the early stages especially.

Each droplet of monomer is about 200 μm in diameter. The water acts as a

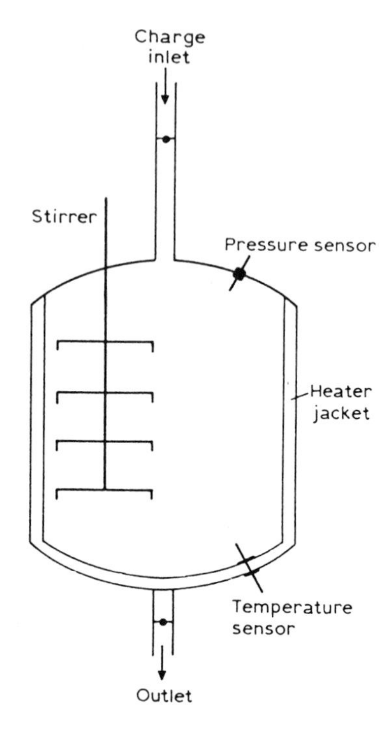

Fig. 2.15 Typical suspension reactor for PVC production.

heat transfer medium and allows easy stirring. The addition polymerisation proceeds within each particle as it would in a bulk process and so the kinetics are similar and the descriptions given in earlier sections for free radical addition processes are applicable. However, certain disadvantages of solvent and bulk processes are avoided. For bulk processes these are:

i) problems of heat transfer,
ii) increasing viscosity which limits stirring
and iii) limited polymerisation rate due to high viscosities reducing diffusion rates (gel effect).

For solvent processes the major problem is that of choosing an inert solvent which does not participate in the polymerisation by chain transfer reactions. This problem is not significant in ionic polymerisations and so industrial ionic polymerisations are often carried out in solution.

Emulsion Polymerisation

This process is at first glance similar to suspension polymerisation but the droplet size is in fact much smaller, in the range 0.05 to 5 μm. The initiator is soluble in the aqueous phase and so can diffuse through it. The monomer droplets are held in a stable emulsion by the use of surfactants, i.e. emulsifiers. This produces, after agitation, an emulsion rather like milk. The mechanism of polymerisation is complex and unlike that previously described.

When surfactant molecules, which are similar to soaps, are first added to water they move around and migrate to the surface and lower the surface tension. When more are added, above a certain concentration they begin to form aggregates or micelles in the bulk of the water. These surfactant molecules are for example the sodium or potassium salts of organic acids. They have hydrophilic heads and hydrophobic tails. In the micelles which are about 6 nm diameter, the tails are drawn together in the micelle centres. If an organic monomer is added it is attracted to the centre of the micelles and the micelles are swollen until an equilibrium is attained with the surface tension forces. However, most of the monomer resides in droplets about 1 μm in diameter. Once formed this system which is an emulsion is energetically stable and so does not suffer from the tendencies to coalesce as found in suspension polymerisation. Because the monomer swollen micelles are so small and numerous they exhibit a much larger surface area to volume ratio than do the droplets. In order to polymerise the monomer, a water soluble initiator is used and this migrates through the water and, because of their large target area, the free radicals generated by initiator are far more likely to enter the swollen micelles.

Suitable radicals can be produced by redox systems or persulphate initiators. A persulphate ion breaks down into suphate ion radicals which act as initiators:

$$\underset{\text{persulphate}}{S_2O_8^{-2}} = \underset{\text{sulphate ion (free radical)}}{2SO_4^{-}}$$

As soon as a radical enters a micelle it initiates polymerisation. As the monomer is consumed it is replaced by monomer diffusing from other uninitiated micelles, which shrink and disappear, and from the droplets. Thus the system, as shown in Fig. 2.16, soon stabilises at a fairly constant number of particles. The monomer concentration in the growing particles tends to reach equilibrium.

A monomer swollen micelle which has been targeted by one initiator molecule contains one growing chain. The chain cannot terminate until a second radical enters the particle when termination occurs immediately. This cycle can then begin again.

A stable number of monomer swollen polymer particles are present after the very early stages of polymerisation. At any time a particle contains either one growing chain or no growing chains. If there are N particles per unit volume there must be $N/2$ growing chains per unit volume. For free radical addition polymerisation the rate is given by equation (2.5):

$$r_p = K_p [M] [RM_i^*]$$

where $[RM_i^*]$ = the concentration of growing chains

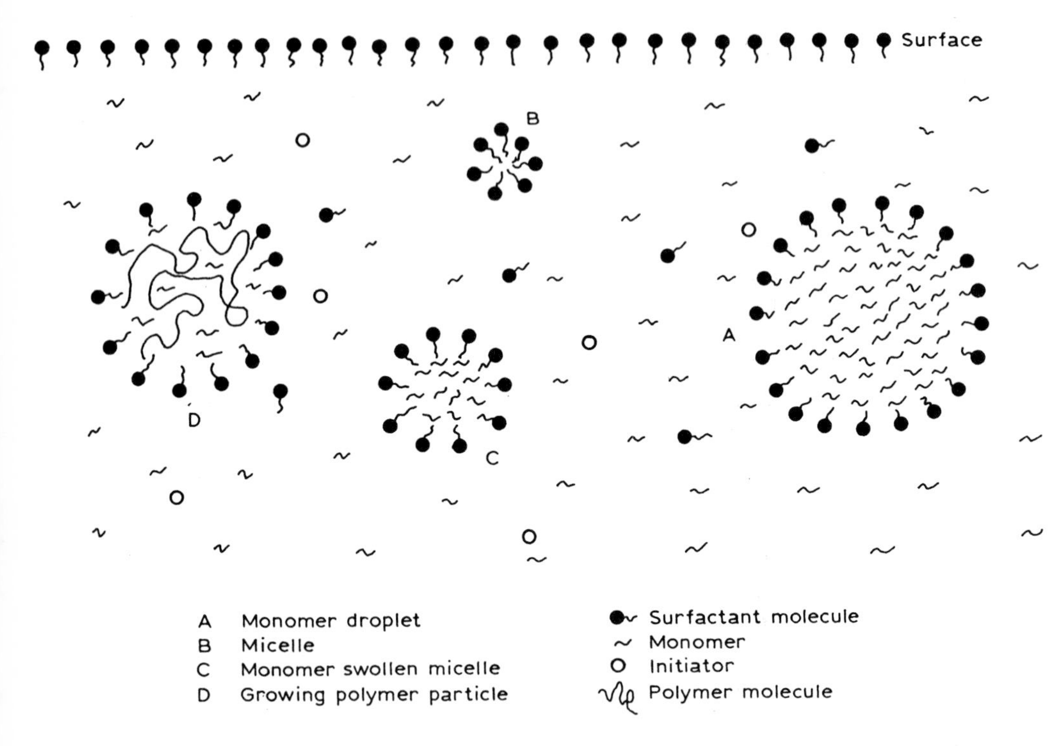

Fig. 2.16 An emulsion polymerisation system.

$[M]$ = the monomer concentration

and K_p = the rate constant.

But the concentration of growing chains is $N/2$ per unit volume. Thus

$$r_p = K_p [M] N/2 \tag{2.71}$$

Thus the rate is independent of initiator concentration (more or less), and, because N remains constant throughout most of the reaction, the rate is also constant. In general the rate can be increased by increasing the number of particles, N. N increases as the soap concentration increases. However, high molar mass polymer can be produced at these high rates unlike other free radical addition polymerisation processes.

Emulsion polymerisations can be carried out in CSTRs because the emulsion is completely stable. This contrasts with suspension polymerisation where batch reactors are used, because flow reactors invariably have stagnant regions where coalescence would occur.

Polymers such as PVC and styrene butadiene rubber copolymer (SBR) can be produced by emulsion polymerisation which is an important commercial process.

REFERENCES

J.M.G. COWIE: *Polymers: Chemistry and Physics of Modern Materials*, International Textbook Company Ltd, Aylesbury, (UK), 1973.

P.J. FLORY: *Principles of Polymer Chemistry*. Cornell University Press, Ithaca, NY, 1953.

M.P. STEVENS: *Polymer Chemistry – An Introduction*, Addison Wesley Inc., Reading, MA, 1975.

R.J. YOUNG: *Introduction to Polymers*, Chapter 2, Chapman Hall, London, 1983.

3 *Polymer Blends and Composites*

INTRODUCTION

Plastics materials are being used in place of metals and ceramics in many applications, but there are three main drawbacks associated with certain deficiencies in mechanical properties namely, low modulus, low strength and poor creep resistance.

The tensile moduli of homopolymers lie in the range 0.3–20 GN m^{-2}, with tensile strength values of between 12–140 MN m^{-2}. For steel and aluminium these values are 200 GN m^{-2} and 200 MN m^{-2} (steel) and 70 GNm^{-2} and 200 MN m^{-2} (aluminium).

This chapter, therefore, will deal with ways in which the mechanical properties of polymeric materials may be improved. This naturally leads to a greater versatility and diversity of application for polymer composites over the homopolymers.

HOMOPOLYMERS

The obvious way of improving the stiffness of homopolymers is to alter the molecular structure in order to make the molecules stiffer. Polyethylene, for instance, is not an engineering thermoplastic because the ease of rotation about the carbon bonds in the backbone chain leads to a soft but tough material. Clearly, if such rotation could be minimised a stiffer polymer would result.

Stiffening of polymer molecules leads to an increase in T_g, and for partly-crystalline materials an increase in T_m. This was discussed in Chapter 1.

Considerable stiffening is achieved by incorporating aromatic units in the backbone chain, but this can be carried too far. An example of this is shown in

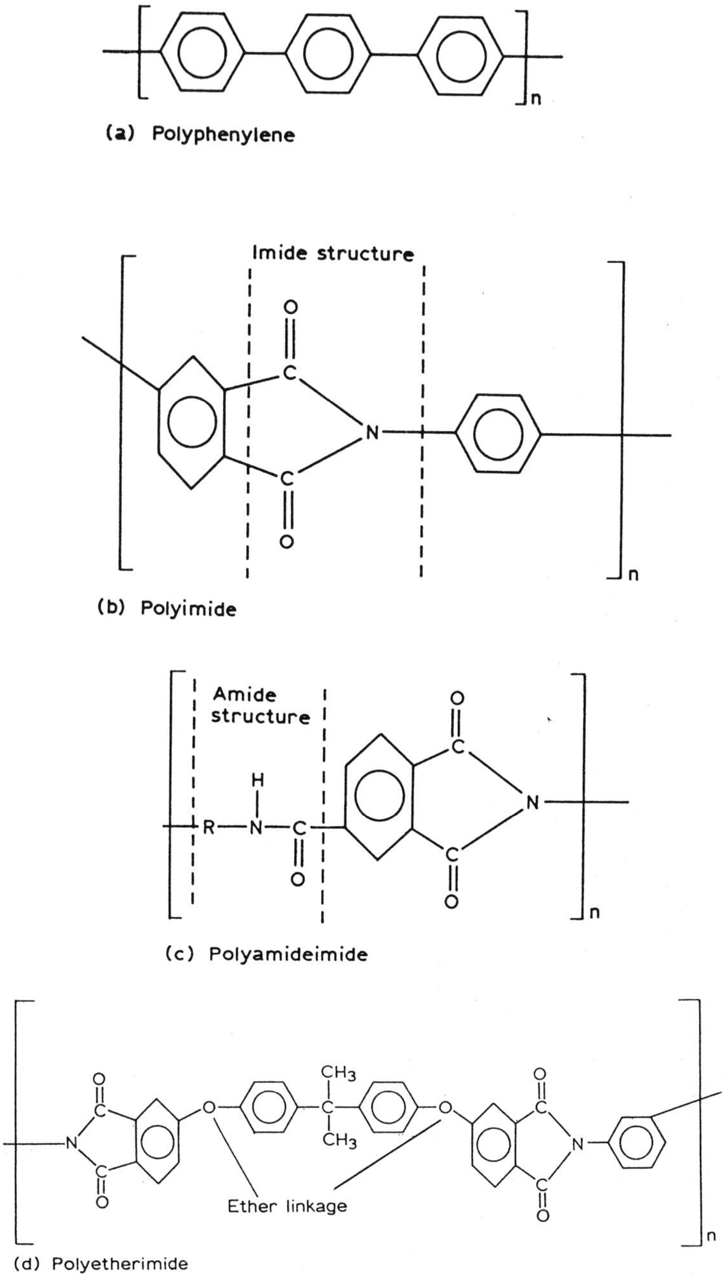

(a) Polyphenylene

Imide structure

(b) Polyimide

Amide structure

(c) Polyamideimide

Ether linkage

(d) Polyetherimide

Fig. 3.1 The repeat units in some high performance polymers.

Fig. 3.1a, which shows the structure of polyphenylene. The decomposition temperature, T_d, of this material is lower than T_m, and therefore it cannot be melt processed. More flexible bonds must, therefore, be placed between the aromatic rings in order to lower T_m.

A similar situation exists with polyimide, whose chemical structure is shown in Fig. 3.1b. This is a thermoplastic that cannot be melt processed because $T_m > T_d$. In order to obtain a melt processable polyimide the structure must be made more flexible, and this is achieved in several ways:

(1) By the incorporation of an amide structure as a part of the repeat unit as shown in Fig. 3.1c — polyamideimide.
(2) By the incorporation of ether linkages (–0–) in the repeat unit at various places — polyetherimide (Fig. 3.3d).

These latter materials are termed speciality thermoplastics and are used in high temperature engineering applications.

CRYSTALLINITY

Another way of improving stiffness and creep resistance is to manipulate the molecular architecture to induce or increase crystallinity, as described in Chapter 1.

An example of crystallinity improving high temperature stiffness is evidenced by polyethersulphone (PES) and polyetheretherketone (PEEK). PES is an amorphous polymer with T_g = 221°C, whereas the T_g of PEEK is 143°C. PEEK, however, is partly crystalline with a T_m of 343°C, and it can be used continuously at temperatures around 250°C. PES cannot be used at these high temperatures despite its having the higher T_g.

CROSSLINKING

Thermosets owe their high tensile modulus, tensile strength and creep resistance to their fixed three-dimensional structure caused by chemical crosslinking. Thermosets are polymers which when formed into the final product undergo a second chemical reaction in which a continuation of heat and pressure causes strong chemical bonds to link the polymer molecules. Once these crosslinks have formed the material will not flow again — a bit like a hard-boiled egg. Examples of these thermosetting materials are unsaturated polyesters, epoxies, phenol formaldehyde, melamine formaldehyde and urea formaldehyde. The crosslinking reaction in rubbers is called vulcanisation or curing. In rubbers the long chain molecules are usually linked by sulphur molecules. The degree of crosslinking determines the stiffness of the rubber.

In rubbers, double, unsaturated carbon bonds are present, as shown in Fig. 3.2. These are more reactive than the more prevalent single bonds and so make crosslinking easy, but the crosslinking occurs at the nearest CH single bond to the C=CH bond.

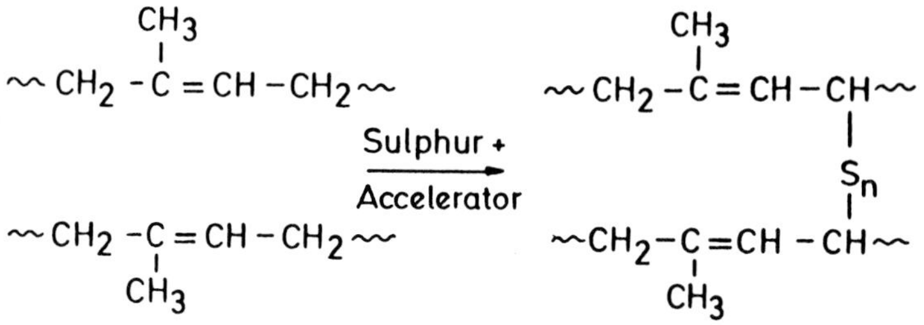

$$\begin{array}{c} \sim CH_2-\underset{}{\overset{CH_3}{\overset{|}{C}}}=CH-CH_2\sim \\ \\ \sim CH_2-\underset{CH_3}{\underset{|}{C}}=CH-CH_2\sim \end{array} \xrightarrow[\text{Accelerator}]{\text{Sulphur +}} \begin{array}{c} \sim CH_2-\overset{CH_3}{\overset{|}{C}}=CH-CH\sim \\ | \\ S_n \\ | \\ \sim CH_2-\underset{CH_3}{\underset{|}{C}}=CH-CH\sim \end{array}$$

Fig. 3.2 The crosslinking reaction in natural rubber.

Rubbers are rarely used in their pure state, but contain fillers and other additives that modify their mechanical properties. This often offers the opportunity of tailoring the mechanical properties to the product requirements. Carbon black is one of the major fillers in rubbers, which improves processability and increases stiffness and strength.

Natural rubber (NR) possesses strength and resilience and in cyclic loading situations there is little heat build up, because of its low hysterisis. Moreover, it bonds well to metal. For these reasons it is ideal for applications such as anti-vibration mountings, and heavy vehicle, aeroplane and racing car tyres.

Synthetic rubbers are copolymerised from smaller molecular units, such as styrene and butadiene (SBR). This is the most widely used general purpose rubber, finding application in car tyres and footwear.

Other types of synthetic rubber include nitrile rubber, polychloroprene, polybutadiene and butyl rubber.

Quite often rubbers are referred to as elastomers because of their large reversible extensions (up to 600%). There is another group of elastomers called thermoplastic elastomers and their properties are discussed in section 3.7.

One of the main drawbacks of thermosets is their inability to be reprocessed or recycled, and this meant that at the advent of the new speciality thermoplastics, thermosets were replaced in many applications. Moreover, the rapid injection moulding possible with thermoplastics increased this move away from thermosets.

Recently, however, with better process control, it is possible to injection mould both thermosets and rubbers and as a result the move away from the use of thermosets is now less noticeable.

RADIATION PROCESSING

Apart from crosslinking by the application of heat and pressure, radiation induced crosslinking by γ-irradiation or electron beam irradiation has become more popular. Crosslinking by γ-irradiation is achieved by exposing the product to a radioisotope such as Co^{60} or Cs^{137}. A particle accelerator is used to provide the electron beam for electron beam processing.

These two methods are used to cure thermosets, crosslink thermoplastics (e.g. polyethylene), synthesise graft copolymers and produce wood/plastic and concrete/plastic composites. In the radiation crosslinking of plastics materials, free radicals are liberated, which give rise to the crosslinking action. A particular scheme for polyethylene is given in Fig. 3.3.

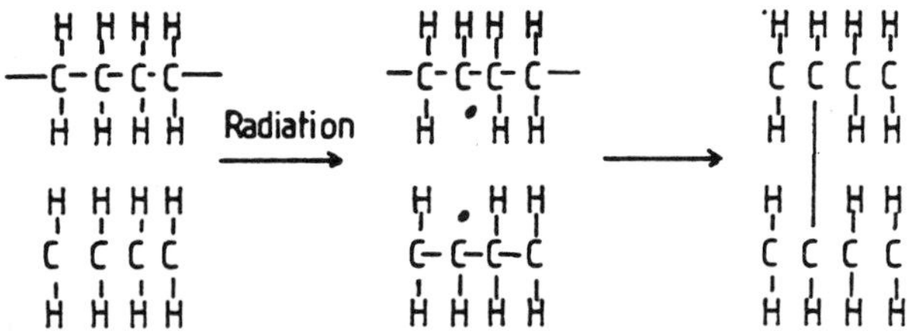

Fig. 3.3 Radiation gives rise to free radicals followed by a subsequent crosslinking reaction.

Radiation processing is used, for example, in wire and cable coatings, shrinkable sleevings and in packaging film. The most widely used polymer is polyethylene. In the last application the radiation causes the polyethylene to shrink when held between 135 and 150°C — shrink wrapping.

POLYMER BLENDS AND SOLUBILITY

Although new homopolymers have been developed, it is an expensive and risky process; the polymer may not have such good mechanical properties as had been hoped, or it may take too long to be accepted in the market place. When one considers that about three quarters of the engineering thermoplastic applications are accounted for by ABS, polyamides and polycarbonate, there is little room for new polymers. Material such as polysulphone (PSU), polyethersulphone (PES) polyetherimide (PEI), polyamideimide (PAI) and polyetheretherketone (PEEK) only account for a few per cent of the total market for polymeric materials.

An obvious way of obtaining a wider range of materials or to obtain synergistic properties is to blend two or more polymers togeher. In practice this Two-phase does not always work very well because, in general, polymers are not miscible. Two phase materials are formed with one phase dispersed as droplets in the other, and the interfacial adhesion is poor. This leads to poor mechanical properties. The miscibility of polymeric materials with each other depends on their solubility parameters. This is explained below.

When a polymer is in contact with a solvent, the polymer molecules are as much attracted to the solvent molecules as they are to their own kind. The resulting mixture will be a single homogeneous phase. Two polymers that are miscible will behave in the same way. If the polymer molecules are immiscible, they will be attracted preferentially by their own kind and will form a two phase blend.

Clearly, the forces of attraction between like molecules are important, and a knowledge of their magnitude may be obained from a consideration of how to prize the molecules apart by thermal means. The energy to separate like molecules may be given by the latent heat of vaporisation L. This, however, overestimates the real value because it includes the energy expended in evaporation. The appropriate energy of vaporisation is given by $(L–RT)$, where R is the universal gas constant.

A better measure would be the energy of vaporisation per molar volume, which is called the cohesive energy density E_c:

$$E_c = \frac{L - RT}{M/\rho} \ \mathrm{Jm^{-3}} \tag{3.1}$$

where MM is the molecular weight and ρ is the density.

The more common term used in solubility studies is the Hildebrand solubility parameter, δ, where

$$\delta = \sqrt{E_c}$$

$$\delta = \left(\frac{L - RT}{M/\rho}\right)^{1/2} \ (\mathrm{Jm^{-3}})^{1/2} \tag{3.2}$$

Values of δ for various polymers, solvents and plasticisers are given in the literature. Some typical values for polymers are given in Table 3.1. These figures are most accurate for $\delta < 19.4\ (\mathrm{MJ\ m^{-3}})^{1/2}$ when the polymers are amorphous and non-polar. A simple method of obtaining the δ of a polymer is to try to dissolve it in a number of non-polar solvents of varying δ and observing which solvent dissolves the polymer best. The δ of the polymer is then the same as that of the solvent.

For non-polar materials miscibility is likely if the difference in the δ's of the two species is less than $0.3\ (\mathrm{MJ\ m^{-3}})^{1/2}$. Values greater than this will lead to phase separation.

It is instructive to examine the thermodynamics of mixing. The first law of thermodynamics gives

$$\begin{aligned} Q &= \Delta E + W \\ W &= Q - \Delta E \end{aligned} \qquad (3.3)$$

where Q is the amount of heat energy absorbed, which contributes to a change in the internal energy ΔE and expends mechanical energy W. In equation (3.3) energy may be expended if $Q > \Delta E$.

Table 3.1 Values of Hildebrand solubility parameter for a number of polymeric materials.

Polymer	δ (MJ/m⁻³)^½	Polymer	δ (MJ/m⁻³)^½
Polytetrafluoroethylene	12.6	Polymethylmethacrylate	18.7
Polydimethylsiloxane	14.9	Polyvinylchloride	19.4
Ethylene-propylene rubber	16.1	Polycarbonate	19.4
Polyethylene	16.3	Polysulphone	21.2
Polypropylene	16.3	Polybutyleneterephthalate	21.5
Polyisoprene (NR)	16.5	Polyethyleneterephthalate	21.8
Polybutadiene	17.1	Polyethersulphone	22.6
Styrene-butadiene	17.1	Nylon 66	27.8
Polymethylphenylsiloxane	18.3	Polyacrylonitrite	28.7
Polystyrene	18.7		

$$\text{Suppose } \Delta A = \Delta E - Q = \Delta E - T\Delta S \qquad (3.4)$$

where ΔS is the entropy change at the Kelvin temperature T. ΔA is called the change in work function.

Equation (3.4) expresses that if more heat is absorbed than is taken up in increasing the internal energy of the system, mechanical energy can be expended.

If the change takes place at constant temperature and pressure P the increase in enthalpy, ΔH, during the process is given by

$$\Delta H = \Delta E + P\Delta V \qquad (3.5)$$

This increase in enthalpy is due to an increase in the internal energy and to the external work done in increasing the volume of the system by ΔV.

Eliminating ΔE between equations (3.4) and (3.5) gives

$$\begin{aligned} \Delta A - \Delta H &= -P\Delta V - T\Delta S \\ \Delta A + P\Delta V &= \Delta H - T\Delta S \end{aligned}$$

$$\text{If} \quad \Delta G = \Delta A + P\Delta V$$

$$\text{Then} \quad \Delta G = \Delta H - T\Delta S \qquad (3.6)$$

where ΔG is called the Gibb's free energy. For mixing to occur $\Delta G < O$.

Equation (3.6) is related to the mixing process in the following way. If $T\Delta S > \Delta H$, the available free-energy is negative, and molecules of the two different species in the mixture will be attracted to one another. In macroscopic terms, if the heat absorbed ($T\Delta S$) during the process is greater than the enthalpy change (ΔH), the Gibbs free energy of mixing will be negative and the two species will be compatible.

The question arises as to whether or not $T\Delta S$ is positive (i.e. is ΔS positive)? The entropy of a system is a measure of the order in that system, and, as the degree of freedom of the molecules increases, the order decreases and ΔS is positive. When mixing occurs between two species of molecule, it is expected that the degree of order is decreased because the species do not order themselves into domains. Their degree of freedom is, thus, increased, giving an increase in entropy. Thus ΔS is then positive during mixing. Mixing will occur, therefore, if the heat of mixing (the enthalpy) $\Delta H, = O$ or if $\Delta H < T\Delta S$.

As a consequence of this:

(1) It is observed that mixing is more likely at raised temperatures.
(2) The greater the increase in entropy, the greater is the likelihood of the species mixing.
(3) Mixing is more likely if the change in enthalpy is small.

It has been shown by Hildebrand that if there is no specific interaction such as hydrogen bonding between the two species

$$\Delta H = V_T (\delta_1 - \delta_2)\, \varnothing_1\, \varnothing_2 \tag{3.7}$$

where V_T is the total volume of the mixture and $\varnothing_1$, $\varnothing_2$ are the volume fractions of each species.

If $\delta_1 = \delta_2$, $\Delta H = 0$ and mixing occurs. If $(\delta_1 - \delta_2)$ is finite, mixing will occur if the value is small such that $T\Delta S > \Delta H$.

The use of the Hildebrand solubility parameter δ in determining whether or not polymers will mix is not its sole application. It may be used in discussing the behaviour of solvents and plasticisers for a given polymer. When searching for good solvents for polymers, tables usually provide a range of values in which solvents would be expected to dissolve the given polymer. The tables often provide the ranges for three types of solvent: non-polar, moderately polar and strongly polar. The size of the range of values of δ is often highly dependent on the type of solvent. This suggests that the simple approach adopted so far for δ may be insufficient. This will be discussed later.

In a good solvent for a given polymer (matching δs) the secondary Van der Waals forces between the polymer segments and solvent molecules are strong, and the polymer molecules spread out in the solvent and occupy a larger hydrodynamic volume. This increases the viscosity of the solution, because the motion of large molecules involves these molecules in more mutual interactions than the motion of small molecules at the same concentration. This viscosity increase is used in a more accurate determination of the solubility parameter of a polymer.

In poor solvents, the polymer molecules curl up and occupy a smaller hydrodynamic volume than in the previous case and this gives rise to lower solution viscosities. The intramolecular forces in the polymer chain are greater than the intermolecular forces between the polymer and the non-solvent. This causes the coiling up of the polymer molecules.

It is easier to get two solvents to mix together than a polymer and a solvent because in the former case the change in entropy ΔS is very large, and ΔG is much more likely to be negative. During solvent polymerisation it is often possible for the monomer and solvent to mix readily but once a critical molecular mass has been reached, the polymer will precipitate out of solution. This kind of behaviour occurs in the synthesis of ABS and high impact polystyrene where it is essential for giving the phase inversion that takes place which disperses the rubber particles.

The reasons for lower values of ΔS in the mixing of polymer molecules is that the disorder is less in long molecules than in short ones because long molecules are more constrained.

During a polymerisation reaction a point may be reached at which the polymer molecular mass is sufficiently large for precipitation to occur. At this point $\Delta G = 0$ and $\Delta H = T\Delta S$. This condition depends on the temperature and the relative molecular mass average. As $\overline{M}_w$ increases, the longer molecules precipitate out. The same will occur if the temperature falls. The limit of solubility for an infinite $\overline{M}_w$, at which ΔS is a minimum is called the Θ condition. Real polymers, in which $\overline{M}_w$ is finite, are soluble at the Θ condition.

Plasticisers are usually high boiling point solvents of the given polymer. They lower the processing temperatures of polymers and render the product flexible and softer, e.g. PVC. Plasticisers should have the following features:

(1) A molecular weight in excess of 300.
(2) A δ similar to that of the polymer.
(3) If the polymer is likely to crystallise the plasticiser must have some specific interaction with the polymer.
(4) The plasticiser may itself be crystalline only if it is capable of a specific interaction with the polymer, e.g. hydrogen bonding.

The plasticiser works by increasing the free volume without eliminating polymer chain interactions. In the case of polar polymers such as PVC an interaction between the polymer and the plasticiser occurs in the form of linkages, the numbers of which depend on temperature. This offsets the spacing between polymer molecules, the greatest effect occurring with the greatest interaction. In some cases interaction is so strong that anti-plasticisation may occur, giving a stiffer product.

From the above it is clear that the Hildebrand solubility parameter δ is an incomplete definition of solubility. For this reason three components of the solubility parameter have been suggested. δ_d is that due to dispersive forces, δ_p is that due to polar forces and δ_h is that due to hydrogen bonding where

$$\delta_2 = \delta_d^2 + \delta_p^2 + \delta_h^2 \qquad (3.8)$$

Values of the three-component δs are available in the literature.

When the solubility parameters of two polymers differ by more than 0.4 $(MJ\ m^{-3})^{1/2}$, a two-phase blend is formed, with the matrix material generally having the higher concentration. If the difference between the two solubility parameters is less than $0.8(MJ\ m^{-3})^{1/2}$ there will be good interfacial adhesion. The range, therefore, in which polymer pairs may have good mechanical properties in blending is very restrictive.

A number of miscible commercial polymer blends are available and possess attractive mechanical properties; e.g. polyphenyleneoxide (PPO)/PS, PVC/N66, PS/PC (up to 60% PC) and PS/tetramethyl PC (MPC). These are exceptions and other ways have to be used to obtain good interfacial adhesion in two phase blends. One method is to employ a chemical bond across the interface, which may be achieved by copolymerisation or by using interpenetrating networks (IPNs) or by a simple compatibilising agent. The blend then becomes a polymer alloy.

COPOLYMERS

In copolymerisation species A and species B are joined together by a chemical bond. There are three main types of copolymer: (a) random, (b) graft and (c) block. The differences between them are shown in Fig. 3.4.

ABBABABAAB

(a) Random Copolymer

AAAAAAAA
B
B
B
B

(b) Graft Copolymer

AAABBBBB

(c) A Diblock Copolymer

AAABBBBAAA

(d) A Triblock Copolymer

AAABBAAABBAAABBAA

(e) An $(AB)_n$ Block Copolymer

Fig. 3.4 Types of copolymer.

In a random copolymer the two species *A* and *B* are irregularly distributed in the polymer chain. This gives a homogeneous material and no phase separation. The lack of regularity prevents any chain alignment and crystallinity.

By altering the proportions of the two species a whole spectrum of grades can be synthesised with properties intermediate between the two homopolymers. Well known examples of this are propylene/ethylene copolymer and ethylene/vinylacetate copolymer (Fig. 3.5).

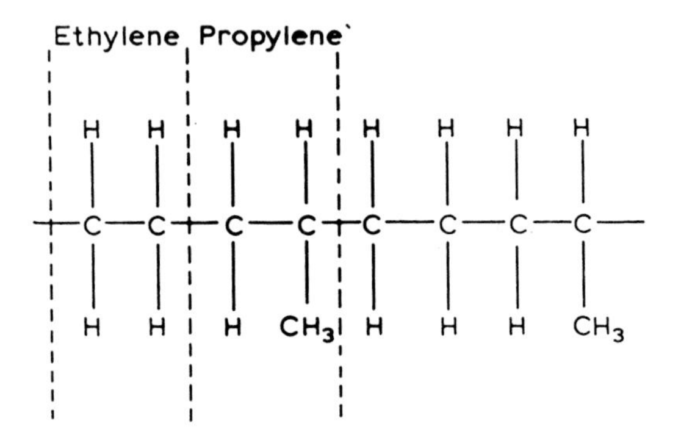

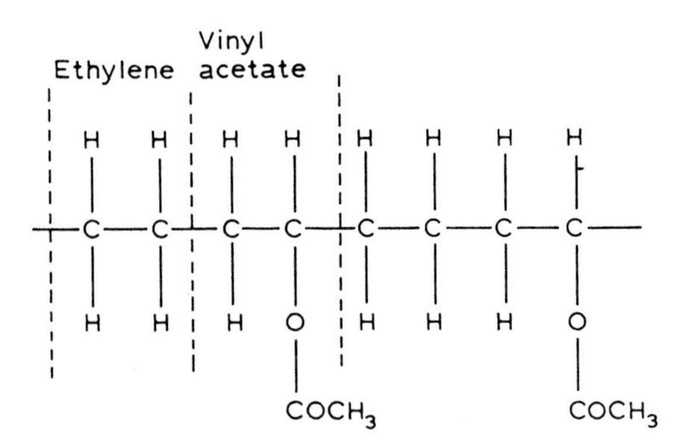

Fig. 3.5 Structures of ethylene/propylene polymer and ethylene/vinyl acetate copolymer.

In the case of graft copolymers, species *B* is attached to the side of the main chain of species *A*. If the two species are miscible, a homogeneous, amorphous material results, as in the previous case, and a whole spectrum of grades is possible with properties intermediate between the two homopolymers. One example of this is modified polyphenyleneoxide (PPO) 'Noryl'. The

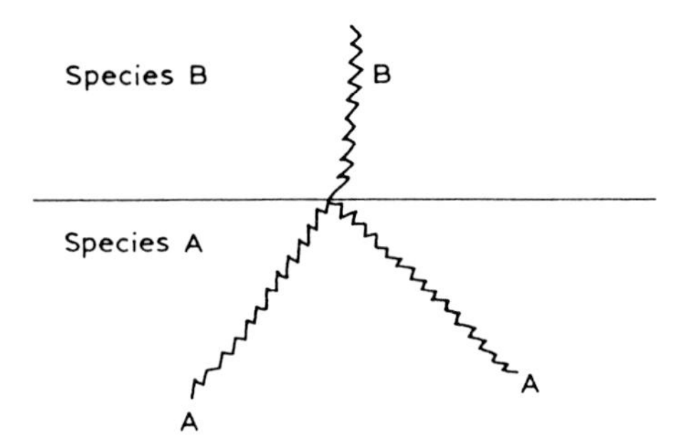

Fig. 3.6 A graft copolymer acting as an interfacial chemical linkage.

homopolymer PPO is difficult to process but has desirable mechanical properties. When PS is grafted on to the backbone chain of PPO, the processability is improved at the expense of only a slight loss in mechanical properties.

In general, polymeric materials are not compatible and in these cases graft copolymerisation leads to phase separation. As this is a pre-requisite to rubber toughening, this route is a good method of suspending the rubber particles in the plastics matrices, both in thermoplastics and in thermosets. The chemical tie across the interface gives good interfacial adhesion, as shown in Fig. 3.6. The particle size achieved is of the order of 1 μm which is ideal for toughening.

This kind of copolymer is used in the toughening of PS (high-impact PS or HIPS), ABS PVC and PMMA and in thermosets such as epoxy resins. Such polymerisation is also carried out with toughened nylons, in which the

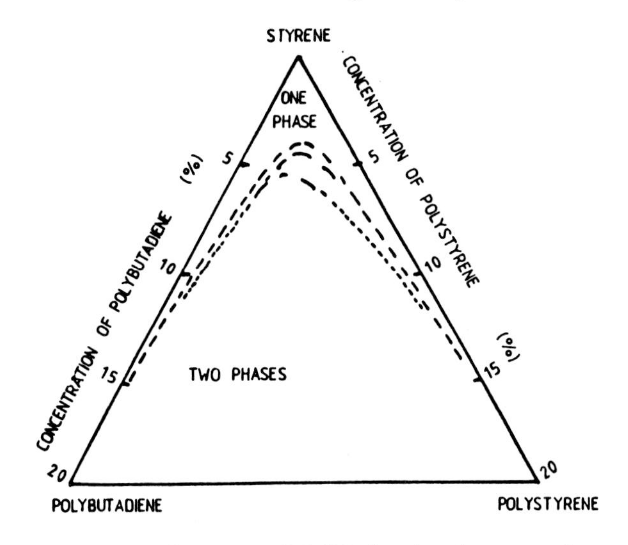

Fig. 3.7 A ternary diagram for styrene/PS/PB showing the range of compatibility of the three species.

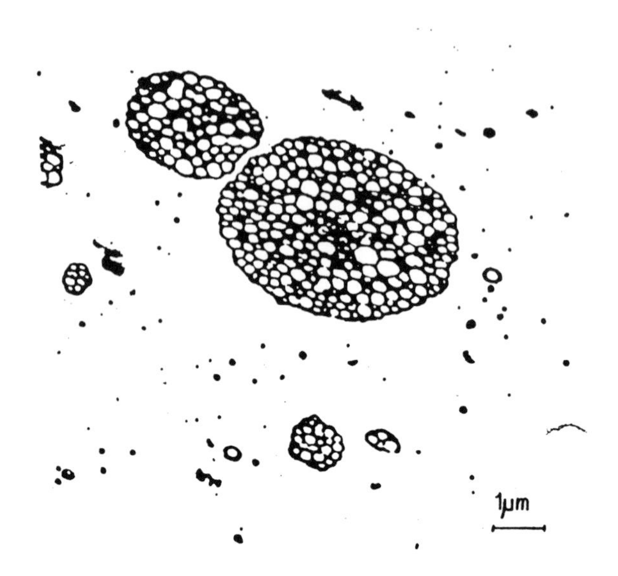

Fig. 3.8 Inclusions of PS in PB particles caused by the phase inversion during the polymerisation of the styrene monomer.

grafted polymer is EPDM, and in toughened polybutyleneterephthalate (PBT), in which the toughening agent is ethylene/propylene copolymer.

When dealing with addition polymers such as PS and PMMA, toughening with polybutadiene is not difficult. HIPS is synthesised in the following way. The polybutadiene (PB) is dissolved in the styrene monomer with which it is compatible. The styrene is then polymerised in the presence of the PB. The polystyrene formed is not compatible with the styrene/PB solution, as shown in the ternary diagram in Fig. 3.7. Phase separation occurs with PS being the dispersed phase. The reason for the incompatibility is that the change in entropy $T\Delta S$ is less with large molecules because they are more ordered and the Gibb's free energy of mixing, from equation (3.6), becomes positive.

At about 2% PS formation there is a phase inversion in which the PB becomes the dispersed phase and styrene/PS the matrix phase. Eventually all the styrene monomer is polymerised.

During the phase inversion, some PS is trapped inside the rubber particles, as shown in Fig. 3.8. This is desirable because for a required particle size of 1 µm, less PB is needed than if the inclusions were absent. Typical toughening values are given for HIPS in Table 3.2. The rubber additive leads to a reduction in tensile strength and flexural modulus.

In the synthesis of HIPS the grafting occurs as a part of the reaction, and it is believed that some of the PB grafts on to the *PS* main chain. Moreover, during the reaction the PB is crosslinked. This maintains the particle size so

Table 3.2 The improvement in toughness in HIPS and SBS block copolymer compared with polystyrene.

Material	Notched Izod impact strength $J m^{-1}$
Polystyrene (PS)	13
High impact polystyrene (HIPS)	90
Styrene–butadiene–styrene (SBS) triblock	400

that future melt processing operations do not break up the particles. Particle size is thus controlled by the chemistry.

In toughened nylons (polyamides) the grafting is deliberately sought, and an appropriately end-capped EPDM is grafted on to the nylon main chain. In this case the rubber is not crosslinked and the desired particle size is achieved by careful processing, which is subject to some secrecy. Typical toughening values are given in Table 3.3.

Table 3.3 Notched Izod impact test values for Nylon 66 and Du Pont super-tough nylon.

Material	Notched Izod impact strength $J m^{-1}$	Tensile strength $MN m^{-2}$	Flexural Modulus $GN m^{-2}$
Nylon 66	53	82	2.8
'Super-tough' nylon	800	53	1.73

Another method of dispersing a rubber phase in a plastics matrix is to use a block copolymer. In these, species A and B are joined end to end to form either: (a) an AB (b) and ABA or (C) an $(AB)_n$ block copolymer. A diblock (AB) copolymer forms a two-phase blend when added to a matrix of either polymer, but the preferred use of a diblock copolymer is as a compatibilising agent across the AB interface, where it works in a similar way to the graft copolymer shown in Fig. 3.6.

With condensation polymers such as polysulphone (PSU) and polyethersulphone (PES) it is not possible to suspend the dispersed rubber phase by the method used for HIPS and ABS. Polybutadiene is, moreover, unsuitable for use with these engineering thermoplastics because of their high processing temperatures and high service temperatures. A block copolymer route involving ABA or $(AB)_n$ blocks with polydimethylsiloxane PDMS has been attempted in PSU.

With PSU there is little choice of synthesis route for rubber toughening, but a copolymer route has also been used with styrene and butadiene to make a triblock SBS copolymer for dispersal in a PS matrix. As seen in Table 3.2, the SBS triblock copolymer gives far better toughening than is present in the traditional HIPS material.

It is believed that the block copolymer method gives greater control over particle size than the graft copolymer route, although only the latter has led to currently available commercial materials. In the synthesis of $(AB)_n$ block copolymers a number of low molecular weight repeat units of species A, called oligomers of A, are terminated with specific end groups. Oligomers of species B are also terminated with functional end groups so that when the two sets of oligomers are reacted together in a common solvent or in the melt phase, the end groups react and disappear, leaving species A and B joined in an alternating sequence. The synthesis routes involved, however, are much more complex than in the manufacture of HIPS.

One disadvantage of rubber toughened polymeric materials that has already been mentioned is the accompanying reduction in tensile strength and elastic modulus. Another disadvantage is an inherent lack of transparency. The reason for this is that generaly the dispersed elastomeric phase will have a different refractive index from the matrix phase. This will cause partial reflection every time a particular ray strikes a particle. Even if the two materials have the same refractive index, transparency may be limited to a narrow temperature range because the temperature coefficient of refracture index may not be the same for the two materials. For this reason safety spectacles are made from inherently tough materials and not from toughened ones.

INTERPENETRATING NETWORKS

As a result of the difficulty of controlling particle size in the rubber toughening of condensation polymers with silicone elastomers, the use of interpenetrating networks (IPNs) seemed a likely method of approach.

An IPN occurs when polymer 1 is crosslinked and allowed to swell in monomer 2 and a crosslinking agent for monomer 2. Monomer 2 is then polymerised and crosslinked when the swelling is complete. If the two polymers are incompatible two phase domains occur, which are intimately mixed and of size of the order of tens of nm. In rubber toughened materials the combination of polymer and elastomeric polymer are used. The variation of the concentration of each species gives grades of behaviour from reinforced rubber (similar to an ABA thermoplastic elastomer) to a rubber toughened plastic.

The term IPN is used because in the limiting case of high compatibility between the two polymers both networks would be continuous and interpenetrating.

Semi-IPNs occur when only one of the components is crosslinked. Such a system is similar to a graft copolymer system in which one of the components is crosslinked. A toughened thermoplastic is made by dissolving polymer 1, which is not crosslinked into monomer 2 and its crosslinking agent. Monomer 2 is then polymerised and crosslinked. The domain size is similar to that produced in graft copolymerisation. IPNs have been used in improved leather goods, piezodialysis membranes, noise damping materials, toughened plastics and pressure sensitive adhesives.

FIBRE REINFORCED PLASTICS MATERIALS

The use of fibre reinforcement in polymer matrices is an excellent and relatively cheap way of improving the stiffness, strength and resistance to temperature and creep of the matrix material. Fibres of glass, carbon, Kevlar or other materials, both metals and ceramics, are used to reinforce both thermoplastics and thermosets. Some typical properties are given in Table 3.4.

Table 3.4 Reinforcing fibres and their mechanical properties.

Fibre	Density g cm^{-3}	Tensile modulus GN m^{-2}	Tensile strength MN m^{-2}
E-glass	2.55	75	2.0
S-glass	2.49	75	5.5
Carbon	2.00	170–200	0.5–1.0
Kevlar	1.45	130	3.0–3.6

In these reinforced materials, the strength and stiffness of the fibres are married to the plastics matrices to give tough composites. The fibres above are very brittle and alone their strength and stiffness are not fully realised. The polymer matrix protects them and transfers the load to them. For this reason good interfacial adhesion is important and much work has been carried out on coupling agents.

As shown in Fig. 3.9, the matrix transfers the load to the fibre as a tensile stress. As the (L/D) ratio of the fibre increases the tensile stress transferred to it increases. At a critical value $(L/D)_c$, the tensile stress in the fibre reaches a maximum value and any further increase in the fibre length leads to no further increase in stress transfer.

The most popular way of reinforcing thermoplastics matrices is to use short glass fibres in the length range 0.125–0.5 mm. During injection moulding these fibres suffer degradation, and the average fibre length in the product is reduced. The degradation occurs mainly in the narrow gate region at the entrance to the mould cavity.

One of the modern trends is to use much longer fibres in the range 6–10 mm. These too suffer from processing degradation but the overall fibre

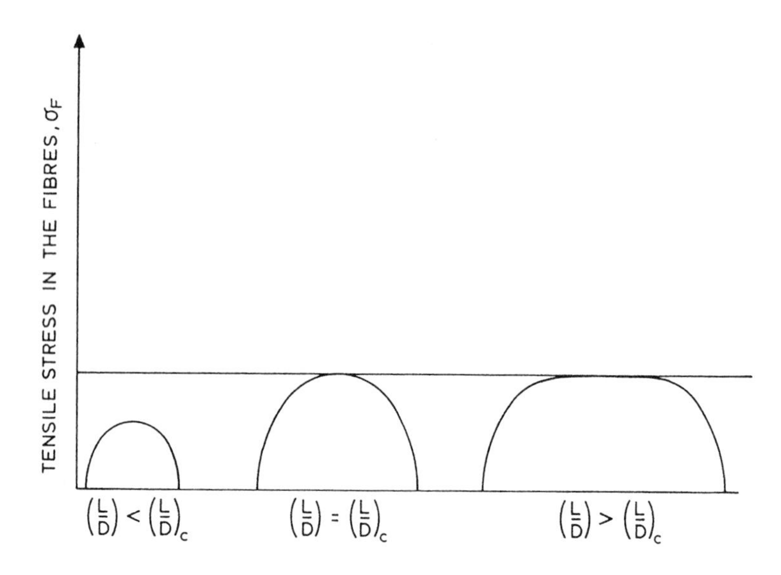

Fig. 3.9 The effect of length/diameter (L/D) ratio of the fibre on the efficiency of the load transfer to the fibre.

Table 3.5 Comparison of the mechanical properties of dry Nylon 66 samples for Maranyl A190 (Unreinforced), Maranyl (A690) (50% short glass fibres) and Verton RF–700–10 (50% long glass fibre reinforced Nylon 66).

Property	A190	A690	RF–700–10
Tensile Strength $MN\,m^{-2}$	83	200	230
Elongation to Break %	60	3	4
Flexural Modulus $GN\,m^{-2}$	2.8	12.0	15.8
Notched Izod Impact Strength $J\,m^{-1}$	53.0	11.0	27.0

length in the product is much larger than for short fibre reinforcement. Table 3.5 shows values of the mechanical properties for Nylon 66 which is unreinforced, short-fibre reinforced and long-fibre reinforced Nylon 66.

Table 3.6 shows the effect of short-fibre reinforcement on the heat deflection temperature of various polymer composites. The values clearly show that partly crystalline matrices show a considerable increase in heat deflection temperature whereas amorphous matrices do not. An explanation may be that the fibres can join the crystalline regions, acting as a rigid bridge across the amorphous regions.

It has been assumed that the benefits of short and long fibre reinforcement are the same in all directions, but this cannot be true because it has also been assumed that the fibres are completely aligned along the direction of tensile stress.

Table 3.6 The effect of short-fibre reinforcement on the heat deflection temperatures of amorphous and partly-crystalline matrices.

Thermoplastic	Heat deflection temperature at 1.81 MNm^{-2}/°C	
	Matrix	Composite
Partly-crystalline		
Nylon 6	75	212
Nylon 66	95	248
PEEK	135–160	286
Amorphous		
PC	130	140
'Noryl'	130	14
PES	201	216

In practice, mechanical properties are improved in the machine direction, along the axis of orientation of the fibres, at the expense of the mechanical properties in the two mutually perpendicular directions. Designing with these anisotropic materials is more specialised.

It is possible to model composite behaviour mathematically, and the simplest method is to assume a rule of mixtures. If the tensile moduli of the matrix, fibre and composite are given by E_m, E_f and E_c and the volume fractions are Θ_m, Θ_f respectively then

$$E_c = \Theta_m E_m + \Theta_f E_f \tag{3.9}$$

This equation assumes a perfect fibre alignment and a perfect transfer of tensile stress from the matrix to the fibre. To test this, assume that the tensile modulus of an *E*-glass fibre is 185 GN m^{-2} and that of Nylon 66 is 2.8 GN m^{-2}. The densities of *E*-glass and Nylon 66 are 2548 kg m^{-3} and 1150 kg m^{-3} respectively.

Consider a 50% glass reinforced Nylon 66, which means 50% by weight. This must be translated into a volume fraction to work with equation (3.9).

This is done as follows. Suppose that the weight fractions of the fibres and the matrix are ϕ_f and ϕ_m and their densities are ρ_f and ρ_m respectively.

Then

$$\Theta_f = \frac{V_f}{V_f + V_m} \tag{3.10}$$

where V_f and V_m are the volumes of the fibres and the matrix.

$$\Theta_f = \frac{M_f/\rho_f}{M_f/\rho_f + M_m/\rho_m} \tag{3.11}$$

Multiply the numerator and denominator by M_c gives

$$= \frac{\phi_f/\rho_f}{\phi_f/\rho_f + \phi_m/\rho_m} \tag{3.12}$$

Multiplying top and bottom by $\sigma_f\sigma_m$ gives:

$$\Theta_f = \frac{\phi_f\rho_m}{\phi_f\rho_m + \phi_m\rho_f} \tag{3.13}$$

In the present example $\Theta_f = 31\%$ and $\Theta_m = 69\%$.

From equation (3.9)

$$E_c = 0.69 \times 2.8 \times 10^9 + 0.31 \times 185 \times 10^9$$

$$= 59.3 \text{ GN m}^{-2}$$

By comparing the theoretical value of 59.3 GN m^{-2} with the values for Maranyl A690 and Verton RF–700–10 of 12.0 and 15.8 GN m^{-2} respectively (Table 3.5), it can be seen that there is a loss of efficiency of the transfer of stress from the Nylon to the glass fibres. This is partly due to imperfect alignment of the fibres. For this reason continuous fibre reinforced thermoplastics have been developed.

This development has been commonplace in thermoset matrices because the impregnation of the continuous fibres with the uncrosslinked resin is easier to achieve than with the higher viscosity thermoplastics resins. Carbon fibre/epoxy composites are a much used composite in aerospace, where strength, stiffness and lightness are important features. One of the newer thermoplastic composites is the carbon fibre/PEEK (APC-2) from ICI and a similar composite is planned with carbon fibre/polyetherimide, which is thought to be easier to process.

These materials are marketed in the form of a very expensive 'plywood'. Continuous collimated fibres make up one ply, and a laminate is made up by joining this ply to others with their continuous fibres at different orientations. This gives useful omni-directional properties.

In order to describe the orientations of the different layers, a control axis is identified and axes parallel to this are referred to as 0. Thus four plies orientated in this direction are denoted by 0_4, and five plies orientated perpendicular to this direction are referred to as 90_5. A set of brackets indicates a repeated sequence, and the suffix outside refers to the number of repetitions. $(0_4,90_5)_2$ means two repetitions of 0_4 and 90_5.

One important constraint is that the lay-up should be symmetrical about the centre to prevent warpage in service. The symmetry is denoted by *S*.

Thus $(0,90)_s$ could be written as 0, 90, 90,0 and $(0,90)_{3s}$ would mean 0, 90, 0, 90, 0, 90, 90, 0. 90., 0. 90. 0. If an odd number of plies is used, the ply that is not repeated is indicated by a bar. Thus (0,90, 0) indicates 0, 90, 0, 90, 0.

A popular lay up form that has uniform properties in all directions in the plane of the sheet is called the quasi-isotropic laminate. Examples of this include $(+45,90,-45,0)_{ns}$ and $(0,90,+45,-45)_{ns}$ where *n* is an integer.

APC-2 contains 62% by volume carbon fibres in PEEK. The theoretical modulus is given by equation (3.9).

$$= 0.38 \times 1.1 + 0.62 \times 185$$
$$= 115 \text{ GN m}^{-2}$$

In practice a ply has a value of modulus which is less than the value derived from the rule of mixtures.

SELF-REINFORCING POLYMERS

Self-reinforcing polymers or thermotropic liquid crystal polymers (LCP) are a new class of materials that combine high strength and stiffness with easy processing. This is due to the liquid crystalline phenomenon or mesomorphism.

Liquid crystal molecules are subject to both long range order, as in solids, and short range order, as in liquids. The individual molecules are aligned with respect to each other within domains — like magnetic domains in iron. When subject to stresses these domains readily align. On leaving dies or gates the domains remain aligned, resulting in a high orientation of polymer molecules, which gives high strength and high modulus. In addition, the frozen-in-stress is low and environmental stress cracking is reduced. Table 3.7 shows mechanical properties of two LCPs.

The viscosity of the materials is reduced because of the alignment of the individual molecules and the domains. The viscous drag between domains moving past each other is less than in the isotropic phase when the entangled randomly orientated polymer molecules are involved in motion. The viscosity rises when the temperature is increased so that mesomorphism is destroyed.

Table 3.7 The mechanical properties of 'Vectra' (Hoechst–Celanese) and 'Xydar' (Amoco).

	Xydar	Xydar/ glass fibre	Vectra	Vectra/ glass fibre
Specific gravity	1.35–1.40	1.6	1.4	1.57
Tensile Strength/MN^{-2}	80–123	136	140–180	200
Elongation/%	3.3–4.9	1.7	1.3–6.9	2.2
Tensile Modulus/GNm^{-2}	–	15.6	10–40	17
Notched Izod				
Impact Strength/Jm^{-1}	75–210	106	53–530	135
Heat deflection temperature @ 1.81MNm^{-2}/°C^{-1}	316–355	346	180–240	230

The name 'self-reinforcing polymer' stems from the fact that the aligned liquid crystalline regions are fibrous and so the material is reinforced by fibres of itself. Under the electron microscope the fibrous structure appears like that of timber.

Liquid crystallinity exists in two forms: lyotropic, in which mesomorphism depends on the concentration of a liquid crystal component in a solvent, and thermotropic, in which mesomorphism occurs in a well-defined temperature range. Both types of liquid crystallinity may occur in heating from the solid to liquid and back again (enantiotropic) or it may happen only on supercooling (monotropic). In the former case the mesophase is thermodynamically stable, whereas in the latter it is only metastable.

On heating a crystalline material that is not a liquid crystal, the solid phase changes to an isotropic liquid in which the Van der Waals forces act between molecules, giving short range order. In liquid crystalline materials a change in conditions causes one or more transitions through mesophases before an isotropic fluid is achieved. In the mesophase long range forces act between groups of molecules and Van der Waals forces act between the domains.

There are three kinds of mesophase, of which two are important to LCPs. (1) the nematic mesophase, (2) the smectic mesophase and (3) the cholesteric mesophase. The molecular configurations in the first two are shown in Fig. 3.10.

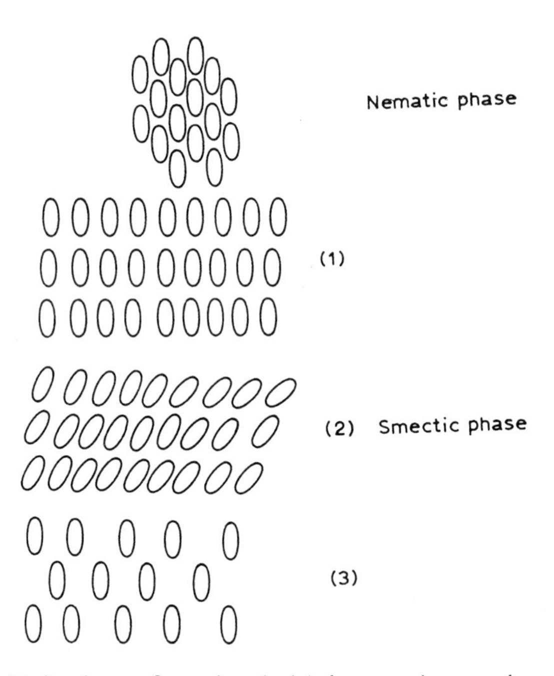

Fig. 3.10 Molecular configurations in (a) the nematic mesophase and (b) the smectic mesophases.

The nematic mesophase represents a one-dimensional long range order in which the molecules are aligned but their positions relative to each other are uncorrelated. This gives a low viscosity fluid and easy processing.

The smectic mesophase shows a two-dimensional order in which alignment and correlation with adjacent molecules are important. This mesophase may occur in three different forms, depending on the degree of order, tilt or positional correlation. This mesophase gives rise to high viscosities in fluids and is not of great interest to the polymer processor. The same may be said of the cholesteric mesophase, which comprises a helical nematic or smectic mesophase.

On changing conditions a liquid crystalline material does not have to pass through all of these mesophases before becoming an isotropic liquid. Materials that exhibit a nematic mesophase suffer an increase in viscosity in becoming an isotropic fluid so that processing temperatures must be well controlled.

Lyotropic liquid crystal polymers were important originally because it is from such a system that 'Kevlar' aramid fibre is made. Here the mesomorphic material is dissolved in a solvent and the well-aligned fibres are spun and dried. The high degree of orientation that results gives the great strength and modulus of 'Kevlar' Thermotropic LCPs are of great interest to manufacturers because these may be melt processed into products.

LCPs obtain their properties by incorporating in them liquid crystal-forming units called mesogenic moieties, which are long and stiff.

The polyester architecture is conducive to acting as a mesogenic moiety. These moieties may be incorporated into the side chain to give comb polymers or in the main chain. The latter is used for engineering LCPs. Fig. 3.11 shows both side chain and main chain LCPs in the nematic mesophase.

In side chain LCPs, the mesogenic moiety is connected to the backbone chain by a flexible spacer group — often a methylene group (CH_2). Its function is to decouple the side-chain motion from that of the stiff main chain so that the mesogenic units may align.

Of interest to polymer engineers are the LCPs with mesogenic moieties in the main chain. There are two types: (a) where rigid segments are separated by flexible segments and (b) where there are totally rigid segments with very little main chain movement. Such materials have a high solid to mesophase transition temperature, as with 'Xydar', and are highly resistant to chemicals.

LCPs may be reinforced as other thermoplastics and processed in similar ways. They are intended for high temperature applications, where resistances to chemicals, UV or ionising radiations are important. Low combustibility and smoke emission are other creditable properties. The main drawback of LCPs is the low weld line strengths where two different flows meet in a mould cavity.

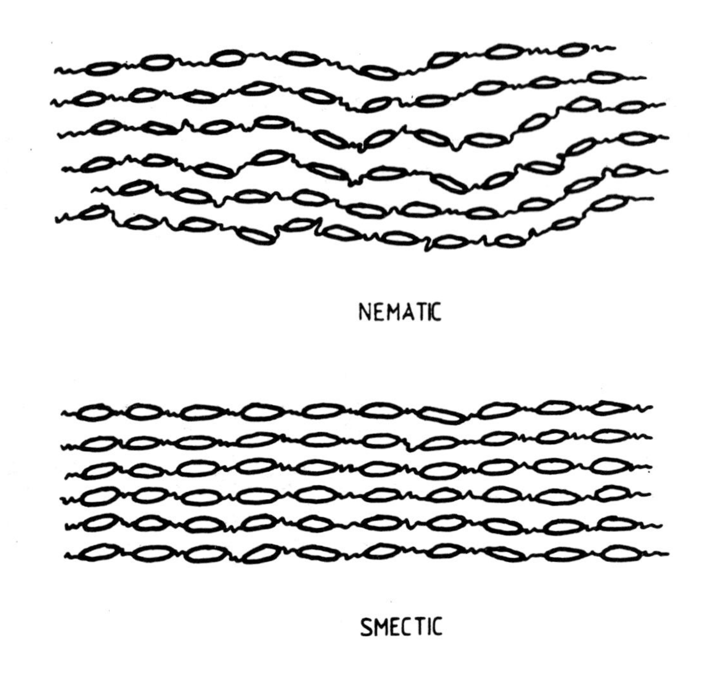

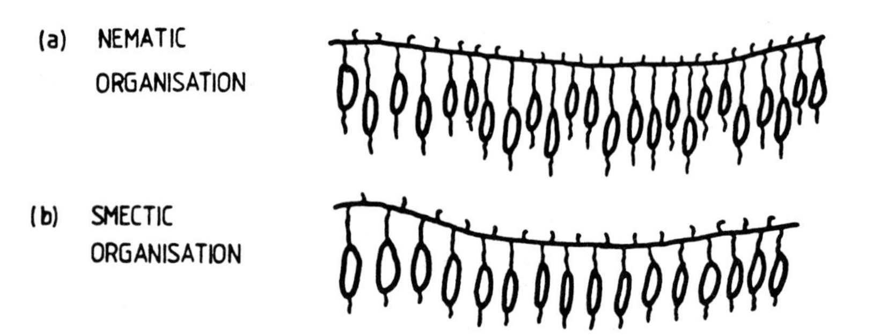

Fig. 3.11 Side chain and main chain liquid crystal polymer structures.

CELLULAR PLASTICS

Foamed or cellular plastics consist of a polymer matrix and a dispersed gas phase. The different kinds are characterised by their cellular morphology, mechanical properties and composition.

The morphology can either be open or closed cell. In the former, the voids coalesce to give continuous solid and gas phases, in the latter, discrete bubbles are formed and dispersed in the solid matrix. Structural foams consist of a solid skin and a cellular core, and often the skin material is different from the core material. These foams have better load-bearing properties than the former types. Other cellular morphologies involve a third component such as glass fibres to give a reinforced plastic form, or hollow-glass, ceramic or plastic microspheres to give a syntactic foam.

Cellular plastics may be divided into rigid or flexible foams. A rigid foam is formed when the matrix polymer is used below T_g or T_m, whereas a flexible foam is used above these temperatures. Most polyolefins, polystyrene, polycarbonate, 'Noryl' and some polyurethane foams are rigid, while rubbers, elastomeric polyurethane and plasticised polyvinylchloride form flexible foams. Semi-rigid foams form an intermediate class.

The main reasons for foaming products are to combine lightness with strength, and to save material. The latter can be demonstrated as follows. Consider the beam under flexure, as shown in Fig. 3.12. The material at the centre, at the neutral axis, is neither compressed nor extended and does nothing to resist bending. The internal bending moment, which opposes flexure, is generated by the outermost layers. In making a uniformly cellular composite beam, a thicker beam of the same mass results. This effectively pushes the solid material away from the neutral axis, thereby increasing the stiffness of the beam.

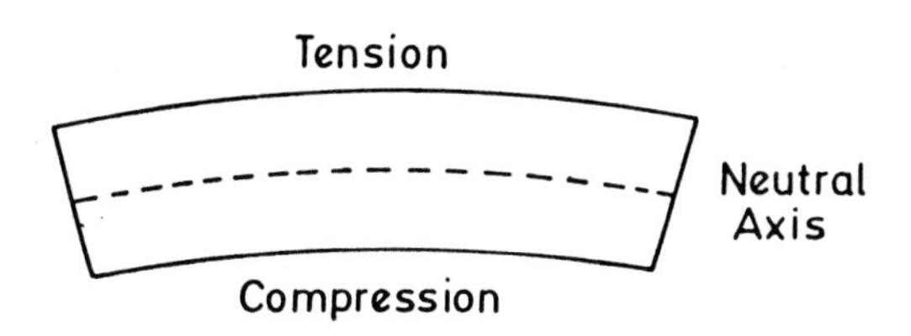

Fig. 3.12 A beam under flexure showing the neutral axis.

There is one disadvantage, the modulus of the composite is reduced but not in the way expressed by the rule of mixtures. It has been shown that

$$E_f = E_m(\rho_f/\rho_m)^2 \qquad (3.14)$$

where E_f and E_m are the moduli of the foam and matrix and ρ_f and ρ_m are the densities of the foam composite and matrix respectively. The density of the foamed composite depends on the foaming fraction *f*, which is the fractional density reduction.

$$f = \frac{\rho_m - \rho_f}{\rho_m} = (1 - \rho_f/\rho_m) \qquad (3.15)$$

A foamed beam is much lighter than a solid beam of equal bending stiffness. The bending stiffness, B for a beam of thickness t, breadth b and modulus E is given by

$$B = \frac{b\,t^3\,E}{12} \tag{3.16}$$

Using the subscript f for the foamed composite and m for the matrix material, and keeping the bending stiffness of both beams the same gives,

$$\frac{b\,t_f^3\,E_f}{12} = \frac{b\,t^3{}_m E_m}{12} \tag{3.17}$$

therefore

$$\frac{t_f}{t_m} = \left(\frac{E_m}{E_f}\right)^{1/3} \tag{3.18}$$

and

$$E_f = \left(\frac{\rho_f}{\rho_m}\right)^2 E_m \tag{3.19}$$

The percentage of material saved $\Delta M = 100\left(\frac{M_m - M_f}{M_m}\right)$ (3.20)

where M_m is the mass of the solid beam.

$$\Delta M = 100\left(1 - \frac{M_f}{M_m}\right) \tag{3.21}$$

For unit length $M_f = t_f\rho_f$, $M_m = t_m\rho_m$

therefore

$$\frac{M_f}{M_m} = \frac{t_f\rho_f}{t_m\rho_m}$$

and so

$$\Delta M = 100\left(1 - \frac{t_f\rho_f}{t_m\rho_m}\right) \tag{3.22}$$

In order to demonstrate the weight saving in a foamed composite product, consider a beam of thickness 20 mm, width 50 mm and length 1 m that is made from N66. It is to be replaced by (a) a 30% glass fibre reinforced N66, (b) an N66 foam of foaming fraction 0.3 and (c) a 30% glass fibre reinforced N66 with a foaming fraction of 0.3. In each case the bending stiffness must be the same.

	Nylon 6,6	Nylon 6,6/30% glass fibre
Flexural modulus/GN m^{-2}	2.8	9.1
Density/kg m^{-3}	1150	1390

Calculation for glass reinforced Nylon 66

$$\frac{t_c}{t_m} = \left(\frac{E_m}{E_c}\right)^{1/3} \tag{3.23}$$

where the subscript c is for the reinforced composite

$$t_c = 20\left(\frac{2.8}{9.1}\right)^{1/3}$$

$$= 13.5mm$$

The percentage mass saving ΔM is given by

$$\Delta M = 100\left(1 - \frac{t_c\rho_c}{t_m\rho_m}\right)$$

$$= \left(1 - \frac{13.5 \times 10^{-3} \times 1390}{20 \times 10^{-3} \times 1150}\right) \times 100$$

$$= 18.5\%$$

Calculation for the foamed Nylon 66

The foaming fraction f is given by equation (3.15)

$$f = \frac{1 - \rho_f}{\rho_m}$$

Therefore

$$\frac{\rho_f}{\rho_m} = 1 - f = 0.7$$

for $f = 0.3$

Using equations (3.18) and (3.19) gives

$$\frac{t_f}{t_m} = \left(\frac{E_m}{E_f}\right)^{1/3} = \left(\frac{E_m}{(\rho_f/\rho_m)^2 E_m}\right)^{1/3} = \left(\frac{\rho_m}{\rho_f}\right)^{2/3}$$

$$t_f = 20\left(\frac{1}{0.7}\right)^{2/3} = 25.4mm$$

$$\Delta M = 100\left(1 - \frac{25.4 \times 0.7}{20}\right) = 11\%$$

Calculation for the foamed and reinforced Nylon 66

$$t_{f'} = 13.5\left(\frac{1}{0.7}\right)^{2/3}$$

$$= 17.1mm$$

The percentage mass saving compared with the Nylon 66 beam is given by

$$\Delta M = 100\left(1 - \frac{17.1 \times 0.7 \times 1390}{20 \times 1150}\right)$$

$$= 27.7\%$$

Table 3.8 The thickness and mass of the beams of equal bending stiffness.

	N66	N66/30% glass	N66/foam	N66/30% foam
Thickness/ mm	20	13.5	25.4	17.1
Mass/kg	1.15	0.94	1.03	0.83

Table 3.8 shows the thickness and masses of the beams for equal bending stiffness.

From the calculations it can be seen that the greatest saving in mass is achieved by using a foaming fraction of 0.7 and regaining the modulus of the matrix material by using glass fibre reinforcement. These kind of considerations are very useful to the product designer.

REFERENCES

D.W. CLEGG and A.A. COLLYER, Eds: *Mechanical Properties of Reinforced Thermoplastics*, Elsevier Applied Science. London, 1986.

D.W. CLEGG and A.A. COLLYER, Eds: *Irrodiation Effects on Polymers*, Elsevier Applied Science, London, 1991.

A.A. COLLYER: *A Practical Guide to the Selection of High Temperature Engineering Thermoplastics*. Elsevier Advanced Technology, Oxford, 1990.

R.J. CRAWFORD: *Plastics Engineering*. Pergamon Press Ltd, Oxford, 1983.

R.J. CRAWFORD: *Plastics and Rubber*. Mechanical Engineering Publications, London, 1985.

R.W. DYSON: *Speciality Polymers*. Blackie, Glasgow, 1987.

D.H. MORTON-JONES and J.W. ELLIS: *Polymer Products*. Chapman and Hall, London, 1986.

4 *The Viscoelastic Nature of Polymers*

INTRODUCTION

Polymers are unusual in that the vast majority of them simultaneously exhibit the properties of both elastic solids and viscous liquids, under a wide range of conditions. They are thus said to be viscoelastic in nature. As a result their response to stress is time dependent and their mechanical properties depend on the rate of deformation. One result of this is that creep rates in thermoplastics are relatively high compared with metals and ceramics at equivalent temperatures. This chapter will explain the molecular mechanisms responsible for deformation in polymers and relate this to the response of polymers to imposed stresses. The property of primary interest will be stiffness as measured from, for example, the slope of a stress strain plot — although this is not necessarily a straightforward measurement as will be seen.

THE ROLE OF MOLECULAR STRUCTURE

The stiffness and other mechanical properties of a polymeric material are governed by the mobility of the constituent polymer chains. The mobility is in turn dependent on the inherent stiffness of the chain, the degree of chain entanglement, the extent of crosslinking and the degree of crystallinity.

We begin by considering an individual molecule in an amorphous polymer. In the absence of stress the chains take up a randomly twisted or convoluted form. When a stress is applied deformation of a molecule can take place by two processes:

i) bond stretching and bond angle opening, and
ii) rotation of segments of chain about the chain backbone.

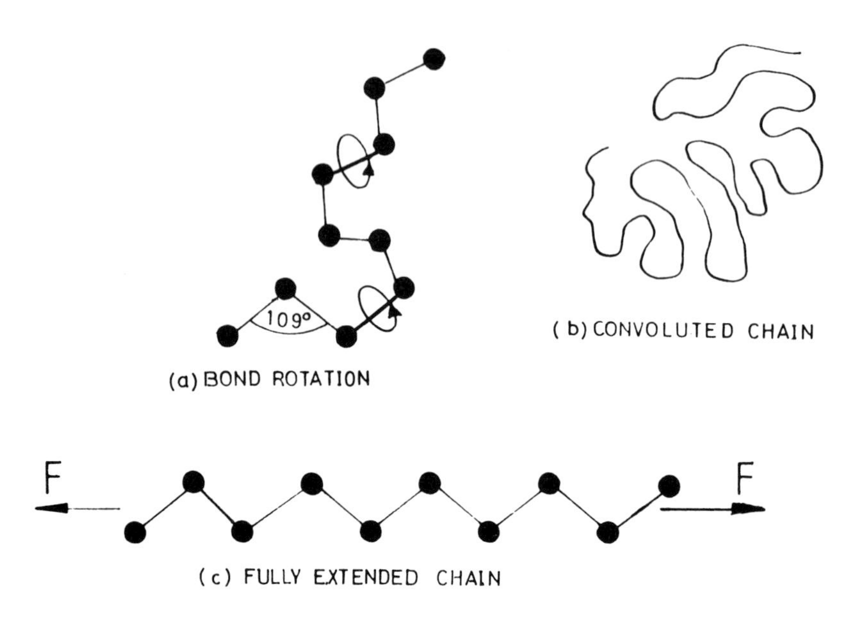

Fig. 4.1 (a) chain segment rotation. (b) a convoluted chain (c) a fully extended chain.

Below the glass transition temperature bond stretching and bond angle opening are the main deformation mechanisms. However, as the temperature is increased through the glass transition temperature individual backbone bonds are able to rotate. Not every backbone bond rotates but perhaps one in every 6 to 16 carbon atoms in a polymer with a carbon chain backbone. This means that the polymer chain is then made up of segments linked by flexible bonds. Each segment may be in the region of 1–2 nm in length. Because the carbon–carbon bonds in the backbone of a typical polymer are at an angle of 109° 28′ to one another, this ability to rotate can bring about enormous shape changes in the polymer chain.

Under stress the chain can stretch out into a linear zig-zag. This process is shown in Fig. 4.1 and is reversible. It gives rise to considerable extensibility and what is termed high-entropy or rubber elasticity. If the polymer is not crosslinked the individual chains will become disentangled and flow past each other as the stress is increased or time passes giving rise to viscous flow. Thus the viscoelastic nature of polymers can be appreciated. However, it must be appreciated that viscoelasticity is strongly affected by molecular structure and by crystallinity and crosslinking.

DEFORMATION PROCESSES

Three basic deformation responses can be identified and they will now be examined in turn.

Normal Elasticity

This is the result of bond stretching and bond angle opening and gives rise to linear or Hookean elasticity with associated high values of modulus. Sometimes the polymers are brittle in this state but this is not always the case. It is easier to explain the normal elastic behaviour of a crystalline polymer such as polyethylene where the packing of molecules is regular within a unit cell. The unit cell of polyethylene is orthorhombic and within each cell the molecules are extended in a planar zig-zag, this makes the analysis easier. The cell is shown in Fig. 1.14.

If the cell is stressed along the c axis then the molecules are stretched along their chain backbones and for simplicity we will neglect bond opening. Each unit cell contains 2 molecules on average. When a stress is applied the molecules are displaced from their equilibrium positions and the potential energy can be described by the following function:

$$U(r) = \frac{a}{2}\left(\frac{-2}{r^6} + \frac{r_o^6}{r^{12}}\right) \tag{4.1}$$

where r_o is the equilibrium interatomic separation and a is a constant.

The first term represents an attractive potential and the second a repulsive potential. The interatomic force is given by

$$F(r) = \frac{-\mathrm{d}U(r)}{\mathrm{d}r} \tag{4.2}$$

These functions are shown in Fig. 4.2.

The breaking stress is represented by the minimum in the $F(r)$ plot and the modulus by the slope. It can be shown that

$$\sigma_B = \frac{2.7\ U_o}{r_o\ A_o} \tag{4.3}$$

and

$$E = \frac{72\ U_o}{r_o\ A_o} \tag{4.4}$$

where σ_B = breaking stress, U_o = bond energy, A_o = section area of one molecule and r_o = equilibrium separation.

The bond energy of the carbon–carbon bond is approximately 335 kJ mol^{-1} = 5.6×10^{-19} J per bond and r_o is approximately 1.5×10^{-10} m. The unit cell has a section area of 36.5×10^{-20}m^2 and each cell contains 2 chains. Thus

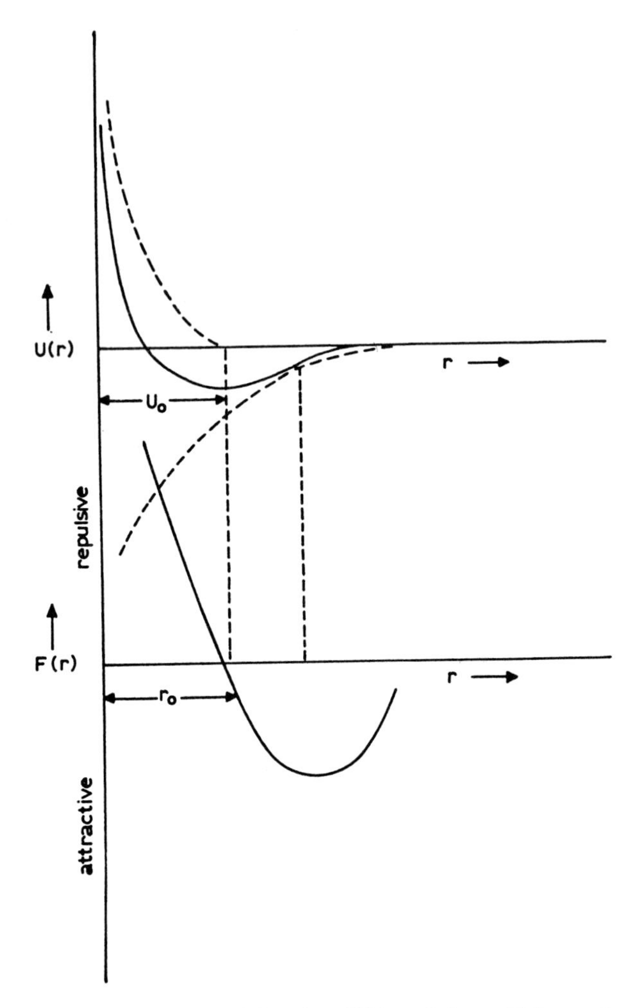

Fig. 4.2 Interatomic potential function $U(r)$ and force $F(r)$ as a function of atomic separation.

equations (4.3) and (4.4) give $\sigma_B = 5.5 \times 10^{10} \mathrm{Nm^{-2}}$ and $E = 1.46 \times 10^{12} \mathrm{Nm^{-2}}$. The value for E is rather too large as it ignores bond angle opening but it is a reasonable estimate for rigid chains. The value of σ_B is about 50 times too large as it ignores defects which give rise to premature fracture or mechanisms which give rise to plastic flow. However, the high values of modulus and low extensibilities of chains in a fully extended form can be appreciated and a similar state of affairs can be expected for amorphous chains below their glass transition temperature. However, in the latter case modulus values may well be lower because bond angle opening plays a larger part in the deformation and also a much larger degree of free volume is present allowing small chain segments and any branches present to show chain motions associated with high elasticity. Considerable scope has been demonstrated for highly oriented polymers.

High, Entropy or Rubber Elasticity

An effective theory of high elasticity has to explain various interesting properties of high elastic materials. The materials which show predominantly high elasticity are rubbers above their glass transition temperatures. Rubbers exhibit increasing elastic moduli as the temperature is increased, unlike metals. They have non-linear load-extension curves and, when crosslinked, show reversible extensions of around 600%. In addition, a stressed rubber band contracts when heated gently whereas unstressed rubber expands as expected. Most polymeric materials show some aspects of these properties under certain conditions and so it is important to have sensible theories to explain them.

A useful analogy can be made with an 'ideal gas'. Thus an ideal elastomer can be thought of as a three dimensional network of polymer chains in which the internal energy U and the volume V are independent of the deformation, and the temperature T and external pressure P are constant. The first law of thermodynamics states that:

$$dU = dQ - dW \tag{4.5}$$

where dU = the change in energy of the system, dQ = the heat absorbed by the system and dW = the work done by the system.

If a sample of rubber is stretched by a force f by an amount dx then work fdx will be done on the system. As rubbers deform at approximately constant volume the contribution pdV is negligible and so the work done by the system when stretched is $-fdx$.

We can assume that the deformation is fully reversible and so the second law of thermodynamics tells us that:

$$dQ = TdS \tag{4.6}$$

where T = temperature in Kelvin and dS = change in entropy of the system.

thus

$$fdl = dU - TdS \tag{4.7}$$

It can then be shown that

$$f = \left(\frac{dU}{dl}\right)_T - \left(\frac{TdS}{dl}\right)_T \tag{4.8}$$

In most materials the first term dominates this expression but in rubbers it is very small and the largest contribution to the force is from the change in entropy, i.e. the second term. The entropy is related to the degree of molecular order and increases as the disorder increases.

The above conventional thermodynamic approach is informative but limited. By adopting a statistical approach it becomes possible to relate the molecular structure changes during deformation to changes in entropy and to derive theoretical expressions for the stress strain response of a rubber-like material. The entropy, S, depends on the number of ways or conformations, Ω, in which a polymer chain can be arranged. This is given by the Boltzman's relationship

$$S = k \ln\Omega \qquad (4.9)$$

where k = Boltzman's constant.

When the material is unstressed a molecule in the network can take up a large number of random, coiled conformations switching from one to another bond rotation. If the material is stretched the chain is extended so reducing the number of conformations so that the state of order and the entropy decreases. This entropy change produces a force which counteracts the applied force and gives rise to what may be called entropy elasticity. When the applied force is removed a rubber contracts so that energy of the system can be minimised by increasing the number of chain conformations and hence the entropy. This is quite unlike normal crystalline solids which we may say exhibit energy elasticity. For the case of uniaxial tension the entropy change S is related to the instantaneous length by

$$S = \frac{Nk}{2}\left(\frac{x^2}{x_o} + \frac{2x_o}{x} - 3\right) \qquad (4.10)$$

where N is the number of network chains per unit volume, x_o is the undeformed length and $\frac{x}{x_o}$ is called the extension ratio.

If equations (4.8) and (4.10) are combined the following expression is obtained:

$$\sigma = G\left(\frac{\lambda - 1}{\lambda^2}\right) \qquad (4.11)$$

where G is the shear modulus (= NKT)

The stress extension ratio response is shown in Fig. 4.3 for cross-linked rubber for both extension and compression. The match with experimental values is good at lower extension ratios.

Clearly the elastic modulus varies with deformation but perhaps the initial modulus is of greatest interest. Network chains are anchored by two cross-links and assuming that defects such as loops and chain ends are absent the density can be given by

$$\rho = \frac{N M_c}{N_A} \qquad (4.12)$$

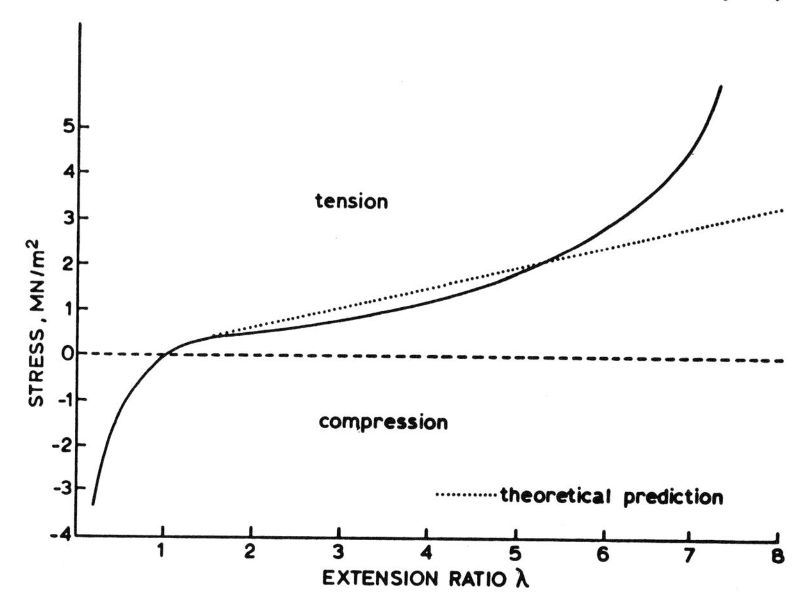

Fig. 4.3 Stress versus extension ratio for a rubber showing experimental and predicted curves for tension and compression.

where N_A is the Avogadro number and M_c is the number average molar mass of network chains.

The shear modulus then becomes

$$G = \frac{\rho R T}{M_c} \tag{4.13}$$

For an incompressible material the tensile modulus, E, is equal to 3G, assuming that the Poisson ratio, υ, is 0.5.

Equation (4.13) yields some very interesting results. It predicts that as the length of network chain is increased (M_c increasing) the modulus reduces. Therefore an increase in crosslink density increases the modulus. Also the increase in modulus with temperature is explained.

Although the above brief outline of the theory of entropy elasticity applies to rubbers (elastomers) it is worth emphasising again that many polymers under certain conditions exhibit rubber-like properties and so elements of the above behaviour can be expected.

Viscous Flow

This is dealt with elsewhere (Chapter 7) but essentially occurs when polymer molecules are able to move relative to one another under the application of a stress field. This process involves considerable molecular flexibility as molecules need to disentangle themselves from one another and to stretch out in the direction of the flow. Thus flow occurs most readily at temperatures significantly above the glass transition temperature of the polymer and is suppressed by crystallisation and crosslinking. Some viscous flow may be possible below the glass transition temperature as limited segmental rotation in the backbone or in side chains may take place. These segmental motions do not occur instantaneously but require a finite time depending on the inherent segmental mobility, dictated by the molecular architecture and the free volume. The free volume decreases with decreasing temperature.

VISCOELASTICITY

We assumed in the previous section that an elastomer or rubber exhibits 'ideal' behaviour in that it always returns to its original dimensions after the removal of an imposed stress and that equilibrium deformations are always achieved. Crosslinked elastomers may come close to this but other polymeric materials, such as thermoplastics, behave differently. Thus the majority of polymeric materials are non-reversible and do not achieve equilibrium behaviour even at very long times. This is a result of viscous flow processes mentioned above and is most significant, as stated in the previous section, above T_g. Thus the response of a polymer to stress is a combination of normal elastic, entropy elastic and viscous flow processes. The contributions of each of these is controlled by the temperature. The viscous and entropy elastic processes which involve extensive molecular rearrangements are time dependent and the overall response is said to be viscoelastic. This is illustrated in Fig. 4.4a, which shows the overall strain response of a linear polymer with time for a constant stress. Fig. 4.4b shows the components of the strain which are the instantaneous elastic e_i the retarded (time dependent) entropy elastic e_r and the flow component e_f.

The overall strain responses at different temperatures are shown in Fig. 4.5. At temperatures above T_g viscous flow dominates and at temperatures below T_g instantaneous or normal elasticity is the dominant process.

It can be seen that viscoelasticity gives rise to creep, stress relaxation and damping. It is also responsible for the observation that the response of a polymeric material in a tensile or similar test is likely to be dependent on the rate of testing. For example the tensile modulus would increase as the strain rate increases.

Creep is the time dependent deformation of a material under constant stress. Stress relaxation is the time dependent decay of stress when under constant deformation. Creep is very important in stressed engineering components such as beams and cables as excessive deflections and failure can

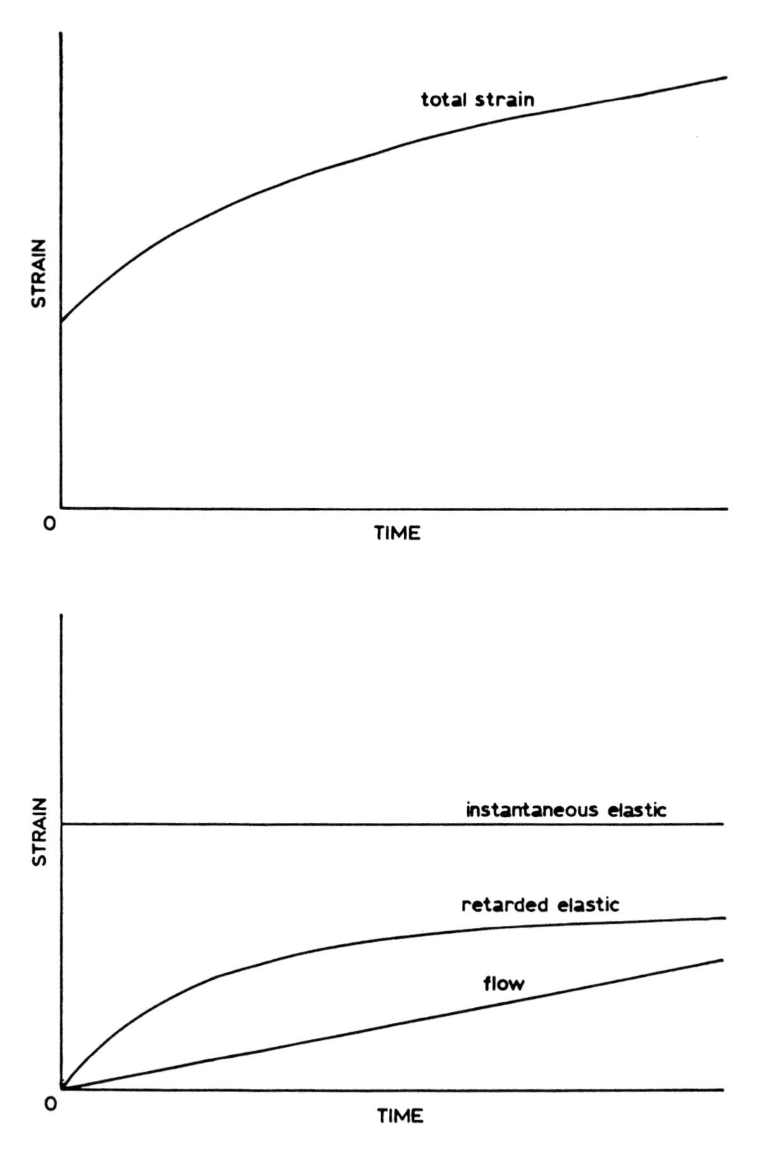

Fig. 4.4 (a) Total strain versus time for a constant stress applied at zero time. (b) Components of the total strain.

occur. Stress relaxation is important in gaskets and seals of all types because as the stress in the sealing material decays the effectiveness of the seal is reduced. These manifestations of viscoelastic behaviour can be described by phenomenological models made up of Hookean springs (σ = Ee) and Newtonian

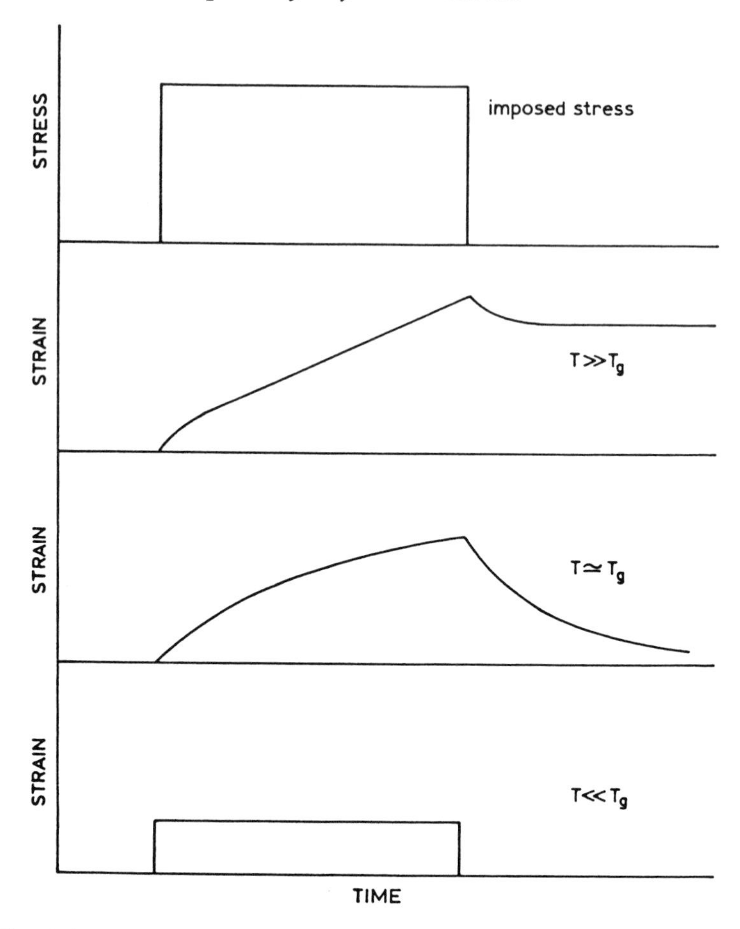

Fig. 4.5 Strain responses to an imposed stress cycle for a linear polymer at various temperatures.

dashpots ($\sigma = \eta\dot{e}$, where η is the tensile viscosity and $\dot{e}$ is the strain rate). They are shown in Fig. 4.6.

The idea is that the springs and dashpots can be used to represent the deformations of long, flexible molecules as they respond in a time dependent way to imposed stresses. Obviously the springs contribute the elastic component and the dashpots the flow and time. However, the springs and dashpots have no real physical significance as regards polymer molecules. These mechanical models can only be used if it assumed that polymers are linear elastic. This means that the elastic modulus (stress/strain) is assumed to be a function of time and not of strain. This is true at reasonably small strains but not at high strains. Rigid polymers can be assumed to be linear elastic up to strains of about 0.5%. More advanced treatments are required for non-linear elastic materials.

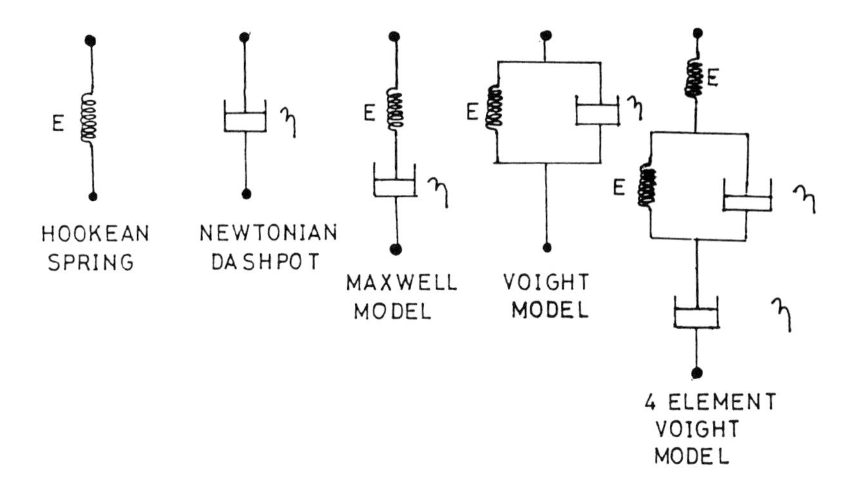

Fig. 4.6 Mechanical models representing viscoelastic behaviour.

Stress Relaxation

Stress relaxation may be modelled by a single Maxwell element. The total strain is the sum of the strains of the elastic spring and of the viscous dashpot. After the element is initially deformed to a fixed strain the subsequent strain rate is zero, i.e.

$$\frac{de}{dt} = \frac{1}{E}\frac{d\sigma}{dt} + \frac{\sigma}{\eta} = 0 \tag{4.14}$$

where $e = \frac{\sigma}{E}$ for the spring (E is the spring constant), σ is the stress at time t, $\tau = \frac{\eta}{E}$ and is the characteristic relaxation time of the system and determines how rapidly the modulus decays with time.

Fig. 4.7 shows equation (4.14) plotted on a logarithmic time scale for a relaxation time of 1 second.

All of the initial contraction takes place in the spring while the dashpot relaxes slowly and enables the spring to expand slowly transferring the initial deformation from the spring to the dashot. Most of the deformation takes place within one decade of time on both sides of the relaxation time. On a log scale the curve has a maximum slope of e^{-1} at time $t = \tau$.

A stress relaxation modulus E(t) $(= \frac{\sigma_t}{e})$ may be defined;

$$E(t) = \frac{\sigma_t}{e} = \frac{\sigma_o}{e}\, e^{-t/\tau} \tag{4.15}$$

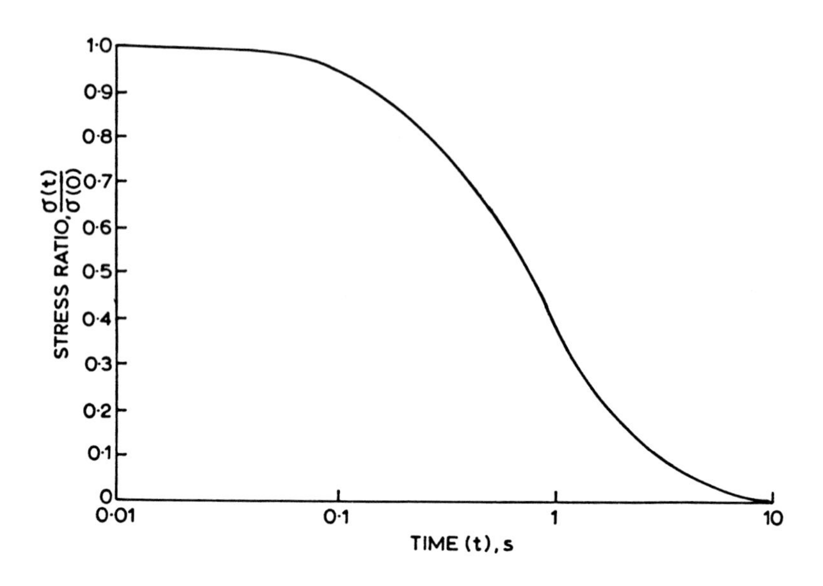

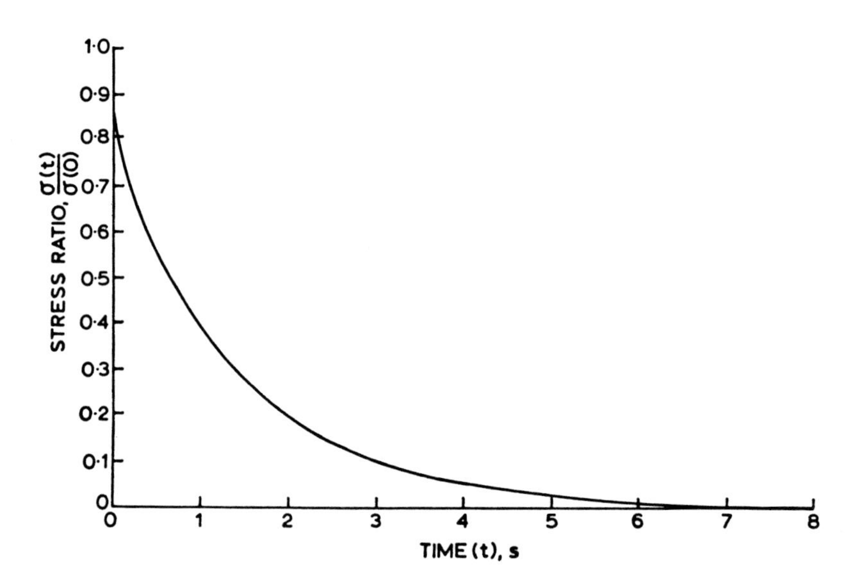

Fig. 4.7 (a) Stress relaxation of a Maxwell element plotted on a log scale. $\tau = 1$ second. (b) The same data plotted on a linear scale.

This should be used instead of the instantaneous elastic modulus in engineering design calculations based on stress relaxation data for polymers and elastomers.

Creep

Creep may be modelled by a 4 element Voigt model. When a constant stress is applied the initial elongation occurs in the single spring (E_1). Later elongation occurs in the other spring (E_2) and the parallel dashpot (η_2) and also from the dashpot in series (η_3). The total elongation is the sum of the individual components

$$e = \frac{\sigma_o}{E_1} + \frac{\sigma_o}{E_2}(1 - e^{-t/\tau}) + \frac{\sigma_o t}{\eta_3} \qquad (4.16)$$

where σ_o is the applied stress and $\tau = \frac{\eta_2}{E_2}$ is the characteristic relaxation time of the system.

If the stress is removed at some time the creep is completely recovered apart from the viscous flow component originating in the dashpot η_3. Examples of creep and creep recovery are given in Fig. 4.8 for a single 4 element Voigt model. This model predicts that most creep occurs within one decade on either side of the relaxation time.

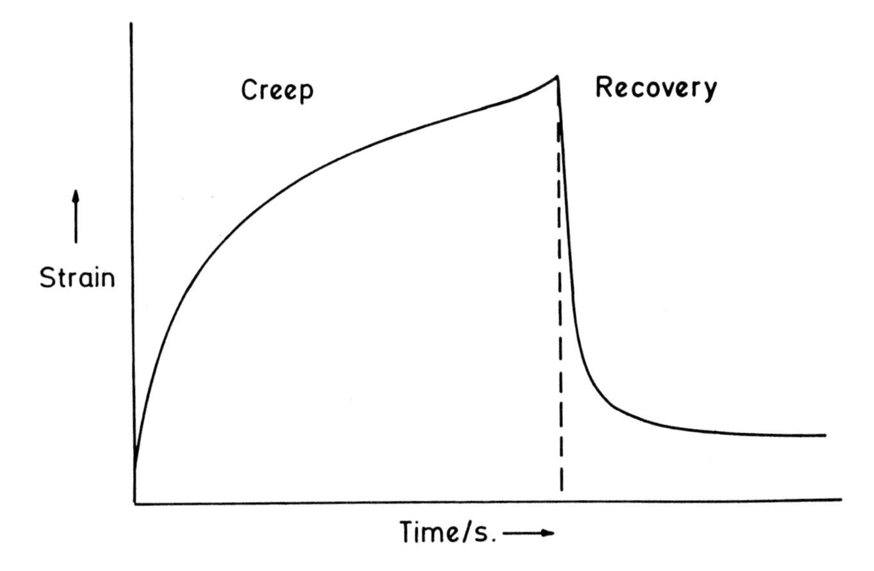

Fig. 4.8 Creep and recovery curves modelled by a 4 element Voigt model.

Distribution of Relaxation and Retardation Times

Real polymers creep and relax over much broader time scales than suggested by the simple models described above. In other words they do not have single relaxation or retardation times. Therefore, the phenomenological models should consist of many Maxwell or Voigt elements in parallel each element having different characteristics. If the distributions are chosen correctly then experimental data can be fitted with reasonable precision. Distributions of relaxation and retardation times can be derived from experimental data and can be used to calculate other viscoelastic data, e.g. dynamic mechanical data from stress relaxation data. This topic is covered in more advanced books.

SUPERPOSITION

There are two superposition theories which are found to be useful in the application of the viscoelastic theory of polymers. They are the Boltzman principle and the time–temperature (W–L–F) theory.

Boltzman Principal

This states that the response of a material to a given stress is independent of the response of the material to any stress already imposed on the material. The effects of different stresses is simply assumed to be additive. It can be applied to both creep and stress relaxation equally well. For example, take the case of stress relaxation. Consider the effect of imposing an initial strain e_o on a material at time t_o followed by increasing the strain to e_1 at a later time t_1. Finally at time t_2 the strain is increased to e_2. This sequence of events and the associated stress relaxations are illustrated in Fig. 4.9.

The stress at any time is found by simple linear superposition and is given by

$$\sigma(t) = e_o E_r(t_o) + (e_1 - e_o)\, E_r\,(t_1 - t_o) + (e_2 - e_1)\, E_r\,(t_1 - t_o) \quad (4.17)$$

for $t_2 > t_1 > t_o$.
A similar expression would apply for the case of creep.

Time – Temperature Superposition

In the case of amorphous polymers it is found that a plot of relaxation modulus at constant time versus temperature yields the same shape curve as the plot of modulus versus log time for the same material as shown in Figs 4.10a and b.

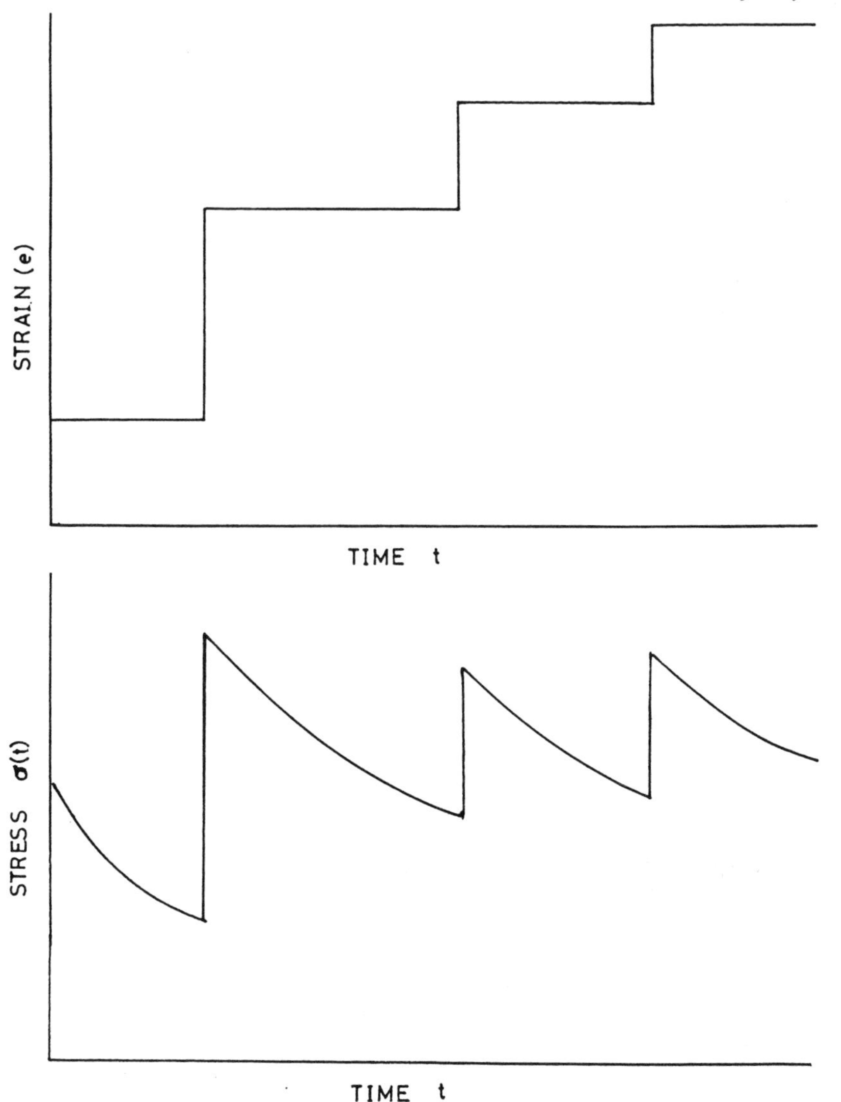

Fig. 4.9 Illustration of Boltzman's superposition principle.

This implies that there is an equivalence between temperature and time. The most notable contributions to this idea have been made by Tobolsky and by Williams, Landel and Ferry. The latter developed the important method of shifting data horizontally along the log time scale by amounts related to temperature differences (W–L–F method).

Consider the stress relaxation data obtained for an amorphous polymer at different temperatures. Each separate curve is derived over a relatively few

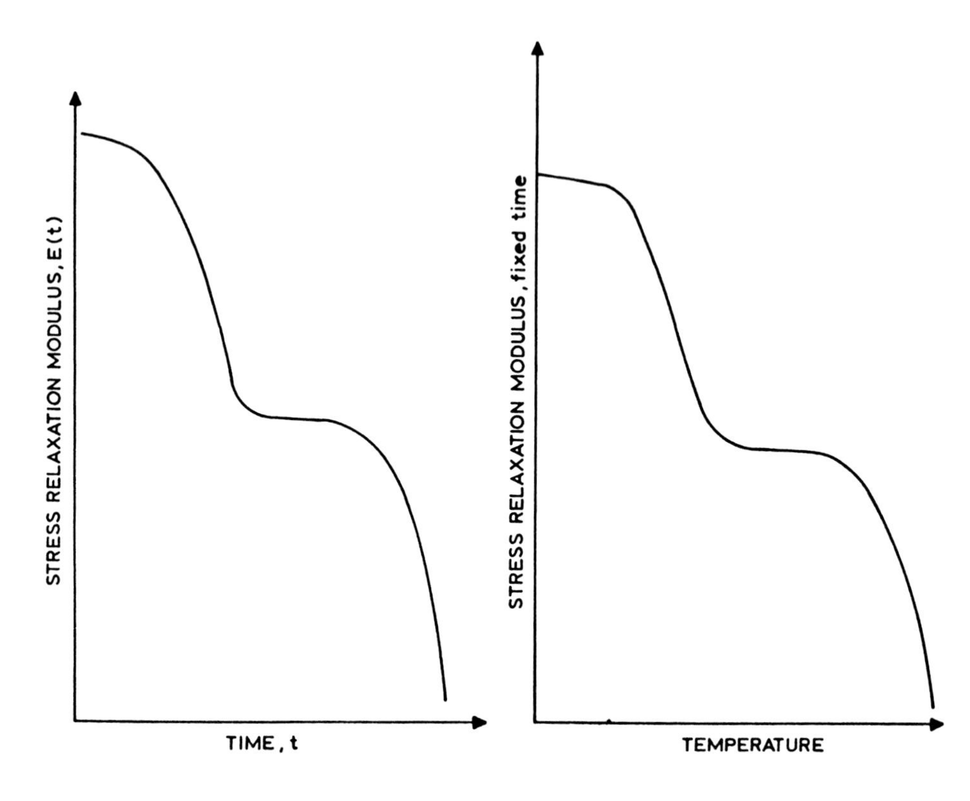

Fig. 4.10 (a) Isothermal stress relaxation behaviour for a high molar mass linear polymer. Stress relaxation modulus plotted as a function of time (b) Stress relaxation modulus (constant time) plotted against temperature for same polymer as in (a).

decades of time but by shifting each curve by a predetermined shift factor, a_T, to a reference temperature a master curve spanning many decades of time can be constructed. Data well outside the range of normal experiments can be obtained in this way. The reference temperature chosen is often the glass transition temperature T_g. This principal is illustrated in Fig. 4.11.

The temperature dependence of the shift factor cannot be satisfactorily described by an Arrhenius relationship over a wide temperature range. The viscoelastic behaviour can only be modelled in this way at higher temperatures. However, Williams, Landel and Ferry derived the following well known empirical relationship for a_T

$$\log a_T = \frac{17.44\,(T - T_o)}{51.6 + T - T_o} \tag{4.18}$$

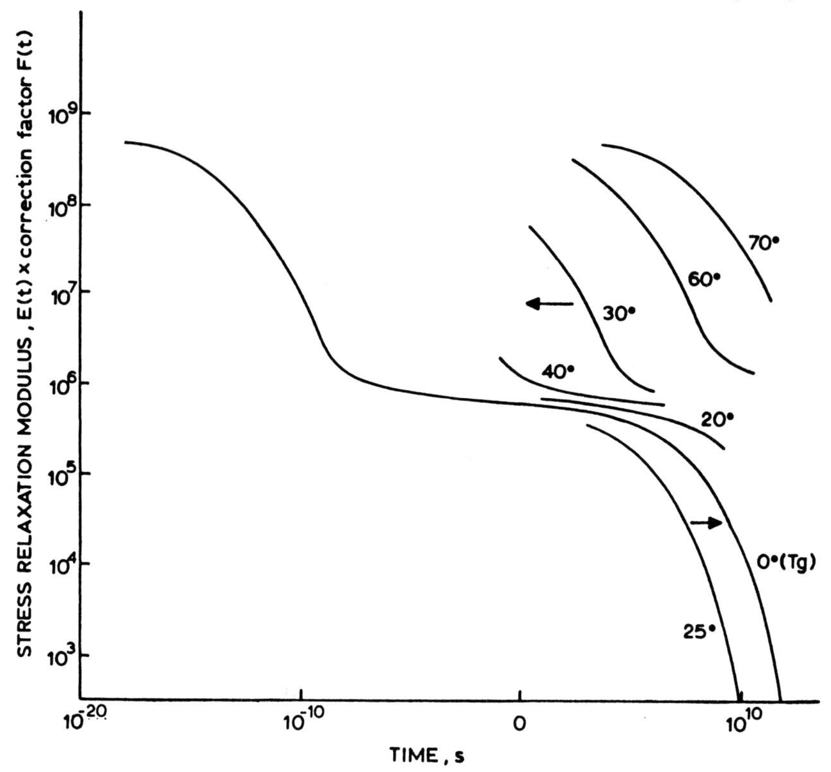

Fig. 4.11 W–L–F time–temperature superposition applied to typical stress relaxation data for an amorphous linear high molar mass polymer. Relaxation data was obtained at various temperatures to obtain a master curve.

where T_o is the reference temperature (often T_g) and T is the temperature at which the data was obtained. (NB The temperatures are in degrees Kelvin.)

It can be seen in Fig. 4.11 that the vertical stress relaxation modulus (E_t) values are multiplied by a correction factor, $F(t)$. This factor is T_g/T above T_g and approximately unity below T_g. It arises from the kinetic theory of rubber elasticity and as a result is only of great significance above T_g where the polymer is behaving in a rubber-like manner.

The stress relaxation curves obtained above T_g are shifted one by one to the right along the log time axis until they superimpose. Curves obtained at temperatures below T_g are shifted in the same way to the left. The long plateau in the master curve is due to chain entanglements acting as temporary crosslinks in high molar mass polymers. Since time and temperature are equivalent the temperature scale can be replaced by a time scale as indicated in Fig. 4.11.

The above approach works best with amorphous polymers. For semi-crystalline polymers problems are encountered at temperatures near the melting temperatue. This is because structural changes take place as the melting temperature is approached. For instance crystalline lamellae become thinner and the rigidity of the material is reduced. This can be taken into account by using a vertical shift as well as the horizontal shift factor. It follows that other temperature dependent structural changes will also make application of the W–L–F method more difficult.

DYNAMIC BEHAVIOUR

The dynamic behaviour of polymers is of especial interest because in many cases the applied stresses or strains vary in a periodic fashion and so it is necessary to know how polymers behave under these conditions. Damping is a particularly interesting property in this respect. In addition dynamic methods of testing polymers yield more information than simple creep or stress relaxation tests because they can be carried out over a much wider range of temperature and frequency. The results of these tests are very sensitive to the molecular structure of polymers and provide very sensitive means of detecting the glass transition, crystalline melting temperature and other changes.

If a sinusoidal stress is applied to a polymer the strain lags behind the stress because of the viscoelastic nature of the material. This can be thought of as a damping process where the strain also varies sinusoidally but behind the stress (*see* Fig. 4.12).

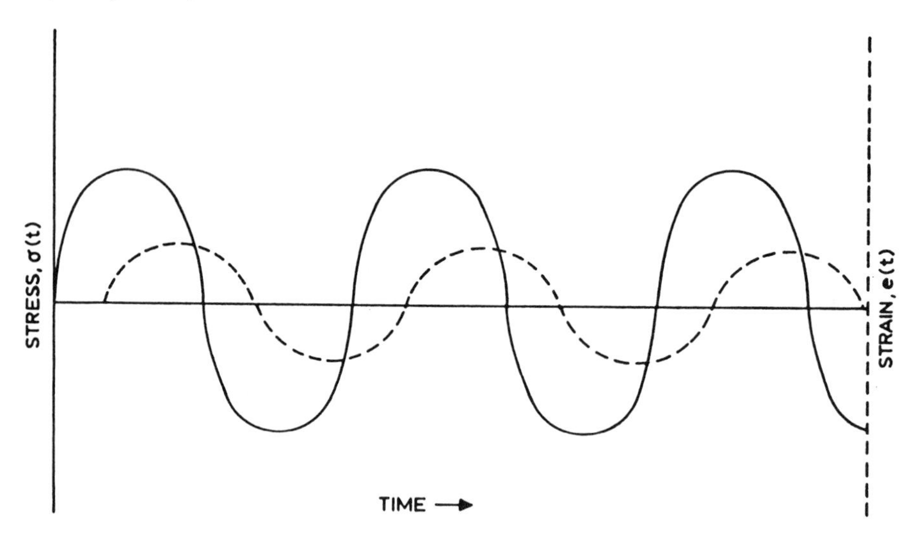

Fig. 4.12 The variation of strain with imposed stress cycling for a viscoelastic material.

Dynamic mechanical quantities are usually given in complex notation but a further explanation of their derivation is given by Young. The stress σ and strain e are given by

$$e = e_o \exp i\omega t$$

and

$$\sigma = o_\sigma \exp i(\omega t + \delta) \qquad (4.19)$$

Complex moduli are defined by

$$E^* = E' + iE'' \qquad (4.20)$$

where $i = (-1)^{1/2}$, E' is the real part of the modulus and E'' is the imaginary part of the modulus.

These expressions can be written in exactly the same way in terms of the shear modulus, i.e. G^* etc.

E'' can be shown to derive from a stress component which lags behind the main stress component, which gives rise to E', by a phase lag angle δ. This phase angle is given by

$$\tan \delta = \frac{E''}{E'} \qquad (4.21)$$

E' is often referred to as the storage modulus because energy is stored by the elastic component which is in phase with the imposed stress cycle. E'' is called the loss modulus because it is related to the energy dissipated during each cycle.

Dynamic mechanical properties of solid polymers can be measured using a torsion pendulum or using more sophisticated instruments which offer close control over frequency and temperature. They go under the general description of dynamic mechanical analysers. Dynamic mechanical analysis is a powerful tool for the investigation of transitions in polymer structure.

Temperature Dependence of Dynamic Properties

The temperature dependence of dynamic properties is investigated by keeping the frequency constant and varying the temperature. Results obtained for natural rubber are shown in Fig. 4.13.

As expected the storage and loss moduli show large drops at the glass transition temperature, with the loss modulus E'' showing a small peak just below T_g. However, the tan δ or damping curve shows a large peak just above T_g, the exact position depending on the frequency. Little damping occurs below T_g as nearly all the energy stored in deformation is recovered very quickly. In the glass transition region some of the molecular chain segments are able to move by bond rotations while others are fixed in position. When a stress is applied some of the segments move quickly. The fixed segments may

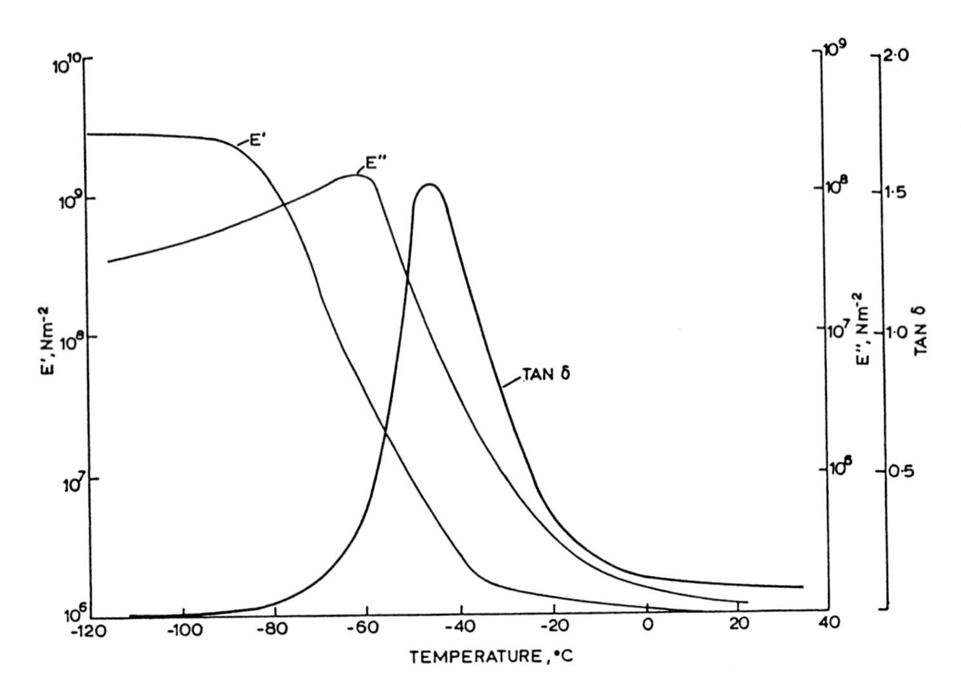

Fig. 4.13 The dynamic mechanical properties of rubber as a function of temperature.

begin to move after some delay but after moving they have less stored energy because the stress has reduced. The excess energy is dissipated as heat. The delayed response causes the deformation to lag behind the imposed stress by the phase angle δ. Thus the damping given by tan δ rises to a peak. Above T_g tan δ drops.

The 4 element mechanical model used to simulate creep behaviour can be used to show how damping changes with temperature. The dynamic mechanical properties of this model are shown plotted against temperature in Fig. 4.14. The viscosities of the dash-pots are assumed to reduce with temperature. At low temperatures the dash-pots will be effectively rigid and the spring E_1 will take up the deformation. The damping will be very low. Assuming that $\eta_1 > \eta_2$ then as the temperature is raised the parallel spring and dash-pot combination will deform but lag behind the deformation of E_1. The stress and strain are out of phase. The dash-pot η_1 dissipates energy as heat and the damping is thus high. At higher temperatures the viscosity of η_2 is still high but η_1 is now so low that it responds rapidly to the stress. The modulus is now lower and the damping small because η_1 is small. The damping increases again at still higher temperatures as the dash-pot η_2 begins to

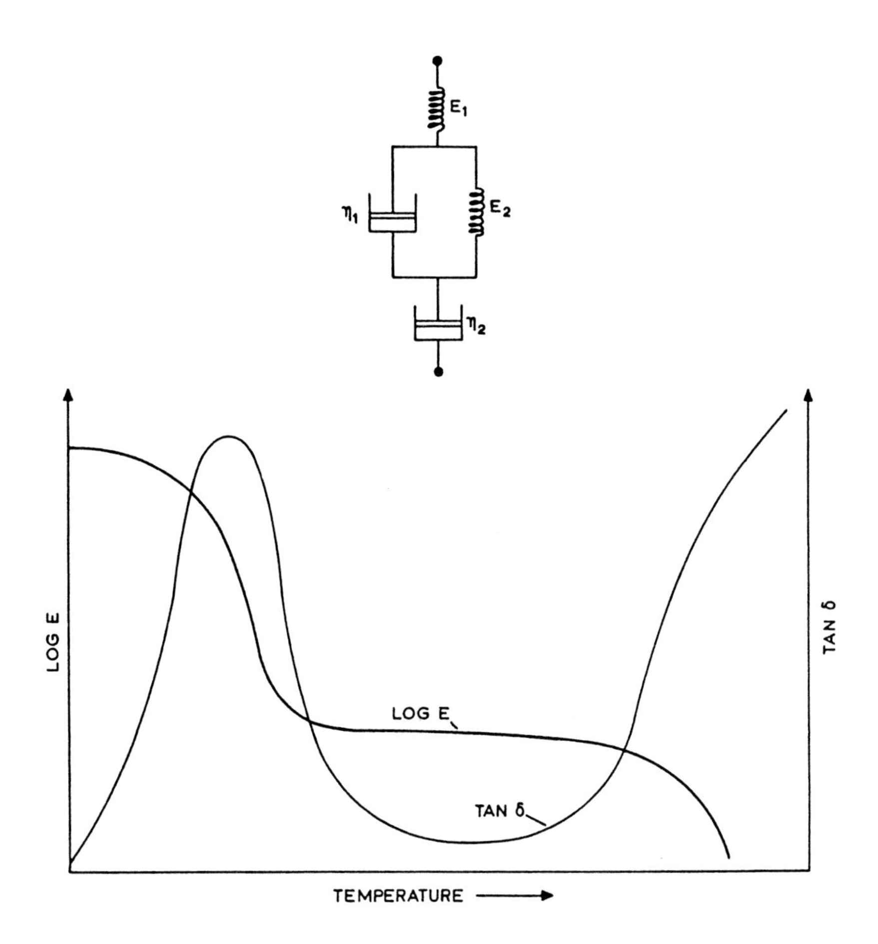

Fig. 4.14 Dynamic mechanical response of a 4-element model as a function of temperature (after Nielsen).

operate and dissipate energy. This would be similar to the behaviour of a molten polymer where the flow is the result of molecular slippage.

Frequency Dependence of Dynamic Properties

If the temperature is held constant and the test frequency varied then the values of E', E'' and tan are found to vary with frequency as shown in Fig. 4.15. These curves can be explained in a similar way to the temperature

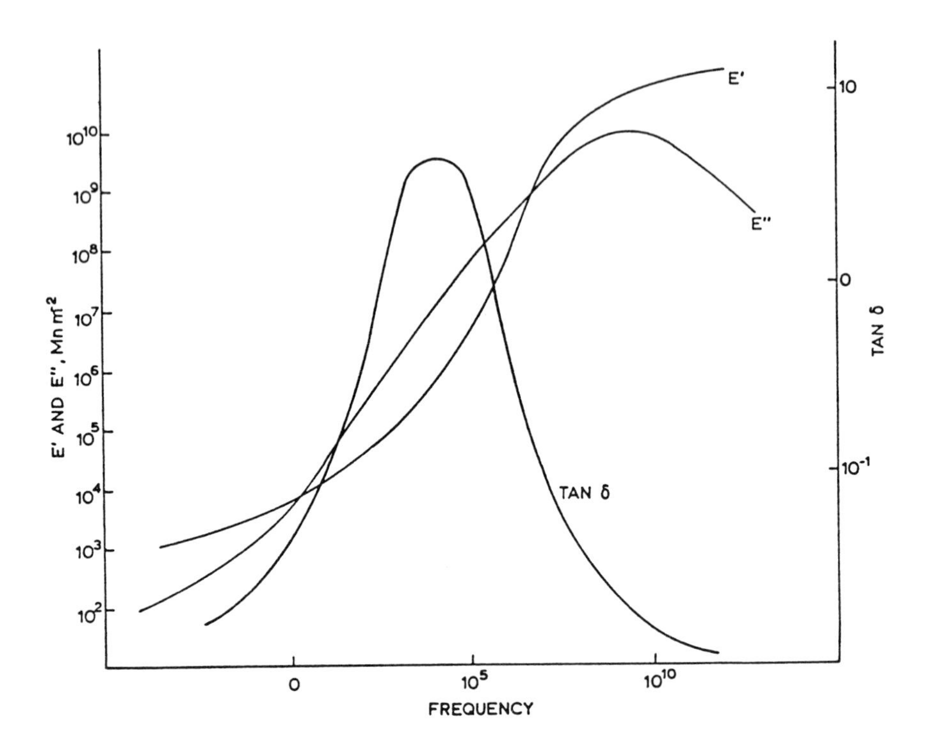

Fig. 4.15 Variation of dynamic properties with frequency for a polymeric material.

dependence above and it is possible to relate temperature and frequency data by the time–temperature superposition principle described previously. The frequency dependence of dynamic properties is of particular interest because damping peaks occur at characteristic frequencies of particular molecular motions within the polymer structure. Thus considerable information on molecular motions in polymers can be obtained. For instance a major peak in E'' occurs at T_g when the frequency equals the natural frequency for main chain rotation. In other words the time necessary for chain movement is matched by the time scale of the applied stress cycle. Above this frequency the material appears stiff as insufficient time is allowed for movement. Below this frequency the material is rubbery and of low modulus because there is ample time in every stress cycle for movement to occur. Melting transitions and secondary transitions can also be detected.

EFFECTS OF STRUCTURE ON VISCOELASTIC PROPERTIES

Viscoelastic properties are affected by a wide range of structural parameters. They include molar mass and its distribution, degree of crystallinity, orientation, degree of crosslinking, copolymer type, physical blend constitution and morphology and the presence of plasticisers. A full account of these effects are given by Nielsen.

BIBLIOGRAPHY

L.R.G. TRELOAR: *The Physics of Rubber elasticity*, Clarendon Press, Oxford, 1958.
J.D. FERRY: *Viscoelastic Properties of Polymers*, Wiley, New York, 1970.
A.G. FREDRICKSON: *Principles and Applications of Rheology*, Prentice Hall, Englewood Cliffs, NJ, 1964.
A.V. TOBOLSKY: *Properties and Structure of Polymers*, Wiley, New York, 1960.
L.E. NIELSEN: *Mechanical Properties of Polymers and Composites*, Marcel Dekker, New York, 1974.
R.J. YOUNG: *Introduction to Polymers*, Chapter 5, pp. 211–256, Chapman Hall, London, 1983.

5 *The Mechanical Properties of Polymers*

INTRODUCTION

The viscoelastic behaviour of polymers was explained in Chapter 4 and the results in terms of creep and stress relaxation were demonstrated. The latter become important over relatively long times and are primarily the result of high elasticity and viscous flow processes. In this chapter we will look at mechanical deformation behaviour over relatively short time scales and at higher stresses leading to yielding and/or fracture. It will then be seen that some polymers have very useful mechanical strengths which can be enhanced by careful processing.

These strengths are especially attractive when assessed relative to the density of a polymeric material, i.e. specific strength. Hence the advertising slogan 'pound for pound stronger than steel', which was, at one time used to advertise the merits of nylon. In this case the strength of nylon is boosted by cold drawing the nylon fibres after spinning. This results in orientation of molecules and crystallites in the draw direction (the long axis of the fibre). A fibre with exceptionally good strength results. Orientation is an extremely useful method of enhancing strength in selected directions and research in this area is being actively pursued to produce ultra high strength polyethylenes and other polymers.

Some of the mechanical properties of polymers can be determined by carrying out stress–strain tests using methods similar to those employed for testing metals. However, one significant and very important difference is found and that is the observation that polymers are generally rate sensitive as a direct result of their viscoelastic nature. For instance, in many cases, if a test is carried out at a higher strain rate a higher tensile modulus will be recorded. Hence it is very important to standardise test conditions.

Because the stress–strain test is easy to carry out, it is very popular and may be used to give information about the tensile (Young's) modulus, yield stress,

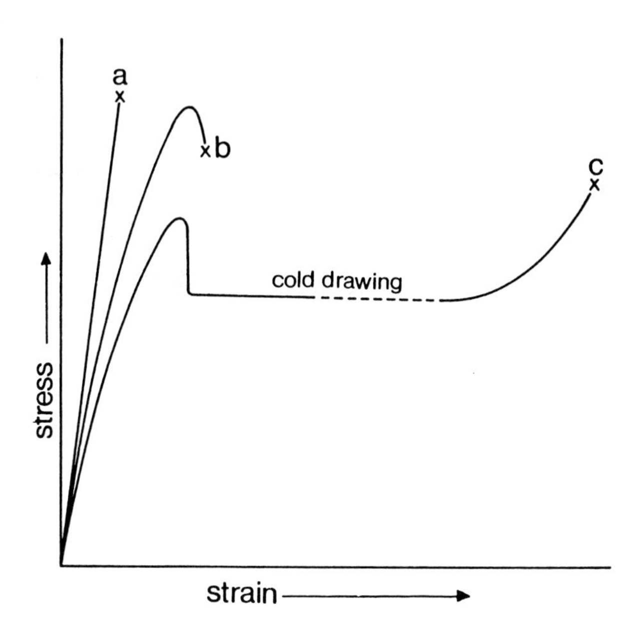

Fig. 5.1 Typical stress–strain responses for different polymeric materials (a) brittle behaviour, e.g. polystyrene (amorphous) below its T_g (b) tough behaviour, e.g. polycarbonate (amorphous) below its T_g and (c) tough behaviour with cold drawing, e.g. Polypropylene (partially crystalline).

tensile strength, fracture stress, toughness and ductility of a polymer. Fig. 5.1 shows several observed behaviours for different types of polymeric materials. While some materials exhibit brittle behaviour, others yield in a ductile manner. Exactly what happens depends on many factors such as temperature and strain rate. Thus it is possible to change from ductile to brittle behaviour in a polymer by a fairly small change in for example temperature. Polystyrene in its untoughened form is brittle below its T_g with near linear elastic behaviour. However, as the temperature is raised above T_g its behaviour changes to viscoelastic with ductile behaviour giving unstable necking characteristics. Care must be exercised in that other glassy polymers may behave differently. For example, polycarbonate is tough and ductile below its T_g. The reasons for this will be explained later.

Many partially crystalline polymers, such as polyethylene, polypropylene and polyamide show rather different behaviour, as shown in Fig. 5.1(c). A distinctly non-linear pre-yield behaviour is evident followed by a yield point. After this the stress drops but stabilises during further straining until a final rise in stress and failure. During the region of constant stress a stable neck is formed which extends until all available material is exhausted. During this cold drawing process molecules are being extended in the direction of stress to cause severe orientation. Molecules may originate from the amorphous regions of the microstructure and/or from the crystallites. However, it is assumed that some orientation of crystallites occurs first, followed by intense general molecular orientation. The result is a highly oriented necked region which is much stronger than the un-necked material. The necked material is highly oriented with molecules running predominantly in the tensile axis. Thus only weak secondary bonds attract the molecules at right angles to the tensile axis. As a result there is a tendency for the drawn region to separate into fibrils and this fibrillation process precedes failure.

The question as to whether a material forms a stable neck or not is very important with regard to the production of fibres, filaments, wires and so on. It is possible to form fibres from some polymers by drawing without a die but this is not possible with glasses or metals. Apart from cold drawing many polymer 'melts' form stable necks at temperatures above their T_g for amorphous polymers or above T_m for partially crystalline polymers.

A little thought about the yield process will help to explain the reasons behind the above observations. In a stress–strain test the stress is usually measured as engineering stress, i.e. force/original cross-sectional area. However, if true stress σ_r is used, i.e. force divided by true instantaneous cross-sectional area A a clearer picture of what is happening emerges. It should also be noted that during deformation the volume remains constant, unless extensive crazing occurs. Hence the linear strain is equal to the negative of the fractional reduction in cross-sectional area. A negative sign is necessary because as the length increases, the area decreases.

$$\frac{dl}{l} = -\frac{dA}{A} \tag{5.1}$$

The strain rate with respect to time can be written as:

$$\frac{de}{dt} = -\frac{1}{A} \times \frac{dA}{dt} \tag{5.2}$$

It is observed that a power law relationship is observed relating strain rate to true stress such that

$$\frac{de}{dt} = k \times \sigma_\tau^{\,n} \tag{5.3}$$

where k is a constant

Combining equations (5.1) and (5.2) and rearranging, the following relationship is obtained:

$$\frac{dA}{dt} = -k \times \frac{F^n}{A^{n-1}} \tag{5.4}$$

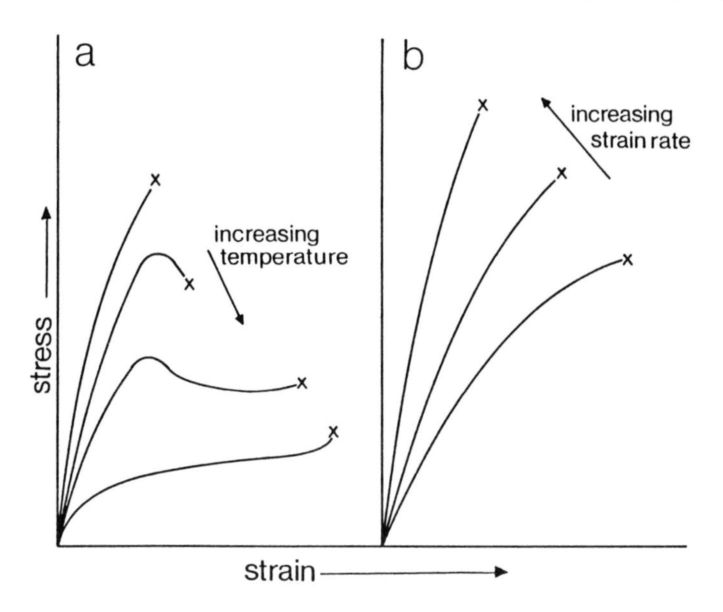

Fig. 5.2 (a) the effect of temperature and (b) strain rate on the stress–strain behaviour of a polymer.

In the case of some thermoplastics when in the 'molten' condition *n* is approximately 1. This means that the rate of reduction of area does not depend on the area. Thus all cross-sections shrink at the same rate and localised or unstable necking does not occur. The same situation also applies to inorganic glasses. Thus polymers and glasses can be drawn down into fibres without the use of dies. However, metals have a value of *n* of about 4. Thus local constrictions in the cross-section shrink faster than the surrounding regions and necking is unstable. The result is that metals cannot be drawn down into wires without the use of dies.

Stress–strain tests on polymers are very dependent on the rate and temperature of testing, mainly because test temperatures are usually close to major transition temperatures. Remember that time and temperature are interchangeable parameters (Chapter 4). Fig. 5.2 shows the effects of changing the temperature and strain rate on the stress–strain response of a polymer.

Whereas it is easy to measure the Young's modulus and yield stress of a metal sample, it is often considerably more difficult in the case of polymeric materials. For instance polymers are usually non-linear in their elastic responses, i.e. their stress–strain response in the elastic region is not a straight line from which the slope can be easily measured. This can be overcome by using a secant modulus as shown in Fig. 5.3. An alternative possibility is to use a tangent modulus, as shown. Clearly the strain at which the modulus is

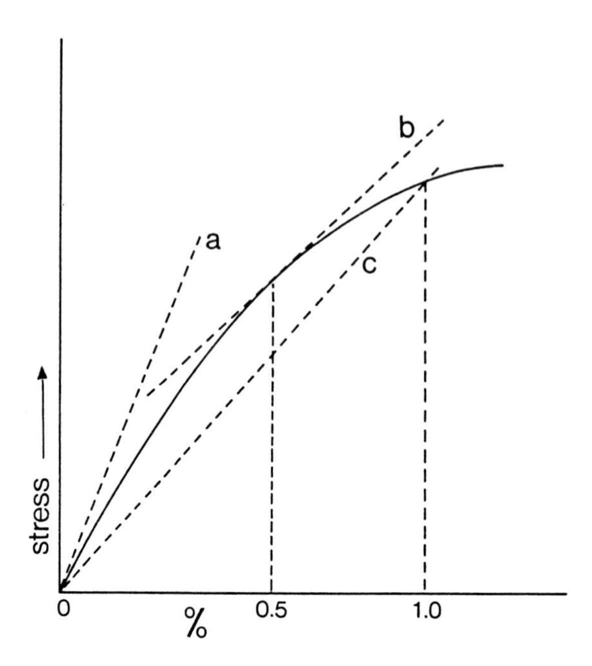

Fig. 5.3 The measurement of the various tensile moduli of a polymer (a) initial modulus at 0% strain (b) tangent modulus at 0.5% strain and (c) secant modulus at 1.0% strain. The secant modulus at 0.2% or 1.0% strain is the one normally used.

to be measured must be specified and the value of the measured modulus depends on the strain at which it is measured. It also depends on the temperature and testing rate as will be apparent from Fig. 5.2. Thus it is very important to specify the testing rate and temperature when measuring modulus and other properties of polymers.

Another problem is the specification of yield stress, beyond which deformation is permanent. In practice this is a difficult property to specify for a polymeric material and so the yield stress is usually taken as the maximum stress, which may well be the fracture stress. The corresponding yield strains may be in the region of 10% which is very much higher than that encountered in metals and ceramics.

YIELDING OF POLYMERS

Yielding of a material is usually assumed to involve a uniform plastic deformation throughout the bulk of the material and with no change in volume. This is the case in some polymers, both crystalline and amorphous. However, an alternative mechanism of yielding called crazing is also observed in some polymers. It is more likely to be found in thermoplastics which are in the glassy state.

Shear Yielding

Polymers which show this type of behaviour possess considerable ductility and toughness even below their glass transition temperatures. They also show strain softening behaviour. This is especially true when stresses are compressive and hydrostatic as in the case of epoxy resins. Epoxy resins may well be brittle when loaded in tension. However, some other polymers, for instance polycarbonate, show yielding behaviour when subjected to tensile stresses.

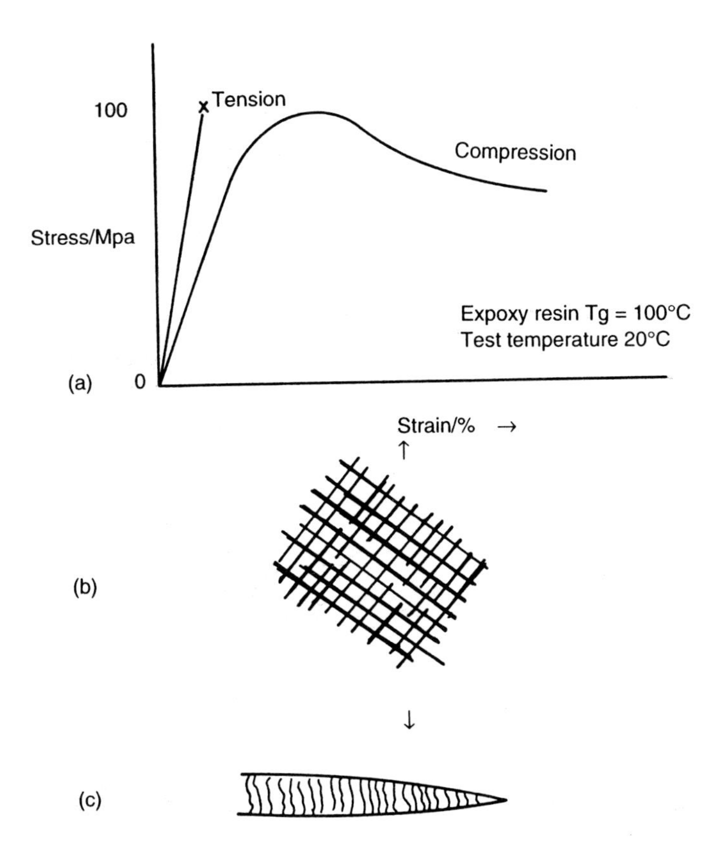

Fig. 5.4 (a) Stress–strain behaviour of an epoxy resin (b) appearance of shear bands in for example polystyrene after compression (c) structure of a craze.

The tensile and compressive response of an epoxy resin is illustrated in Fig. 5.4(a). In the case of epoxy resins deformation is homogeneous with non localised yielding. However, this is rarely the case. For example, polystyrene when loaded in compression shows fine deformation bands (*see* Fig. 5.4b). Deformation is highly localised within each band. Once a shear band is initiated it will continue to operate because it has a lower flow stress than its surroundings. Hence shear bands are associated with strain softening. Shear bands are of lower volume than the surrounding material.

Crazing

If a sample of a glassy plastic such as polymethyl methacrylate is subjected to a high tensile stress below its glass transition temperature the small crack-like features, called crazes, may be seen. They can be seen when some types of plastic ruler are bent too enthusiastically and can easily become true cracks which cause complete fracture. However, crazes are not cracks but very slim regions of polymer which are full of voids. For this reason crazing results in an overall increase in volume of a polymer which yields in this way. Crazing tends to occur under tensile stresses in polymers such as polystyrene, where a bulky group (a phenyl group in this case) is attached to the repeat unit. On the other hand polymers which show shear yielding have their bulky groups incorporated into the backbone as can be seen in the structure of polycarbonate.

Crazing occurs in the presence of hydrostatic tensile stresses. Crazes initiate on the surface of stressed polymers. They are regions of cavitated material. Each craze tapers gradually and is bridged by drawn and oriented polymer fibrils of approximately 20 μm diameter. The nucleation of crazes is thought to occur by a local yielding and cold drawing process which takes place in a constrained region of the material. Stresses are concentrated at the craze tips. Crazes grow slowly at first and absorb considerable amounts of energy. Hence crazing can contribute to toughening. However, crazes subsequently turn into true cracks which propagate rapidly. Hence polymers which craze are normally brittle unless the growth of crazes can be restricted. Crazes cause a stress whitening effect when present through large volumes of a polymer. The volume of the polymer increases. The structure of a craze is shown in Fig. 5.4(c).

Control of Crazing

A useful improvement in toughness is achievable if craze growth is controlled. If crazes are prevented from becoming true cracks and the overall deformation mechanism can be restricted to nucleation and limited growth of crazes, then considerable energy can be absorbed. Hence the material would be tough. This can be achieved by dispersing rubber particles through the polymer matrix. These particles assist in nucleation of crazes but also block craze growth before cracks are formed. However, the microstructure must be carefully controlled to achieve the required rubber particle size and separation. This is the principle behind high impact polystyrene (HIPS) and acrylonitrile butadiene styrene (ABS).

Fracture

Fracture in a brittle manner is often observed in plastics. It is difficult to predict and occurs at low strains with small energy inputs. In common with other materials brittle fracture originates from defects which have a stress concentrating effect. These defects can be of many types originating in processing, manufacturing processes, damage sustained in service or even poor design.

Several simple tests are available to assess fracture behaviour and these will be briefly reviewed later. However, analysis of fracture is properly addressed via the methods of fracture mechanics which have been well developed for metals over the last forty years. However, the theory is only well developed for linear elastic materials. Clearly the majority of plastics are far from linear in their elastic behaviour and so care must be taken in applying linear elastic fracture mechanics to plastics. Even so fracture mechanics has been successfully applied in many instances. For example the 1986 revision of BS3505 for PVC pressure pipe includes a fracture toughness requirement.

Fracture mechanics concentrates on the stress intensity generated by the presence of flaws. However, it is worth pointing out that flaws in plastics may not be as well defined as in a ceramic where flaws can usually be readily identified as cracks or voids. In plastics defects may be of a nebulous nature arising from localised heterogeneities, such as poor additive dispersions, degraded regions, lack of fusion and agglomeration of fillers. This is in addition to defects such as moulding defects and design errors, such as sharp corners.

Fracture Mechanics Theories

Modern fracture mechanics is derived from the work of A.A. Griffith published in the 1920s. This work assumed that brittle fracture originated from a critical crack or defect and that the energy required to increase the length of such a defect is provided to create the new surfaces and their associated surface energies. This energy must be supplied by a decrease in the stored elastic strain energy in the material. This is the basis of the energy balance approach.

In addition to the non-linear nature of plastics another problem invariably arises in that plastics exhibit some degree of local yielding at the tip of the growing crack. Thus a simple surface energy term γ cannot be used in a mathematical analysis. A solution is to replace γ with G, the strain energy release rate, which includes both the surface energy and plastic energy components which constitute the total energy which has to be expended. This becomes G_c (J m^{-2}) when the critical value to produce fracture is reached. Thus Griffith's original approach which was deduced for a truly brittle solid, i.e. glass, was modified slightly to give

$$\sigma_f = \left(\frac{E\, G_c}{a} \right)^{1/2} \tag{5.5}$$

where E is the tensile modulus and a is the crack length.

Later work by Irwin in the 1950s approached the problem of fracture from a different direction. He related the fracture to the intensity of the stress K produced at the crack tip. Fracture occurs when this stress becomes critical, i.e. K_c.

For the simple situation of an infinite flat sheet sample containing a defect of length 2a at its centre the following applies:

$$K_c = \sigma\, (\pi\, a)^{1/2} = (E\, G_c)^{1/2} \tag{5.6}$$

Because many geometries vary from the simple one assumed above, a calibration factor, Y, must be included. Y has been derived for many different geometrical cases. The expression for K_c now becomes:

$$K_c = \sigma\, Y\, (\pi a)^{1/2} \tag{5.7}$$

K_c is usually referred to as fracture toughness (N m$^{-3/2}$). It is measured in a tensile type sheet sample where an edge crack is made to propagate by the application of a stress applied at right angles to it. This is referred to as the Mode I crack growth condition. The measured value of K is then referred to as K_{IC}.

Thus it can be seen that sensible methods of measuring K_{IC} can be designed. Other related parameters can be measured. For instance the crack opening displacement (COD) is important in relation to plastics. The critical value of COD for Mode I geometry is the vertical displacement at the crack tip face just as crack propagation occurs. However, it should be realised that crack growth may not be unstable and sudden but may be time dependent. It may occur where K_I is less than K_{IC}. A simple model which can be applied to plastics takes the following form:

$$\frac{\mathrm{d}a}{\mathrm{d}t} = C K_I^{m} \tag{5.8}$$

where C and m are parameters (assumed constant) for specific plastics.

Because of the conflicts between the assumptions of linear elastic fracture mechanics and the properties of most plastics, it should be realised that plastics which have a high yield strength and are relatively brittle or have been embrittled by degradation for example lend themselves best to analysis by this technique. An associated problem is the size of the plastic zone. K_{IC} is usually measured under plane strain conditions using thick samples. The assumption is that deformation in the sample thickness direction is zero. This means that the thickness of the sample must be significantly greater than the plastic zone size. The problem with plastics is that it is difficult to fabricate good quality and representative thick samples. Having established plane strain conditions where deformation is restricted in the sample thickness direction, the stress system at the crack tip becomes triaxial and the expression for fracture toughness must be modified

$$K_c = \frac{EGc}{1 - \gamma^2} \tag{5.9}$$

where γ = Poisson's ratio.

Impact Tests

Failure due to sudden loads is very common in service conditions and is more likely to occur than failure under slowly applied forces. This happens because impact loads are applied too quickly to allow relaxations of molecular structure and the result is fracture resulting from chain breakage and/or separation of interfaces between phases.

Various test methods have evolved to measure the susceptibility of plastics to impact loads. In general these aim to measure the energy required to fracture standard test specimens under standardised high rates of loading. Clearly there are problems associated from producing standard geometry specimens from a wide range of mouldings extrusions and so on. For this reason several types of impact test have evolved, some suitable for thick mouldings and some for thin sheet, etc.

Impact tests can be categorised into pendulum type tests and falling weight tests. Pendulum tests include Charpy and Izod tests. These tests use the potential energy of a pendulum to fracture a standard sample. The sample contains a sharp notch which produces a triaxial stress state and acts as a stress concentrator. Fracture originates at the notch. A serious problem is that a number of plastics are notch sensitive, e.g. uPVC and the recorded fracture energy will depend on the severity of the notch. It is thus important to ensure that the notch is identical from sample to sample.

An associated problem is that it is difficult to compare the results from different types of plastics which have different notch sensitivities. Care must also be taken in defining the test temperature and loading rate. Plastics are temperature dependent materials and as mentioned in Chapter 4 temperature and time are interchangeable variables. Thus in impact testing it is to be expected that impact velocity and test temperature have interchangeable effects. Many plastics show distinct ductile–brittle temperatures, e.g. polypropylene homopolymer at around 0°C. Clearly the precise position of this transition is sensitive to test temperature, loading rate and notch sharpness.

It is possible to apply fracture mechanics to pendulum tests and to derive the fracture toughness. This has been used successfully in some instances.

Furthermore recent developments have enabled high speed data capture using computers and this has led to instrumented pendulum and falling weight impact tests. To achieve this a force transducer or accelerometer is fitted to the pendulum or impacter and the output sampled during fracture as a function of time. Detailed information about the fracture process can be obtained.

Environmental Stress Cracking

An important cause of embrittlement in plastics is environmental stress cracking (ESC). It occurs when a stressed plastic which would not otherwise be expected to fail does so when exposed to a sensitising environment. The environment is nearly always a liquid or vapour which may well not attack or damage the plastic if the stress were absent. The effect is therefore synergistic in nature and not necessarily associated with other degradation mechanisms. The commonest cases are those involving amorphous polymers in contact with organic solvents, e.g. polycarbonate in contact with low molecular weight hydrocarbons. However, detergents are also fairly potent and cause premature failure when in contact with semi-crystalline plastics such as polythenes. It has been found that, in polythenes, both the basic characteristics of the polymer, e.g. molecular weight, and also process history are controlling factors. In the latter respect, the crystalline morphology plays an important part and this is controlled in part by the thermal experiences of the polymer during processing.

The indications are that aspects of the microstructure which restrict strain when the polymer is under stess reduce the ESC susceptibility. In fact there is a critical strain which can be determined below which ESC does not occur. ESC is assumed to originate from microscopic defects where the active medium interacts with the high stress region at the crack tip. Diffusion into the polymer would be expected to be rapid because the molecules are in a higher than normal energy state than usual due to the high stress. The material then becomes weakened and the crack propagates. Sometimes multiple crazing occurs leading to a whitening effect but without widespread propagation. The stress is relaxed by the multiple crazes without the necessity for any major crack formation. In some respects the mechanism is similar to environmental stress cracking in metals although the mechanisms are quite different.

ESC is usually assessed by empirical standard tests, some of them very simple. For example the Bell Telephone test is one of the simplest and oldest and involves immersing bent strips of plastic containing a line defect in a medium containing alkyl aryl polyethylene glycol. The samples are then examined for signs of defects. An objective of other tests is to determine the critical strain at which ESC occurs. In general a variety of parameters can be measured including the time to initiate visible damage such as crazing.

In practice there are many factors which can contribute to ESC and great care must be taken in particular where organic chemical environments are to be used. The stress system is critical and internal stresses resulting from moulding or welding processes may be sufficient to cause problems.

Fatigue

Fatigue is a major cause of failure in metallic components. It also contributes to premature failure in plastics. Failure follows the application of cyclic stress. Fatigue in plastics varies in detail from that occurring in metals.

Certainly fatigue failure occurs in some cases as a result of the propagation of unstable cracks or defects under the application of high frequency cyclic stresses. This leads to brittle fracture and can be analysed by fracture mechanics.

On the other hand where the plastic flows in a viscous manner under conditions of rapid cyclic stressing, heat builds up as a result of the low conductivity and reduces the yield stress. When the peak applied stress exceeds the now reducing yield stress, ductile failure occurs. This is known as thermal fatigue and in this case fracture mechanics cannot be applied. It is interesting to note that if the frequency of stressing of a normally brittle plastic is increased it is possible to build up heat in certain localities so that the plastic then fails in a ductile manner.

Hardness

Hardness is an important property as it is easily and rapidly measured. However, its interpretation is often complex. Firstly there is the question of exactly what part of the plastic sample is being tested. Clearly it includes the surface but also involves some sub-surface material. It is readily related to yield stress and modulus, and is an easy way of estimating these properties. However, care must be taken because the surface layers of a fabricated polymer may have very different molecular orientation, residual stress levels and other properties compared with the bulk material.

Problems are encountered with plastics which are particularly prone to viscoelastic deformation. Any attempt to probe the surface will then yield time dependent properties. While this can be utilised to investigate time dependency of certain properties it does raise a serious difficulty in terms of standardisation of hardness tests. This problem does not exist in the hardness testing of all but the softest of metals.

There are a wide range of hardness testing techniques as outlined below and comparative hardness values for different methods are given in Table 5.1.

Scratch Tests

The simplest are scratch tests, the principle being that a material can be scratched by another that is marginally harder. It is thus possible to build up a hardness scale on this basis. This is called the Mohs scale. It runs from 1(talc) to 10(diamond). However, the Mohs hardness of most plastics lies in the range of 2 to 3. Thus this scale is of little use for plastics, although it is possible to establish an *ad hoc* scale for plastics based on a modified Mohs scale, using alternative materials for reference.

Indentation Tests

The most popular tests involve probing the surface with a loaded indenter. The depth of penetration or related measurement is then recorded. Common tests are the Brinell, Vickers, Knoop, Rockwell and Barcol tests. These were designed primarily for metals but can be adapted for use with plastics. On the other hand the durometer is designed primarily for plastics and uses a needle type indenter. The depth of penetration under load at a fixed time is recorded. Two common scales are used, Shore A and D, the difference being in the indenter geometry and spring loading force. Portable instruments are available. Special hardness tests of a similar type are also available for rubbery polymers. They use hemispherical indenters and give results based on International Rubber Hardness Degrees. It it worth noting that plasticised PVC is assessed using a hemispherical indenter in exactly the same way but the penetration under load is used to give a BS softness number.

Rebound Hardness

It is also possible to measure hardness by observing the rebound height and/or rebound acceleration of a small missile dropped from a fixed height. This is truly non-destructive and can be adapted electronically to produce useful instruments. The most common instrument is the Shore Sceleroscope. In this method a missile tipped with a diamond cone is dropped onto a surface from a fixed height. The harder the surface the less energy is absorbed and so the

Table 5.1 Hardness values for plastics derived by different methods (after Crawford).

	HDPE	PA	PP	PMMA	PS	uPVC
Vickers	5	5	6	5	7	9
Rockwell	R75	M75	R100	M102	M83	M60
Durometer (Shore D)	70	80	74	90	74	80
Shore Scleroscope				99	70	75

rebound height is greater. This type of test is widely used with metals for testing components such as rolls without inflicting surface damage. It should be of use particularly with the harder plastics.

REFERENCES

H.A. BARNES, J.F. HUTTON and K. WALTERS: *An Introduction to Rheology*, Elsevier Science Publishers, Amsterdam and New York, 1989.

A. BLUMSTEIN: *Polymeric Liquid Crystals*, Plenum Press, 1983.

W. BROSTOW and R.D. CORNELIUSSEN (eds): *Failure of Plastics*, Hanser, Munich (1986).

J.A. BRYDSON: *The flow Properties of Polymer Melts*, Iliffe, 1980.

L.L. CHAPOY: *Recent Advances in Liquid Crystal Polymers*, Elsevier Applied Science, London, 1985.

R.J. CRAWFORD: *Plastics Engineering* 2 edn., Pergamon. Oxford, 1987.

R.W. HERTZBERG and J.A. MANSON: *Fatigue of Engineering Plastics*, Academic Press, New York, 1980.

A.J. KINLOCH and R.J. YOUNG: *Fracture Behaviour of Polymers*. Applied Science Publishers, London, 1983.

P.C. POWELL: *Engineering with Polymers*, Chapman and Hall, London, 1983.

I.M. WARD: *Mechanical Properties of Solid Polymers*, Wiley, New York, 1971.

J.G. WILLIAMS: *Fracture Mechanics of Polymers*, Ellis Horwood, Chichester, UK, 1984.

6 *The Rheology of Polymer Melts*

INTRODUCTION

When polymeric materials are shaped into final articles, they must suffer deformation and flow. The deformation and flow behaviour (known as the rheological properties) are, therefore, of paramount importance. Thorough understanding of the rheological properties of materials helps in:

1) correcting processing faults and obtaining a good production rate; and
2) knowing how to choose the best polymer or polymer combination for use under given process conditions to achieve the desired mechanical properties in the final product.

This chapter deals with the important rheological parameters and how engineers measure, present and use rheological data.

NEWTONIAN FLUIDS

In Chapter 3, the concept of viscosity was introduced with regard to spring and dashpot representations of creep and stress relaxation phenomena. This rheological parameter is now defined more precisely.

Consider the two parallel planes A and B of area A and a distance h apart shown in Fig. 6.1. The planes are separated by a viscous fluid. It is necessary to apply a shearing force F to the upper plane in order to move it with a velocity v. This force F is transmitted by the parallel planes of fluid from the plane A to the plane B, which requires an equal and opposite shearing force F to keep it still. This force F is transmitted by the viscous drag that each plane of fluid has due to its viscosity. Newton showed that the viscous drag or shearing force F transmitted by the fluid between the planes A and B was due

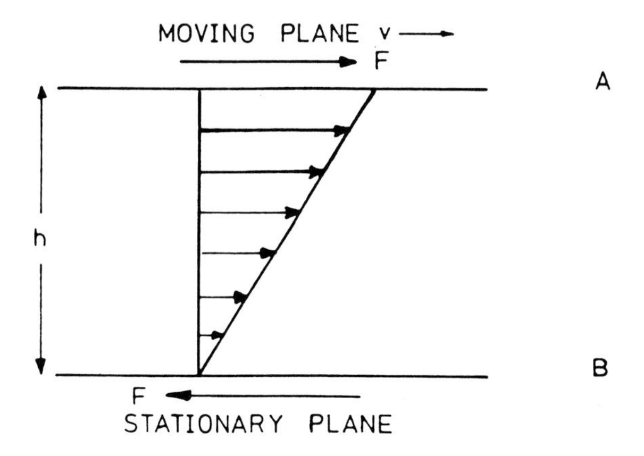

Fig. 6.1 Flow between two parallel planes.

to the relative velocities of the planes of fluid which travel at velocities between zero and v.

$$F = \eta A \times \text{velocity gradient}$$

where the velocity gradient is the difference in velocity of the planes *A* and *B* divided by *h*. The constant η is known as the coefficient of viscosity or viscosity.

$$F = \eta A \frac{v}{h} \tag{6.1}$$

Rearranging this equation gives

$$\eta = \frac{F}{A^{v/h}}$$

from which η is defined as the frictional force in *N* exerted on unit area of a fluid in a region of unit velocity gradient.

The term v/h is quite adequate for describing linear velocity gradients, but as these are generally an exception a more generalised form of velocity gradient is used. This is the shear rate, $\dot{\gamma}$, which is given by

$$\dot{\gamma} = dv/dh$$

Another simplification is achieved by using the shear stress τ, which equals F/A. These changes give

$$\tau = \eta \dot{\gamma} \tag{6.2}$$

which is called Newton's equation and fluids that obey it are known as Newtonian fluids.

Newtonian behaviour is exhibited by all gases and by simple fluids in which the viscous dissipation of energy is due to collisions between small molecules as, for example, in water, mercury and acetone. Some polymer melts, particularly polyamides, behave in a Newtonian way at low shear stresses, but generally this is too simple a model for them.

In equation (6.2), τ is in N m^{-2} or Pa, $\dot{\gamma}$ is in s^{-1} and η is in Nsm^{-2} or PaS. However, often the viscosity η is given in the old CGS units. The conversion is

$$1 \text{ poise} = 10^{-1} \text{ Nsm}^{-2}$$
$$1 \text{ cp} = 10^{-3} \text{ Nsm}^{-2}$$

Typical values of viscosity are: for air at NTP 1.8×10^{-5} Nsm^{-2}, for water at 20°C 10^{-3} Nsm^{-2} and for polymer melts 10–10^{4} Nsm^{-2}.

A SUMMARY OF TYPES OF FLUID FLOW

Fluids in which the viscous dissipation of energy is due to collisions between large molecules, a dispersed second phase or colloidal systems, do not obey Newton's equation. Three main classes of fluid flow behaviour are recognised and are summarised below:

1) Time-independent fluids: fluids in which the shear viscosity is a single valued function of the shear stress.
   Newtonian fluids fall into this class.
2) Time-dependent fluids: fluids in which the shear viscosity depends on the shear stress and the duration of shear.
3) Elastico-viscous fluids: fluids that are predominantly viscous but possess elasticity like solids. Viscoelastic materials are solids that show some flow, such as metals undergoing plastic deformation or plastics above their T_g. Often the two terms are not distinguished and visco elasticity is used to express both.

TIME-INDEPENDENT FLUIDS

The shear stress-shear rate behaviour of this class of fluids is shown in Fig. 6.2. The Bingham plastic fluid is one that will not flow if the applied shear stress, τ, is less than the yield stress of the material, τ_y. The equation that describes this rather idealised behaviour is

$$\tau - \tau_y = k\dot{\gamma} \tag{6.3}$$

where k is called the coefficient of rigidity or the coefficient of plastic viscosity. Like equation (6.2) (Newton's equation), equation (6.3) describes the constitution of the material and is known as a constitutive equation.

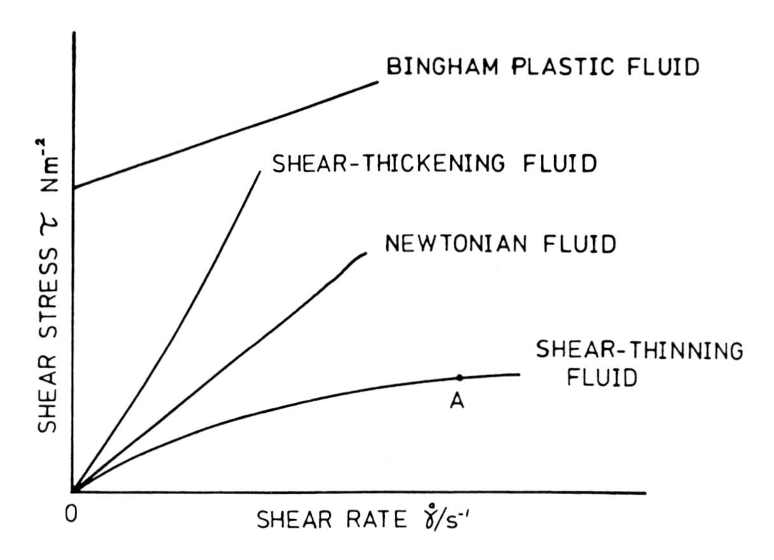

Fig. 6.2 Time-independent Fluids.

Toothpaste is a well-known example of this class, and in polymeric materials this sort of behaviour occurs in plastics matrices with high loadings of elastomer in them. Examples include styrene-acrylonitrile and polyvinylchloride loaded with polyisobutylene.

Shear-thinning fluids, sometimes called pseudoplastic fluids, show a range of shear viscosities, the value of which depends on the shear stress. The term apparent shear viscosity is often used to acknowledge this range of values, because a value quoted in the absence of the shear stress value is meaningless. The flow curve is shown in Fig. 6.2.

The apparent shear viscosity, often shortened to shear viscosity, is not defined by the slope of this graph at any point but by the gradient of the line OA. The apparent viscosity at A is given by

$$\eta_a = \tau / \dot{\gamma} \tag{6.4}$$

Flow curves of log shear stress against log shear rate drawn for these fluids often show straight portions over a decade or so of shear rate (Fig. 6.3). For this reason the fluids are termed power law fluids, and an empirical constitutive equation is used, namely:

$$\tau = k\dot{\gamma}^{n} \tag{6.5}$$

where k is a constant dependent on temperature and n is called the shear-thinning index, which for these fluids is less than unity. For Newtonian fluids $n = 1$.

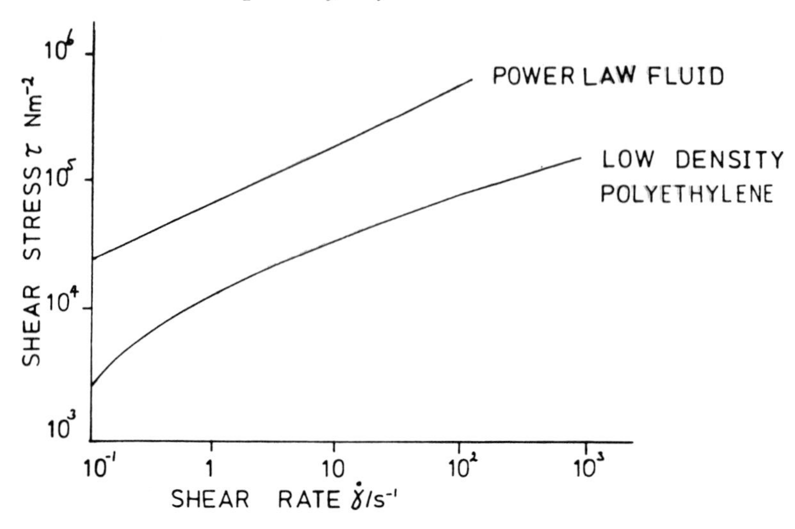

Fig. 6.3 Power law flow and the shear-thinning flow of a polymer melt.

Taking logs of both sides of equation (6.5) gives

$$\log \tau = \log k + n \log \dot{\gamma}$$

which is in the form $y = mx + c$, from which n is obtained from the slope, and k is found by choosing a convenient point on the line and substituting the values of τ, $\dot{\gamma}$ and n in equation (6.5).

As seen from Fig. 6.3, the power law is quite useful, but for the shear-thinning flow of polymer melts is only an approximation.

An expression for the apparent viscosity, η_a, as a function of shear rate, $\dot{\gamma}$, can be obtained by combining equations (6.4) and (6.5).

$$\eta_a = \tau/\dot{\gamma} = \frac{k\dot{\gamma}^n}{\dot{\gamma}} = k\dot{\gamma}^{(n-1)} \quad (6.6)$$

As $n < 1$, η_a decreases as $\dot{\gamma}$ increases and the value of k is the value of the apparent shear viscosity at unit shear rate. The unit shear rate viscosity k is often used as a reference, giving

$$\eta_a = k\dot{\gamma}^{(n-1)} \quad (6.7)$$

The model used to explain shear-thinning behaviour in polymer melts is that under shear the long asymmetric molecules, which are initially entangled and randomly orientated, become less entangled and more aligned along the streamlines. They thus interact less with each other when they are more aligned, and the behaviour becomes more Newtonian as the orientation becomes more complete. Polymer melts behave in a Newtonian manner at low shear stresses when the entanglement and orientation are not much affected by the stress. They behave like shear-thinning fluids at moderate stresses, but at high stresses they behave elastically.

Shear-thickening behaviour or dilatancy is the opposite of the former, η_a increases as $\dot{\gamma}$ increases. It is exhibited by fluids with high solid to liquid contents. In general, this behaviour does not occur in polymer melts but has been recorded when melts tend to crystallise during flow. Often these fluids follow a power law equation in which $n > 1$. More details are given in (1).

TIME DEPENDENT FLUIDS

For these fluids the apparent shear viscosity depends on the shear stress and the duration of shear: the fluids have a memory. There are two types, thixotropic and negative thixotropic or anti-thixotropic.

A thixotropic material exhibits a reversible, isothermal gel–sol–gel transformation. It can be regarded as a time-dependent shear-thinning fluid. In such a fluid, the changes in structure are not taking place immediately or very quickly. A negative-thixotropic fluid shows the opposite behaviour, a reversible, isothermal sol-gel-sol transformation. This can be regarded as time-dependent shear-thickening behaviour.

Time-dependent behaviour is not generally important in polymer processing. Sometimes polymer melts appear to be thixotropic or negative-thixotropic but this is generally due to degradation or crosslinking, and the behaviour is not reversible and is therefore not truly time-dependent as it is defined. Fuller details of the mechanism of time-dependent behaviour are given by (6).

ELASTICOVISCOUS FLUIDS

This group of fluids combines the properties of viscosity and elasticity. Many commonplace fluids have elastic properties. If one applies a tangential force to a bowl of tomato soup, the soup moves in the direction of the rotation, slows down, stops, and then flows back the other way by a small amount. This is a recoil and the amount of this recoil is a measure of the fluid elasticity.

There are a number of models proposed to explain elasticoviscous behaviour in the many varied kinds of fluids that show it. In polymer melts, the elasticity arises from the deformation of the long chain molecules, which can coil, uncoil, align along the streamlines and entangle. The possible conformations of a polyethylene molecule are numerous because of the ease of rotation of the C–C bonds along the backbone chain. The degree of elasticity is highly dependent on the molecular architecture and is discussed fully in the next chapter.

There are a number of phenomena associated with the elasticity of these fluids. These phenomena often give process engineers some problems, which fortunately are not insuperable. One of the phenomena, stress relaxation, has been described earlier in Chapter 4. The others are spinnability, die swell, extrudate distortion and draw resonance.

Spinnability is the ability of an elasticoviscous fluid to be drawn into a fine, stable thread. This is essential to successful extrusion, where the extrudate must maintain its shape until it cools below its melting or softening point.

Die swell or the Barus effect occurs when an elasticoviscous fluid emerges from a die. The extrudate diameter can be several times that of the die. The swelling ratio is the extrudate to die diameter. It is important to understand this phenomenon so that dies can be made to the correct size and shape to give product to the correct specification. Die design is very skilled.

Extrudate distortion occurs when the flow rate through the die is too high, giving a slightly wavy or extremely distorted extrudate. This limits the maximum production rate on a given extrusion machine at that process temperature.

Draw resonance likewise produces an extrudate distortion and occurs when the extrudate is hauled off from the die at too fast a rate.

There are a number of other interesting elastic phenomena, but space does not permit a discussion of them. The above four will be described more fully later.

In the preceding remarks, it was intended to show the types of fluid flow and place polymer melt flow in context. By way of a summary, it can be said that, at low shear stresses, it is sometimes possible to use the Newtonian model to describe polymer melt flow and make calculations with equation (6.2). A higher shear stresses, shear-thinning behaviour occurs and over a few decades of shear rate the power law equation (equation (6.5)) may be applicable. At high shear stresses, the elasticity of the melts becomes important and they must be regarded as elasticoviscous fluids. This is not easy. Often it is not possible to describe the melt flow by one of the above equations with sufficient accuracy, and then a more graphical approach must be used to estimate flow conditions.

SHEARING AND TENSILE FLOWS

The models and behaviours described so far relate only to shear flows. This is a severe limitation because many of the flow fields experienced by melts in travelling through processing equipment are a combination of shear and tensile flows. For example, when a melt leaves a relatively wide heated barrel, as shown in Fig. 6.4, to enter a narrow die, the streamlines converge and conform to a half-angle of convergent flow determined by the characteristics

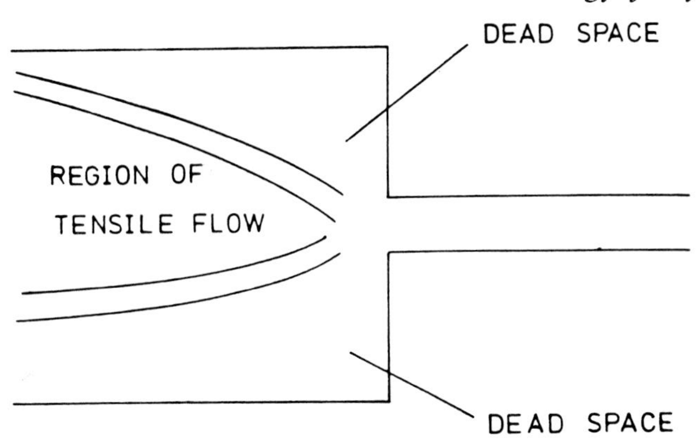

Fig. 6.4 A melt entering a die and encountering a tensile or stretching flow field.

of the melt. This curvature of the streamlines gives rise to an extensional, tensile or stretching flow, which causes tensile stresses in the melt. The viscous and elastic behaviour of the melts must, therefore, be examined under both shear and tensile conditions.

POLYMER MELTS IN SHEAR FLOW

First, it is important to define the shear rate ranges that are of interest to plastics processors. In most extrusion processes, the shear rate range in the die is $10–10^3$ s^{-1}, whereas the injection moulding process can induce shear rates in the gate region leading into the mould of between $10^3–10^5$ s^{-1}. A thermoplastic is deemed suitable for processing if its apparent shear viscosity lies between $10–10^4$ N sm^{-2} in the appropriate shear rate range. Polytetrafluoroethylene (PTFE) has too high a viscosity to be processed like other plastics and is fabricated by a pressure sintering technique.

Fig 6.3 shows a flow curve (log τ versus log $\dot{\gamma}$) for a power law fluid and a typical polymer melt. Manufacturers, however, usually produce data in the form shown in Fig. 6.5, log η versus log τ where η is the apparent shear viscosity. This graph is more useful to engineers because they can easily calculate the wall shear rate involved at a certain output rate from the geometry of the die, and can from the data discover the values of η and τ at the die wall. The shear rate lines are shown in Fig. 6.5, and their positions are calculated in the following manner. Consider the point *X* on the graph; its position is (10^3, 10), which gives $\dot{\gamma} = \tau/\eta = 10^3/10 = 100$ s^{-1}.

The position Y is defined by (10^4, 10^2), which again gives $\dot{\gamma} = 100$ s^{-1}. A line drawn from *A* through *B* thus joins points of shear rate 100 s^{-1}.

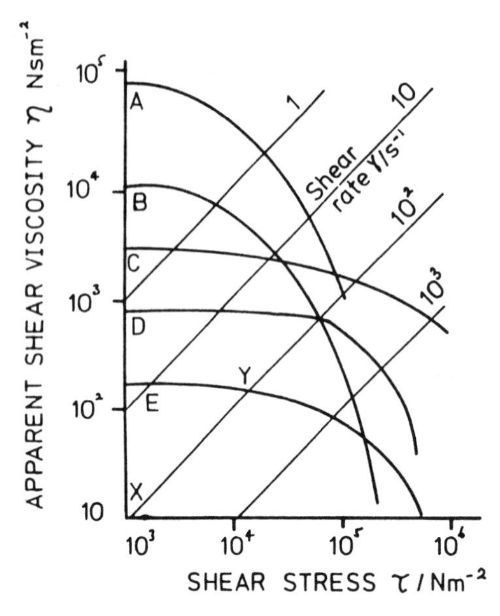

Fig. 6.5 The variation of log apparent shear viscosity with log shear stress for *A* low density polyethylene (170°C) *B* propylene/ethylene copolymer (170°C) *C* polycarbonate (270°C) *D* acetal copolymer (200°C) and *E* nylon 66 (285°C).

From this figure, it is possible to analyse the suitability of melts for various processes. For example, the shear rates involved in compression moulding are very low. A melt of low shear viscosity such as Nylon 66 would fill the mould better than any of the others shown. The most viscous ones would be totally unsuitable.

At a shear rate of 100 s^{-1}, a typical value for an extrusion die, the nylon would be the most fluid but the LDPE could be extruded easily and would keep its shape on leaving the die better than the nylon.

In blow moulding, in which a pipe of material (called a parison) is required to hang vertically for a short time until a mould is closed around it, a highly form-stable melt such as LDPE is essential for the preservation of the parison's integrity. The nylon shown would be useless for this process.

At a shear rate of 10^4 s^{-1}, typical of injection moulding, the easiest melt shown to process is propylene/ethylene copolymer, which at low shear rates has a very high melt viscosity.

The above analysis shows the importance of rheology to polymer processing. The important rheological parameters so far introduced are the shear viscosity (the word 'apparent' will now be dropped), the constant *k* and the shear-thinning index *n*. The next section introduces another rheological parameter, the tensile viscosity.

POLYMER MELTS IN TENSILE FLOWS

Tensile, elongational or stretching flows, as they are often called, occur when:

a) the confining chamber through which a melt is travelling changes size or shape;
b) a melt is hanging and deforms under its own weight; and
c) a bubble of material is expanded like a balloon.

In the last two cases, it can be seen that the stresses involved act at right angles to the surfaces of the melt and not along surfaces of the melt as in flow through a die. The tensile or elongational viscosity, η_E, is a measure of the resistance of the melt to tensile flows. The equation relating tensile stresses to tensile strain rates is similar to that for shear flows, namely:

$$\sigma = \eta_E \dot{\varepsilon} \qquad (6.8)$$

where σ is the tensile stress and $\dot{\varepsilon}$ is the tensile strain rate in s^{-1}. As the strains involved here are large, the Hencky strain is used, where

$$\varepsilon = \ln (l/l_o) \qquad (6.9)$$

where l is the final length and l_o is the initial length.

$$\dot{\varepsilon} = \frac{d}{dt} (\ln l/l_o) \qquad (6.10)$$

Such elongational flows were first studied by Trouton using high viscosity materials such as pitch and wax. For these, he found that

$$\eta_E = 3\eta \qquad (6.11)$$

and such Troutonian behaviour is found to be true for incompressible Newtonian fluids. This is true only for elasticoviscous fluids at low stresses. At high stresses η_E/η becomes much greater than 3 for polymer melts.

The variation of log tensile viscosity with log tensile stress is shown in Fig. 6.6. Three types of behaviour have been noticed:

1) The tensile viscosity is independent of tensile stress, as in Troutonian fluids.
2) The tensile viscosity decreases with increasing tensile stress. This is called tension-thinning and is analogous to necking in metals.
3) The tensile viscosity increases with increasing tensile stress. This is typical of LDPE and is called tension-stiffening. Such melts form very good parisons because at regions of local thinning in the walls, the increase in tensile viscosity arrests further thinning — a kind of self-healing mechanism.

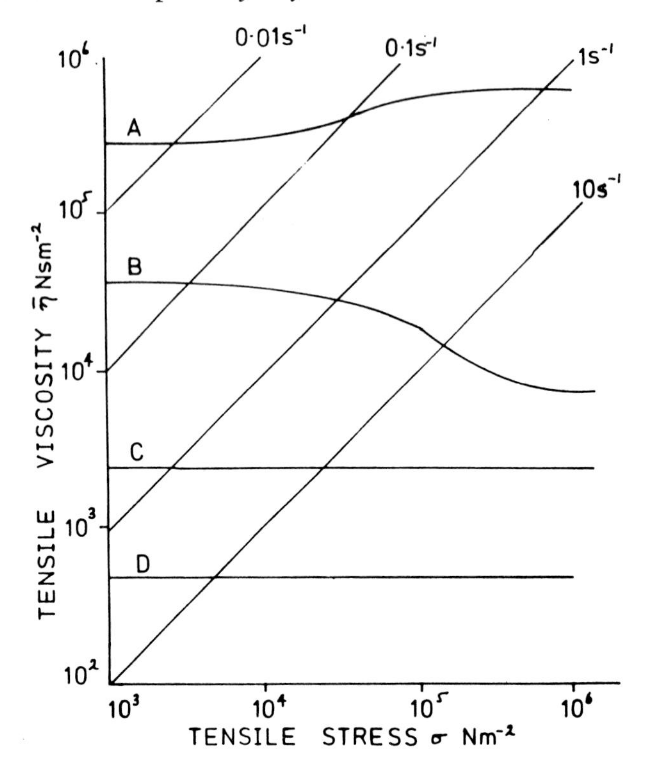

Fig. 6.6 The variation of log tensile viscosity with log tensile stress for *A* low density polyethylene (190°C) *B* propylene/ethylene copolymer (230°C) *C* acetal copolymer (200°C) and *D* nylon 66 (285°C). The ends of the lines represent the onset of extrudate distortion.

SHEAR ELASTICITY

A melt will deform continuously under stress, but once the stress is removed some deformation will be recovered. Hence the recoil in the soup mentioned earlier. At low shear stresses, the ratio of the stress to the recoverable strain is constant but at higher shear stresses this ratio increases to a maximum value of about 6. This ratio of shear stress, τ, to recoverable shear strain, γ_R, is the shear modulus, G, in flow.

$$G = \tau/\gamma_R \qquad (6.12)$$

Fig. 6.7 shows the variation of log shear modulus with log shear stress for several polymer melts. The diagonal lines of equal shear strain are found in the same way as the shear rate lines in Fig. 6.5. From Fig. 6.7 it can be seen how inelastic nylon is, as most of the other melts shown have a maximum recoverable elastic strain of 6.

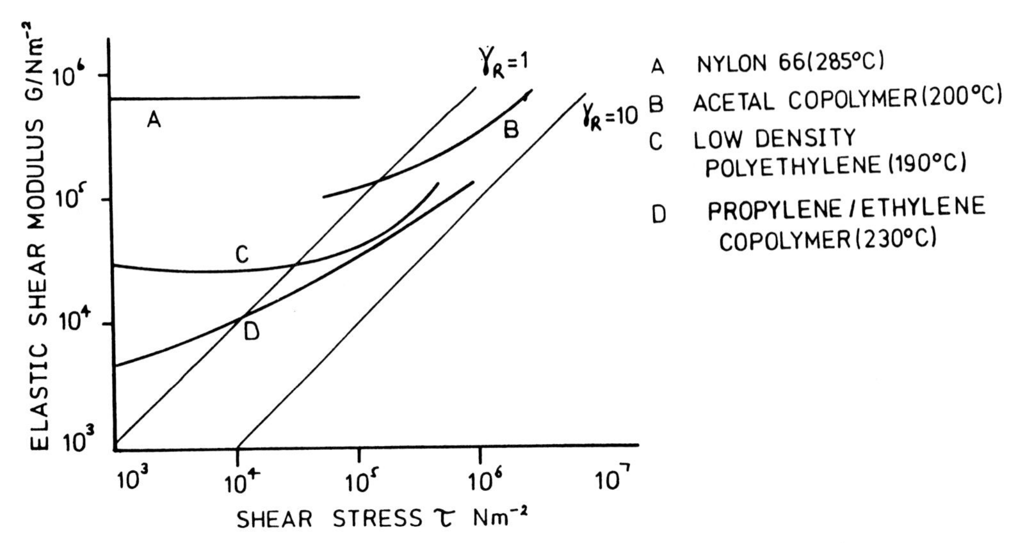

Fig. 6.7 The variation of log shear modulus with log shear stress for *A* nylon 66 (285°C) *B* acetal copolymer (200°C) *C* low density polyethylene (190°C) and *D* propylene/ethylene copolymer (230°C).

Fig. 6.8 shows another representation of the elastic data. This figure illustrates the variation of log recoverable shear strain with log shear stress, and emphasises the inelasticity of Nylon 66.

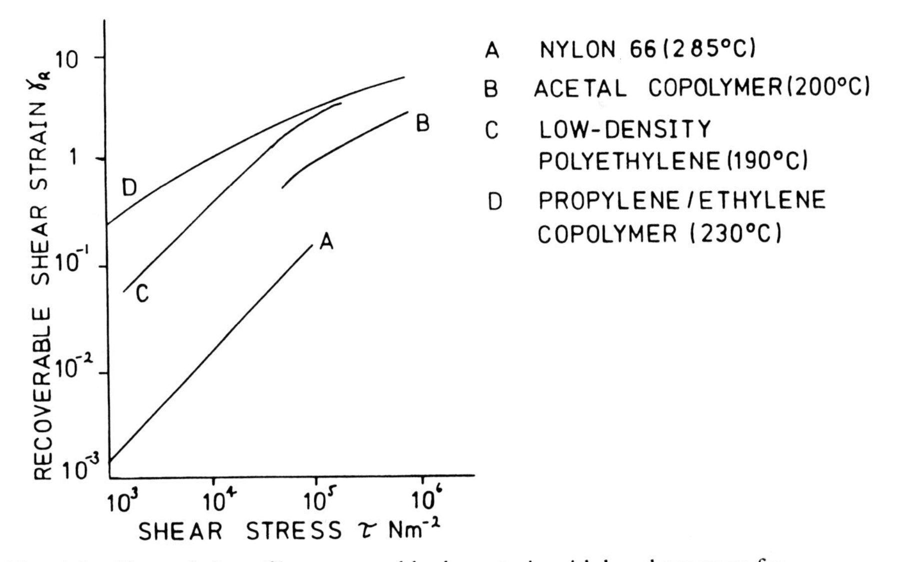

Fig. 6.8 The variation of log recoverable shear strain with log shear stress for *A* nylon 66 (285°) *B* acetal copolymer (200°C) *C* low density polyethylene (190°C) and *D* propylene/ethylene copolymer (230°C).

SHEAR DIE SWELL

Die swell or the Barus effect is shown in Fig. 6.9, and is accounted for in the following way:

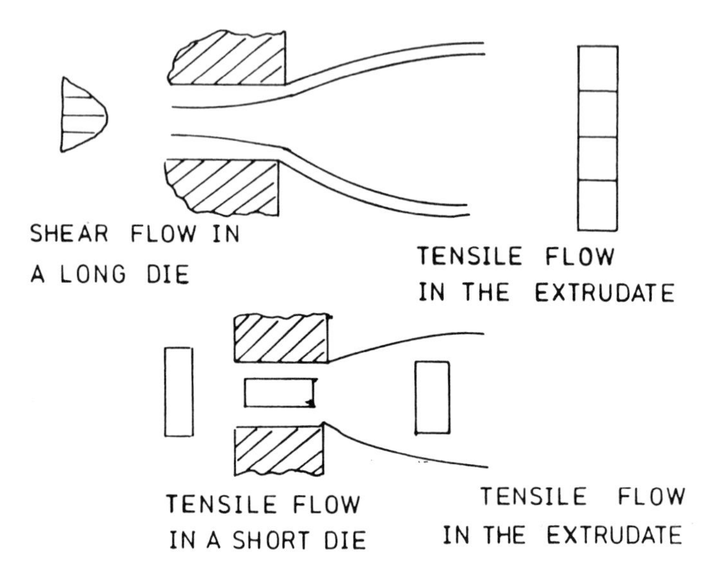

Fig. 6.9 Die swell or the Barus effect from long and short dies.

The polymer molecules in the reservoir are coiled in a random fashion. In the entrance region to the die, the extensional flow uncoils and aligns the polymer molecules along the converging streamlines and the molecules pass along the die. Stress relaxation of the tensile stresses occurs as the molecules pass through the shear field in the die, and the shear stresses only maintain a certain degree of stretching and orientation. If the die is sufficiently long, all the tensile stress will relax out, giving a die swell due to the shear stresses only.

At the outlet, the polymer molecules recoil due to the removal of the shear stress and the continuation of the random Brownian motion. This causes the increase in cross-sectional area and a decrease in the longitudinal direction.

The swelling ratio depends on the shear rate, temperature and pressure as shown in Figs 6.10(a) and (b). The results of this work indicate the following points:

a) Die swell increases with shear rate up to a maximum that occurs just before the onset of melt fracture.

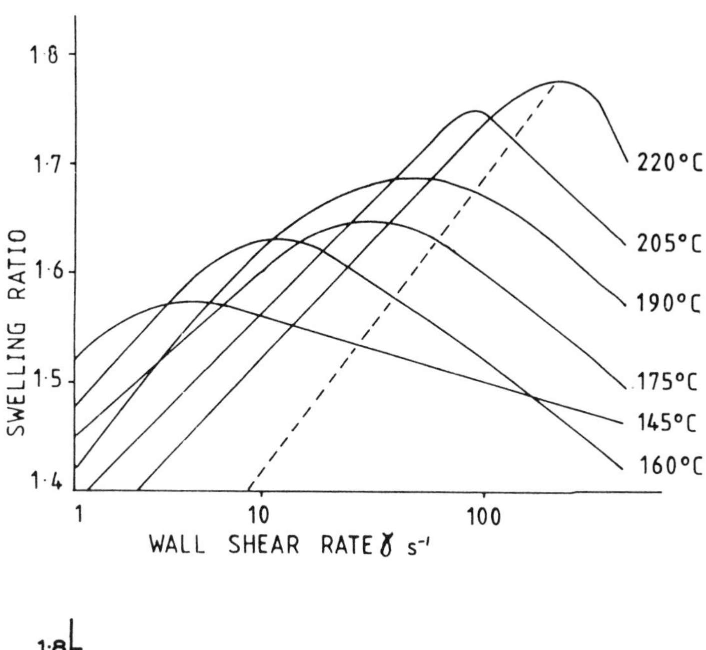

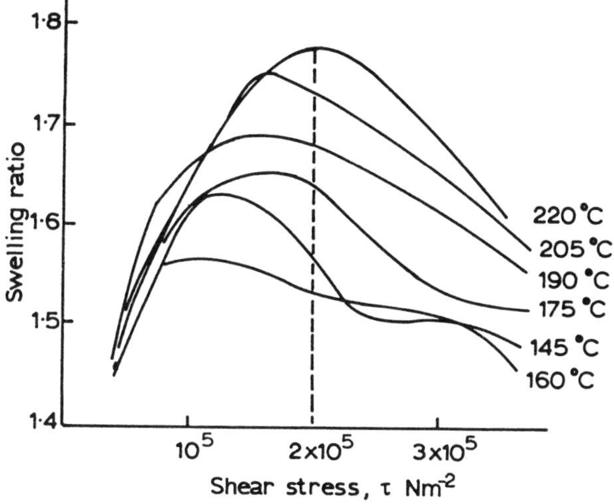

Fig. 6.10 (a) and (b) The effect of shear rate, temperature and pressure on swelling ratio for low-density polyethylene. The dotted line shows the onset of extrudate distortion.

b) At a fixed shear rate, the swelling ratio decreases with increasing temperature, but the maximum value, which occurs at higher shear rates, is greater.
c) At a fixed shear rate, the die swell decreases with increasing die length as the tensile stresses relax.

RELATING SWELLING RATIO TO RECOVERABLE STRAIN

Several equations have been suggested for relating swelling ratio to the shear stress at the wall of the die or to the recoverable shear strain. It is by this sort of method that one obtains values of melt shear modulus as a function of shear stress (Fig. 6.7) or the variation of recoverable shear strain with shear stress (Fig. 6.8). The problem with all the models is that there must be simplifying assumptions, which at times are not justified. This kind of situation is prevalent in the study of elasticoviscous fluids.

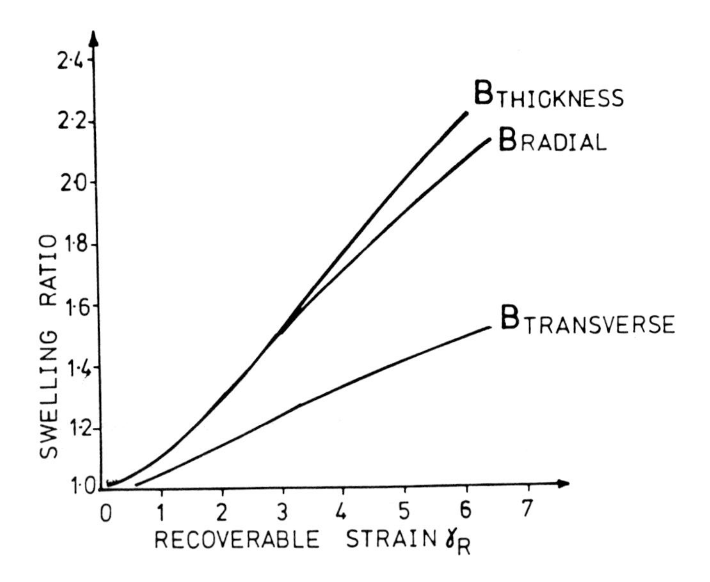

Fig. 6.11 The variation of swelling ratio with recoverable shear strain for capillary and slit dies.

Process engineers tend to use equation (6.13) described by

$$B_R^2 = \tfrac{2}{3}\,\gamma_R\,[(1+\gamma_R^{-2})^{3/2} - \gamma_R^{-3}] \tag{6.13}$$

where B_R is the shear swelling ratio, and γ_R is the recoverable shear strain. This equation is expressed in graphic form in Fig. 6.11. The swelling ratios B_H and B_T represent those in the thickness and transverse directions in a slit die, as shown in Fig. 6.12. An annular die as used in blow moulding or pipe extrusion can be regarded as a slit die of width equal to the average circumference of the annular die and of thickness equal to the gap width of the annual die. It can be shown that $B_H = B_T^2$.

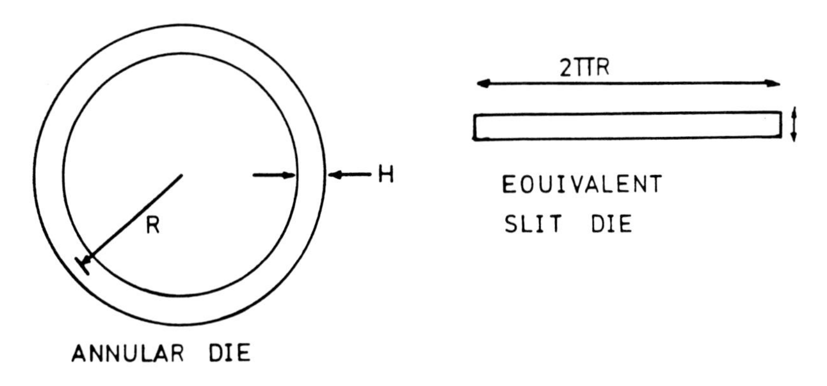

Fig. 6.12 An annular die expressed in terms of a slit die.

Example 6.1

After all this theory, it may be instructive to use the data to predict the swelling ratio for a melt under the following conditions.

An annular die of internal diameter 50 mm and outer diameter 56 mm is used in the extrusion of a propylene/ethylene melt at a flow rate of 2.4×10^{-5} m^3 s^{-1}. Calculate the outside diameter and thickness of the resulting pipe caused by shear die swell.

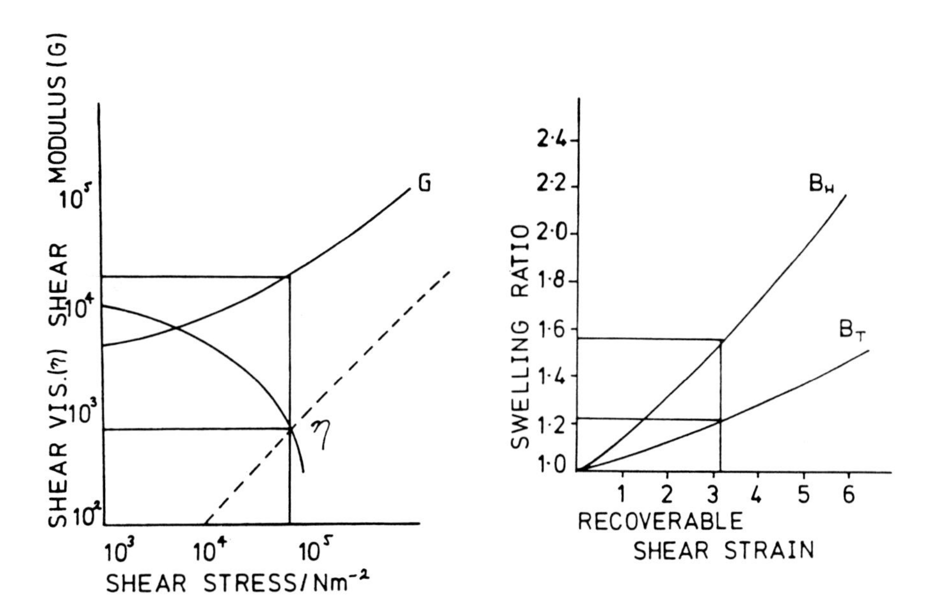

Fig. 6.13 Data for Example 6.1 (a) shear viscosity and shear modulus (b) swelling ratio strain.

Solution

The annular die can be regarded as a slit die of length equal to the average circumference $\pi\left(\frac{50+56}{2}\right)$ mm and of thickness 3 mm.

The equations relating the wall shear rate, $\dot{\gamma}$, to the flow rate Q through (a) a circular die and (b) a slit die are:

$$\text{(a)}\quad \dot{\gamma} = \frac{4Q}{\pi r^3} \qquad (6.14)$$

$$\text{(b)}\quad \dot{\gamma} = \frac{6Q}{Th^2} \qquad (6.15)$$

where r is the die radius for (a) and T and h are the average circumference and thickness respectively for (b). Clearly, we use equation (6.15).

$$\dot{\gamma} = \frac{6 \times 2.4 \times 10^{-5}}{\pi \times 53.0 \times 10^{-3}\,(3 \times 10^{-3})^2}$$

$$\dot{\gamma} = 100\ \text{s}^{-1}$$

On the graph in Fig. 6.13(a), draw a line of shear rate 100 s^{-1}. This cuts the data curve at (7×10^4 Nm^{-2}, 700 Nsm^{-2}). Therefore the value of τ is 7×10^4 Nm^{-2}.

On the graph in Fig. 6.13(a), find the value of the shear modulus for a shear stress τ of 7×10^4 Nm^{-2}.

$$G = 2.2 \times 10^4\ \text{Nm}^{-2}$$

The recoverable strain $\gamma_R = \tau/G$

$$\gamma_R = \frac{7 \times 10^4}{2.2 \times 10^4} = 3.2$$

Use Fig 6.13(b) to find the swelling ratios for a recoverable strain γ_R of 3.2.

$$B_T = 1.22$$
$$\text{and } B_H = 1.56$$

The outside circumference of the pipe

$$= \pi \times 56 \times B_T$$
$$= \pi \times 56 \times 1.22 = 214.6\ \text{mm}$$

Therefore outer diameter = 68.3 mm

The thickness of the pipe

$$= 3 \times B_H$$
$$= 3 \times 1.56 = 4.7\ \text{mm}$$

TENSILE ELASTICITY

In elongational flows, the tensile elasticity of the melt plays an important part in the melt rheology. On approaching the abrupt change in dimensions at the entrance to a die, the melt maintains a minimum energy by conforming to a natural angle of convergence for streamlined flow. Here the shear and tensile stresses occur together. In long dies, the tensile component decays by stress relaxation, but in zero length dies the stress is entirely tensile and gives rise to tensile die swell.

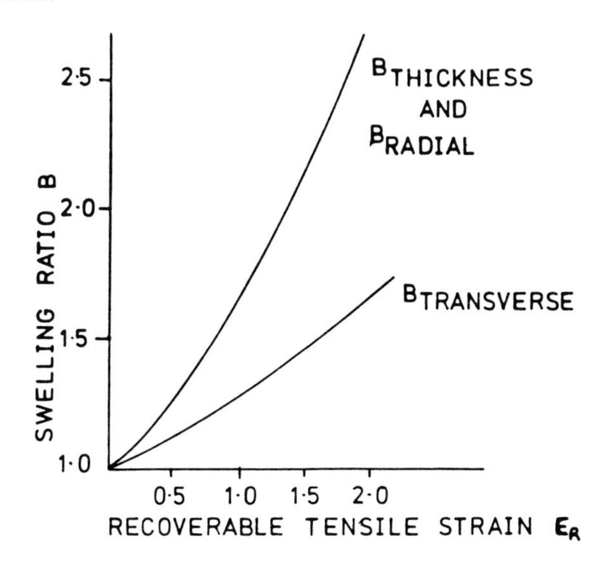

Fig. 6.14 The variation of swelling ratio with recoverable tensile strain for capillary and slit dies.

Fig. 6.14 shows the swelling ratios for capillary and slit dies as a function of recoverable tensile strain. These graphs are the tensile equivalent of those in Fig. 6.11, and likewise the curves are obtained from theoretical relationships (8), namely

$$B_R{}^2 = \exp \varepsilon_R \tag{6.16}$$

$$B_H = B_R = B_T^2 \tag{6.17}$$

The data for the variation of tensile modulus with tensile stress is not often given by manufacturers. Here, one has to make the assumption that $E = 3G$. This is certainly true for low stresses. As mentioned earlier, the maximum recoverable tensile strain is about 2.

With respect to moleculor scale events, three components of deformation are distinguishable when polymer materials are subjected to tensile stresses.

1) Bending and stretching of inter- and intra-molecular bonds occurs as in Hookean solids. This gives an instantaneous elastic deformation and, as seen in Chapter 4, can be represented by a Hookean spring. At room temperature polystyrene behaves like this. In melts, the breaking of bonds gives rise to extrudate distortion.
2) Chain uncoiling with no slippage is the response of an ideal rubber. The rate at which this process takes place depends on the ease with which molecules can coil and uncoil. In most processes, this occurs so rapidly as to be regarded as instantaneous, but when it does not occur rapidly the retarded elastic response is represented by a spring and dashpot in parallel. Well-vulcanised rubbers show this behaviour.
3) Chain clippage occurs in which flow is caused by the irreversible sliding of polymer chains over one another. This is shown by thermoplastic melts and is represented by a dashpot.

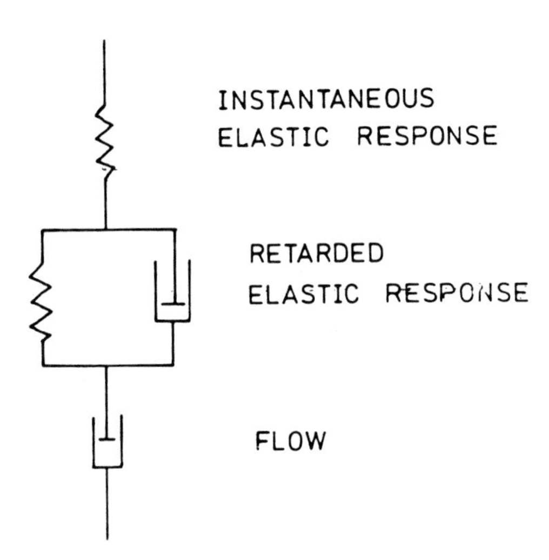

Fig. 6.15 The combined Voigt model for the representation of the tensile behaviour of polymer melts.

Fig. 6.15 shows the combination of all three phenomena as represented by the combined Voigt model.

A SUMMARY OF THE RHEOLOGICAL PARAMETERS

The flow of polymer melts often involves shear and tensile components and viscosity and elasticity effects. At low stress the elasticity is often unimportant but at high stresses all the rheological parameters may be needed. This makes a theoretical approach to the problem almost impossible.

The rheological parameters described are the shear viscosity η, the shear-thinning index n, the constant k, the tensile viscosity η_E, and the shear and tensile moduli.

In the next section the viscous and elastic effects are brought together to determine which, in a given process, predominates.

THE DEBORAH NUMBER N_D

When one moulds 'Silly Putty' it flows like a fluid; when it is subjected to a rapid tensile deformation is breaks like a brittle solid. Clearly the rate at which the process occurs is important in any explanation. With polymer molecules there is a characteristic time called the natural time, or Maxwell relaxation time. It is denoted by λ, where

$$\lambda = \eta/G \text{ shear} \qquad (6.18)$$
$$\text{or } \lambda = \eta_E/E \text{ tensile}$$

Table 6.1 shows some typical values of natural times. It can be seen that the relaxation time for water is very short, and experience shows that water flows rather than bounces. The relaxation time for 'Silly Putty' is large, as the molecules cannot respond to rapid processes, and if the process time t is short and smaller than λ, it behaves elastically and bounces or recoils.

Fluid	Natural Time
Water	10^{13} s
Oils	10^{-5} s
Polymer melts	seconds

A number called the Deborah number N_D is used to express the ratio λ/t. If $N_D << 1$ the behaviour is viscous; if $N_D >> 1$ the behaviour is elastic. Consider the following examples to illustrate the importance of the Deborah number.

Example 6.2

An acrylic is injection moulded at a melt temperature of 230°C (for which data is provided in Fig. 6.16). The maximum shear rate involved is 10^3 s^{-1} and the melt takes 2s to fill the mould. Find the Deborah number and state whether the behaviour is viscous or elastic. We would hope that it will be viscous.

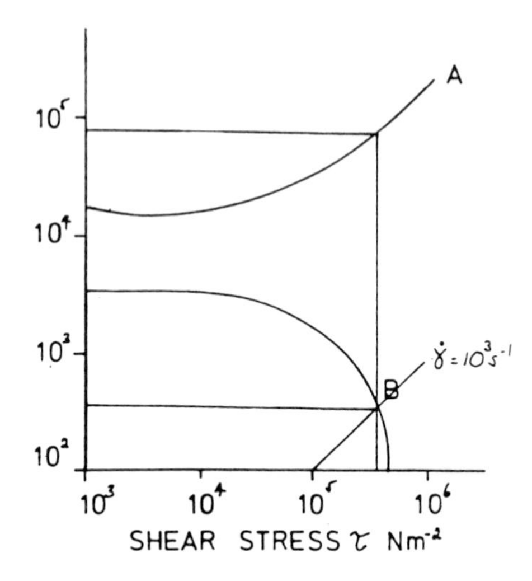

Fig. 6.16 Data for Example 6.2. *A* is the shear modulus curve and *B* is the curve for shear viscosity.

Solution

Find where the curve for acrylic cuts the 10^3 s^{-1} shear rate line in Fig. 6.16 and note the apparent shear viscosity η and the shear stress τ.

$$\tau = 3.5 \times 10^5 \text{ N m}^{-2}$$

$$\eta = 360 \text{ N s m}^{-2}$$

Draw a vertical line on the shear modulus-shear stress graph (Fig. 6.16(b)) to represent $\tau = 3.5 \times 10^5$ N m^{-2} and note where it cuts the graph for acrylic. Note the value of shear modulus.

$$G = 7.5 \times 10^4 \text{ N m}^{-2}$$

$$\text{Therefore} \quad \lambda = \eta/G = \frac{360}{7.5 \times 10^4} = 4.8 \times 10^{-3}\text{s}$$

$$N_D = \lambda/t = \frac{4.8 \times 10^3}{2}$$

$$= \underline{2.4 \times 10^{-3}}$$

The behaviour is viscous as $N_D << 1$.

Example 6.3

An LDPE parison hangs for 2s, during which time its length increases from 150.0 mm to 153.1 mm prior to inflation in a blow moulding process.

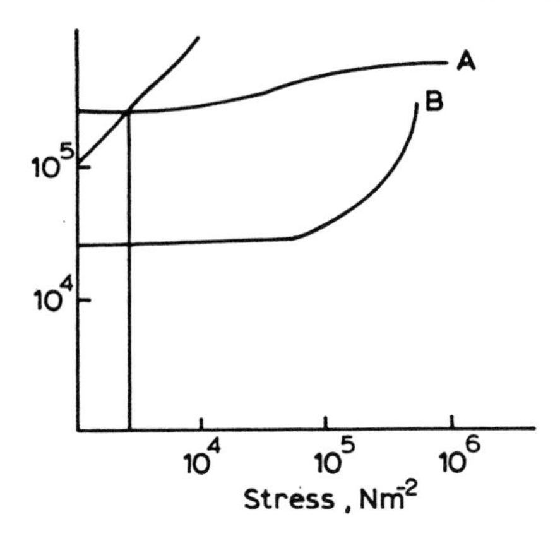

Fig. 6.17 Data for Example 6.3. *A* is the tensile viscosity curve and *B* is the curve for shear modulus.

Calculate the tensile strain rate and using Fig. 6.17, find the Deborah number and state whether the viscous or elastic behaviour of the melt controls the flow. Is the melt strength exceeded?

Solution

The Hencky strain is used in the calculation of tensile strain rate ε.

$$\dot{\varepsilon} = \frac{1}{t} \ln l/l_o = \frac{1}{2} \ln \frac{153.1}{150} = 10^{-2}\ s^{-1}$$

Draw, on Fig. 6.17, a tensile rate line of 10^{-2} s^{-1} in the same way as the shear rate lines are drawn. Find where this line cuts the LDPE graph and note σ and η_E.

$$\text{Tensile stress} \quad \sigma = 2.7 \times 10^3\ N\ m^{-2}$$

$$\text{Tensile visosity}\ \eta_E = 2.7 \times 10^5\ N\ sm^{-2}$$

Often there is no data for tensile modulus versus tensile stress. To overcome this, use the shear modulus-stress graph shown in Fig. 6.17 and imagine that the problem deals with shear flow. Construct a vertical line from $\tau = 2.7 \times 10^3$ N m^{-2} and note where it cuts the *G* graph for LDPE. From this

$$G = 2.7 \times 10^4\ N\ m^{-2}$$

We assume that $E = 3G$ and therefore

$$E = 8.1 \times 10^4\ N\ m^{-2}$$

$$\lambda = \eta_E/E\ \frac{2.7 \times 10^5}{8.1 \times 10}$$

$$N_D = \tau/t = {}^{3.3}/_2 = 1.65$$

The behaviour is predominantly elastic but the viscous component is large.

The LDPE graph ends abruptly at a tensile stress of 10^6 N m^{-2} and this represents the stress at which rupture occurs. In this problem $\sigma = 2.7 \times 10^3$ N m^{-2} and no rupture should occur.

These two examples show that tensile flows are much more likely to involve a higher elastic component than shear flows. The reason is that, as the shear viscosity falls with increasing shear stress and the shear modulus increases with increasing shear stress, the ratio η/G becomes smaller as shear stress increases. The flow is more likely to be viscous than elastic.

In tensile flows, the tensile viscosity decreases slowly with increasing tensile stress, and in some cases increases (LDPE). The tensile modulus increases with tensile strain, so that the ratio η_E/E decreases but not as rapidly as the natural time in shear flows. Thus, the elastic component of the tensile flow predominates at much higher stresses than in shear flow.

EXTRUDATE DISTORTION

When extruding polymeric materials through dies, it is usually desirable to produce smooth surfaces. However, as the rate of extrusion increases there occurs a critical point at which the smooth finish is lost. Under these circumstances the extrudate may become slightly wavy or possess a helical structure, which turns into a gross distortion if the process rate is increased further. This distortion is called elastic turbulence or melt fracture.

Another kind of distortion is associated with a matt finish on the surface of the extrudate. This is known as sharkskin. Both types of distortion owe their origins to different phenomena and their onset marks the upper limit of the process rate under the prevailing conditions.

MELT FRACTURE

The occurrence of extrudate distortion had been observed initially in rubbers, but as extrusion is not a common processing route for rubbers, the phenomenon was not extensively studied until it was observed in thermoplastics.

The properties of melt fracture include:

1) The onset of melt fracture occurs at a critical shear rate $\dot{\gamma}_c$. The corresponding critical shear stress, τ_c, lies in the range 0.1 to 1 MN m^{-2} for most commercial polymers.
2) The extrudate may be in the form of a screwed thread, a regular spiral, an irregular spiral or a bamboo shape.

3) Melt fracture can be delayed to higher shear rates by increasing the temperature (*see* Fig. 6.10) but τ_c varies little with temperature.
4) The product $\tau_c \overline{M}_w$ is approximately constant, but polyvinylchloride is an exception to this rule.
5) The critical shear rate, $\dot{\gamma}_c$ can be increased by tapering the inlet to the die. Process engineering data is always quoted for dies with untapered entrances, but in die design, tapers of less than 10° are used. This can increase $\dot{\gamma}_c$ by a factor of 100.
6) The critical shear rate can be increased by using a die of greater length to diameter (L/D) ratio.

There is some disagreement as to the exact cause of melt fracture. It is believed that it is attributable to a slip–stick mechanism at the die wall or in the melt very close to the wall. Some observations have led to the belief that conditions at the die entry (inlet fracture) as well as in the die itself (land fracture) cause this phenomenon.

The inlet explanation of melt fracture involves the dead zones that occur either side of the die inlet. Melt collects here that has a different thermal history to that issuing directly into the die. At low flow rates, the boundary between the main flow and the dead space is well defined, but at higher flow rates, when the tensile stresses are of the order of the tensile strength of the melt, a swirling rotary motion occurs, and material from the dead space flows into the die. As the flow rate further increases, so the oscillation between melt from the main flow and dead space increases. This causes a mixture of melts with different thermal histories and gives the helical distortion typical of LDPE, PS, PMMA, N66, PVC and PP.

In materials such as HDPE, linear silicones and PB, and size of the dead spaces is small. The irregular type of distortion observed is attributed to very small high frequency oscillations of the flow lines occurring above the inlet.

The latter theory accounts for the different types of melt fracture in LDPE and HDPE, but the whole subject requires much research. Fortunately the ways to minimise this phenomenon are well-known.

SHARKSKIN

Sharkskin is characterised by a series of ridges at right angles to the flow. It varies from a matt finish to a gross distortion.

Sharkskin is distinguished from melt fracture in the following ways:

1) It has a distortion perpendicular to the flow direction rather than a helical or irregular one.
2) It can occur at lower extrusion rates and appears to be a function of the linear output rate ($Q/\pi r^2$) rather than the shear rate ($4Q/\pi r^3$).

3) It is very temperature dependent and its onset can be considerably delayed by employing die tip heating.
4) It is insensitive to die entry angle and L/D ratio, although it may be minimised by using a shorter die.
5) It is insensitive to molecular weight average but is influenced by polydispersity in that monodisperse polymers are more prone to sharkskin.
6) The higher the swelling ratio and the lower the shear-thinning index, the lower the degree of sharkskin.

The sharkskin phenomenon is probably due to a surface effect rather than instabilities at the die entrance or a slip–stick effect. Consider Fig. 6.9 showing die swell. At the die exit the flow profile changes abruptly to a plane profile. This requires an acceleration of the outer lays of the melt. A viscous material can cope with this, but if an element of elasticity is present, tensile stresses build up on the surface that may exceed the tensile strength of the melt. This results in a surface tear followed by stress relaxation and thence a build-up of stress again to repeat the cycle.

Some degree of elasticity is required for this phenomenon, yet it certainly is not more prominent in melts of high elasticity — quite the reverse, in fact.

DRAW RESONANCE

Draw resonance is an elastic phenomenon associated with the haul-off rate from the die. If the melt is hauled away at a faster rate than it leaves the die, wich is usual, the extrudate suffers an extensional flow. This stretching is characterised by a draw ratio or stretch ratio.

$$\text{Draw Ratio} = \frac{\text{Haul-off velocity}}{\text{Die outlet velocity}}$$

As the draw ratio increases, a critical value is attained at which pulsations occur giving a variation in the diameter of the extrudate — a little like melt fracture. The draw resonance becomes worse if the draw ratio is further increased. The manner in which it is thought to occur is explained below.

The drawdown due to the haul-off reduces the die dwell and extrudate distortion (if it is occurring) and increases the orientation in the extrudate. At a critical draw ratio, the melt begins to slip at the die wall in the exit region. Further increases in draw ratio cause further thinning of the extrudate and the region of slip extends towards the die entrance as shown in Fig. 6.18. This effectively reduces the length of the die, until a time is reached when the melt is being pulled directly from the inlet region and scarcely touches the walls of the die. Under these conditions, the extrudate will alternately suffer adhesion to and dehesion from the die wall, and regularly spaced nodules appear on the extrudate.

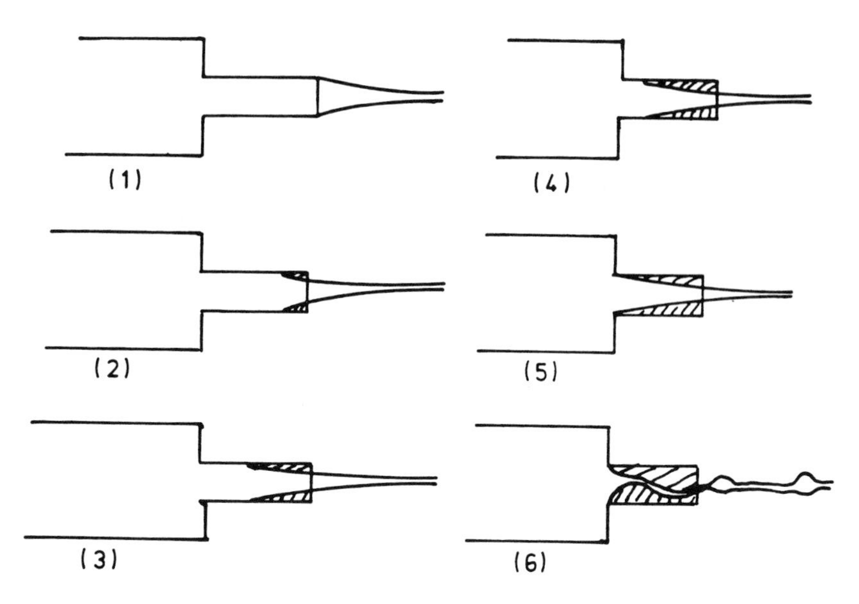

Fig. 6.18 The progress of draw resonance with increasing draw ratio from (1) to (6).

Draw resonance may be reduced by:

a) reducing the draw ratio,
b) increasing the melt temperature,
c) tapering the die inlet, and
d) increasing the L/D ratio of the die.

FILLED POLYMER MELTS

As mentioned in Chapter 3, polymers are often reinforced with fibres to improve their mechanical properties. Such materials must be processed and their rheological behaviour understood.

When small fibres are added to a melt, both the shear and tensile viscosities increase, as shown in Fig. 6.19, for shear flows. This effect is far more noticeable at the low shear or tensile rates. This low rate enhancement becomes greater when:

1) There is an increase in filler concentration.
2) There is an increase in the aspect ratio length/diameter of the filler particles.
3) There is a decrease in the size of the filler particles for a given concentration, causing a greater area of interface between the filler and the melt.

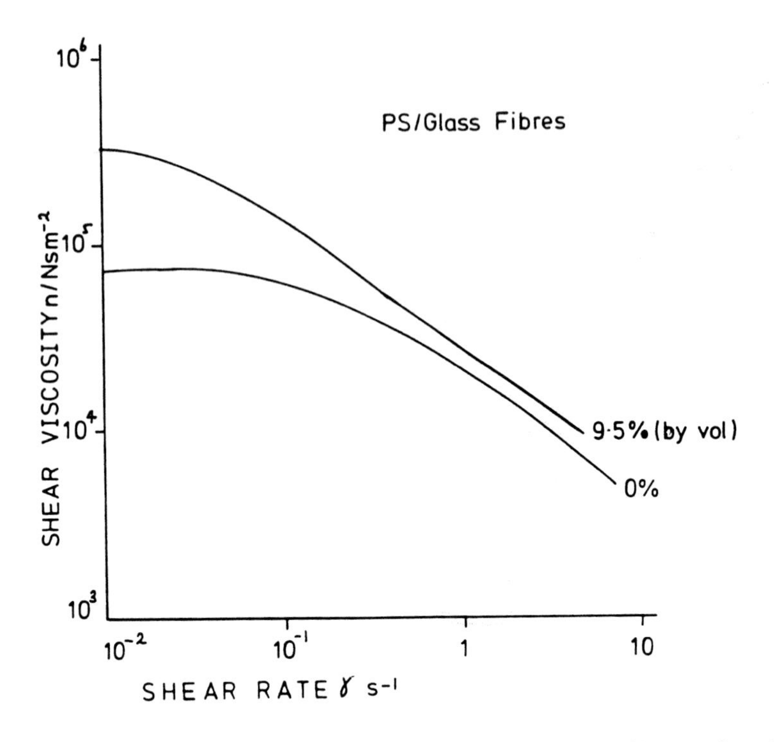

Fig. 6.19 The variation of shear viscosity with shear stress for fibre reinforced thermoplastics.

4) The filler has no compatibilising surface coating.
5) The polar structure of the filler is large.

In addition to these viscosity increases, a yield stress is often apparent and occasionally thixotropy occurs. This is particularly true when the filler is of sub-micron size.

The recoverable elastic strain is generally reduced by the addition of fibres, although if the particles form a network structure by agglomerating in the melt, an anomolous recoverable elastic strain may occur. In general, a fibre additive reduces die swell and delays the onset of extrudate distortion. Clearly, from 2) above, additions of microspheres have a lesser effect than fibres.

The behaviour described is attributed to:

1) the gross behaviour of the particles in the flow, such as alignment along the streamlines which takes up energy, particularly in tensile flows; and
2) the interparticle forces, which if large as with polar fillers such as talc and calcium carbonate, increase the viscosity and yield stress; while if small as

in carbon black (in which only Van der Waals forces predominate) the effect is less noticeable.

LIQUID CRYSTAL POLYMERS

The introduction of liquid crystalline structures or mesogenic moieties into a polymer chain is a new innovation. These mesogenic units are introduced into the main chain and may be separated by flexible spacers or may be completely stiff. Polyesters are popular structures into which liquid crystallinity is introduced. The self-reinforcing polymers are thermotropic, which means that on heating the thermoplastic changes from the sold state into a mesophase. Only after higher temperatures are introduced does the thermoplastic form an isotropic melt phase. In self-reinforcing polymers, rather than those designed for electrical applications, the mesomorphic phase is a nematic one, in which a one-dimensional long range order exists.

Apart from the obvious advantages of self aligning of the molecules, giving a high tensile modulus and high strength, the orientation of the molecules causes a dramatic reduction in shear viscosity. Moreover, after extrusion the molecules, which like to remain aligned under the action of the long-range order, there is little or no die swell.

Table 6.2 Typical Processing Temperatures of Liquid Crystal Polymers.

	'VECTRA'	'XYDAR'
Processing Temperature/°C	285–325	400–425
Melting point/°C	275–330	423

The strange thing about liquid crystalline polymers is that they are processed below their isotropic melting temperatures, as seen in Table 6.2. In addition, the liquid crystallinity enables easy processing in the relatively low shear viscosity range for these high-temperature engineering thermoplastics. Fig. 6.20 shows flow curves for a series of liquid crystalline polymers. The viscosities are surprisingly low for such stiff polymer molecules.

PLASTICISERS AND LUBRICANTS

Plasticisers are often added to polymers to modify their mechanical behaviour. PVC is a well-known example. Sometimes a plasticiser may be added, to imrove the processability or to stabilise the polymer during processing. Occasionally, a plasticiser is present by accident, in the form of an

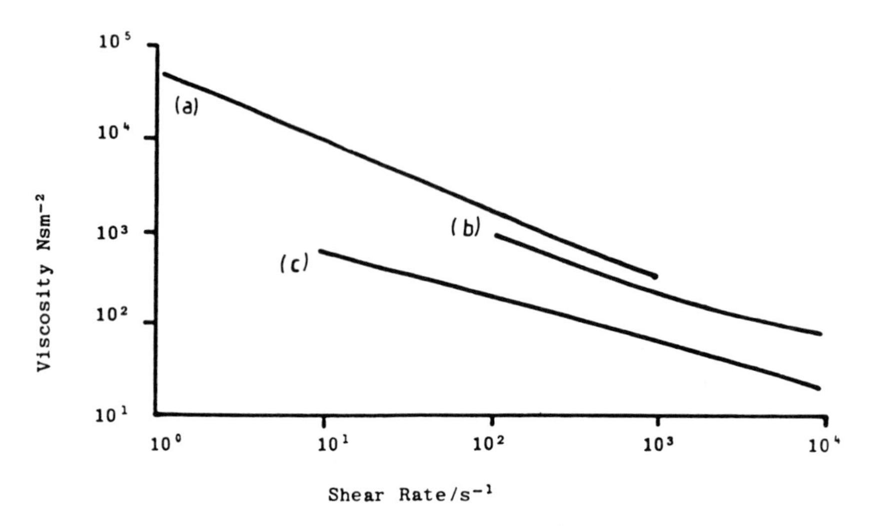

Fig. 6.20 Flow curves for (a) xydar, (b) polyethylene and (c) vectra.

unwanted impurity or unreacted monomer. In each case, the plasticiser modifies rheological behaviour.

The plasticiser is compatible with and hence miscible with the polymer, and is added at the concentration of a few per cent. It lowers the T_g of the blend by creating more free volume by spacing out the macromolecules. This increases the molecular weight average between entanglements and lowers the shear viscosity.

The effectiveness of a plasticiser depends on:

1) its concentration,
2) its compatibility with the polymer, and
3) its viscosity.

The plasticiser lowers the shear viscosity of the blend by:

1) lowering the T_g of the blend,
2) decreasing the density of the entanglements, and
3) acting as a low viscosity diluent.

There is no completely satisfactory method for calculating the blend viscosity as a function of the viscosities of the constituents and their volume fractions. The simplest method (which works only for fluids that have little thermodynamic interaction) of determining the blend viscosity at temperatures well above T_g is to use the mixture rule

$$\log \eta_o = \phi_p \log \eta_{op} + \phi_s \log \eta_{os} \tag{6.19}$$

where η_o, η_{op} and η_{os} are the zero shear rate, or low shear rate, viscosities of the blend, plasticiser and polymer respectively and ϕ_p and ϕ_s are the volume fractions of the plasticiser and polymer.

The above kind of plasticisers are often referred to as internal lubricants. Another way of increasing the output of an extrusion process is to use an external lubricant. Its purpose is to reduce the resistance to flow, and to improve the surface finish of the product.

External lubricants are largely incompatible and hence insoluble in the polymer, and they act in a similar way to a traditional lubricant in reducing mechanical friction. They migrate to the interface between the polymer melt and the walls of the machinery, where the high shear rate regions are found. This induces some slip at the wall of the machine.

Sometimes the extrusion rate for a given pressure is increased dramatically by the addition of a small amount of lubricant. However, the slippage induced reduces the shear flow and induces a 'plug flow' (like toothpaste). This increases the tensile stresses and, as such, increases die swell and causes melt fracture at lower rates. It is important to be aware of these pitfalls when using lubricants.

POLYMER BLENDS

Two or more polymers are sometimes blended together to improve service performance, to give a spectrum of mechanical properties (propylene/ethylene copolymer), to extend the use of expensive engineering thermoplastics, to improve processability (polyphenyleneoxide/polystyrene graft copolymer), to re-use scrap; but by far the most important use is to improve mechanical properties (rubber toughening of thermoplastics).

The rheology of blends depends on their morphology, which is determined by:

1) the concentration of the dispersed phase,
2) the viscosity ratio of the two phases,
3) their elasticity ratio,
4) the size and form of the dispersed phase,
5) the miscibility and presence of compatibilisers,
6) the character and size of the interface domains, and
7) the method of processing, which has some influence on 4).

In the case of blends of partly-crystalline polymers, the situation is more complex. Shear and pressure effects may alter the rate of crystallisation, may modify the nucleation mechanism and possibly alter the type of crystals formed.

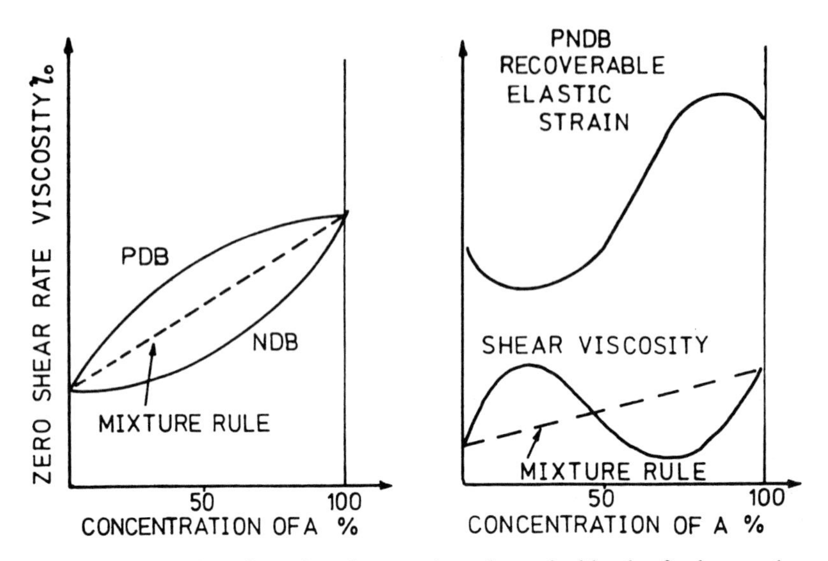

Fig. 6.21 Deviations from the mixture rule as shown by blends of polymers *A* and *B* in a melt.

Fig. 6.21 shows the type of blend behaviour observed:

1) according to the mixture rule, equation (6.19), and described earlier;
2) an increase in viscosity and elasticity over the simple mixture rule as shown by miscible blends such as polyphenyleneoxide/polystyrene and immiscible blends with strong interactions across the interfaces between the phases (called positive deviation blends PDB);
3) a decrease in viscosity and elasticity below the mixture rule predictions, as shown by immiscible blends where the adhesion at the interface is poor and the interaction weak, giving gross delamination between phases (NDB); and
4) both positive and negative deviations, as shown in Fig. 6.21b, with the maximum viscosity coinciding with the minimum elasticity at one concentration and a reversal of this at another (PNDB). This occurs when there is a concentration-dependent change in morphology, such as phase reversal or change from a discrete dispersed phase to an interlocked structure.

BLOCK COPOLYMERS

Block copolymers show unique rheological properties owing to the formation of a discrete domain structure. A model for them would be like that of an emulsion with the addition of chemical bonds across the interface. Random copolymers are PDB and ABA block copolymers have a greater degree of positive deviation than the random copolymers or the homopolymers.

The model proposed for the behaviour of ABA block copolymers is that the ABA macromolecule lies in two A-domains and a B-matrix and the rheological parameters are determined by the properties of the interface. Again insufficient data is available to gain a complete understanding of the observed phenomena. It is known that the rheological behaviour is influenced greatly by the character of the central block and that these melts show non-Newtonian behaviour at low shear rates.

THERMOSETTING MATERIALS

Initially thermosetting resins consist of low molecular weight polymers, fillers, pigments, lubricants and hardness. The viscosity is low. On heating, the viscosity decreases further until the crosslinking reaction commences. The viscosity rapidly increases until flow eventually stops. Fig. 6.22 shows the total flow curve for a thermoset. As temperature increases, the flow rate increases, but the time for the thermoset to gel decreases. This results in a maximum in the total flow curve as shown. In processing these materials, it is of paramount importance to use the correct processing temperatures, particularly if the moulds used are difficult to fill.

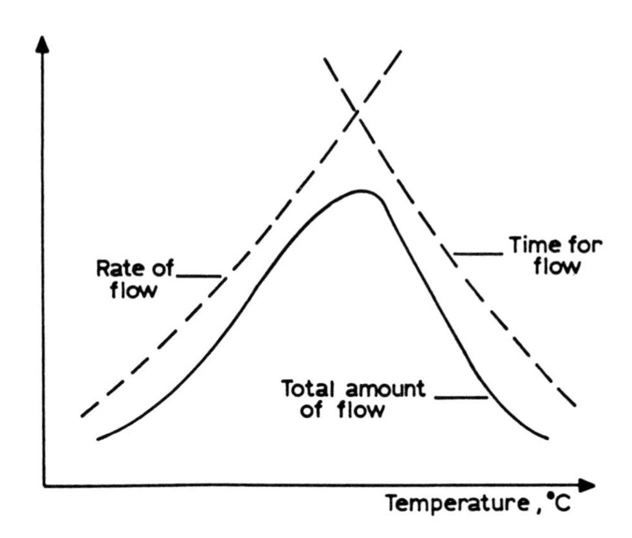

Fig. 6.22 The total flow curve of a thermoset as a function of temperature.

THE MEASUREMENT AND PRESENTATION OF RHEOLOGICAL DATA

There are many ways in which the rheological data can be obtained, but the two simplest methods which are used by process engineers and plastics manufacturers are:

1) the Melt Flow Indexer, and
2) the Capillary Rheometer.

The instrument shown in Fig. 6.23 is called the Melt Flow Indexer. It consists of a heated barrel with a die of L/D ratio just less than 8:1, which is fitted in the base. Polymer is packed into the barrel and a piston is placed above it. This piston is weighed down by a 2.16 kg weight (for polyethylene, for which the instrument was originally developed). This weight pushes the melt through the die, and the amount of polymer in grams collected in 10 minutes at the temperature prescribed is termed the melt flow index. A large value of MFI indicates a very fluid melt. The term MFI should be applied only to polyethylene when the measurement is made according to the standards BS 2782 or ASTM D1238 (Condition E), but the term is often used for some other polymers, or in the shortened form 'melt index'.

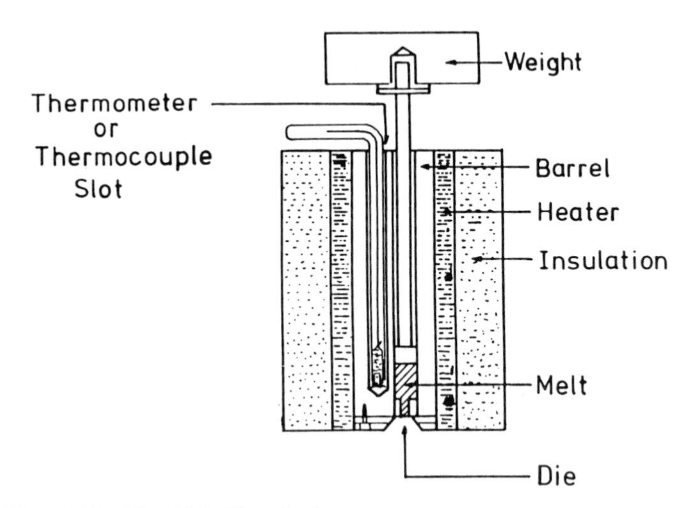

Fig. 6.23 The Melt Flow Indexer.

The advantages of this method are its simplicity and its relative cheapness. It is suitable for testing batch to batch consistency or for assessing thermal or shear degradation of a processed polymer. The main disadvantage is that it is largely a 'one point' test and, as such, gives limited rheological data.

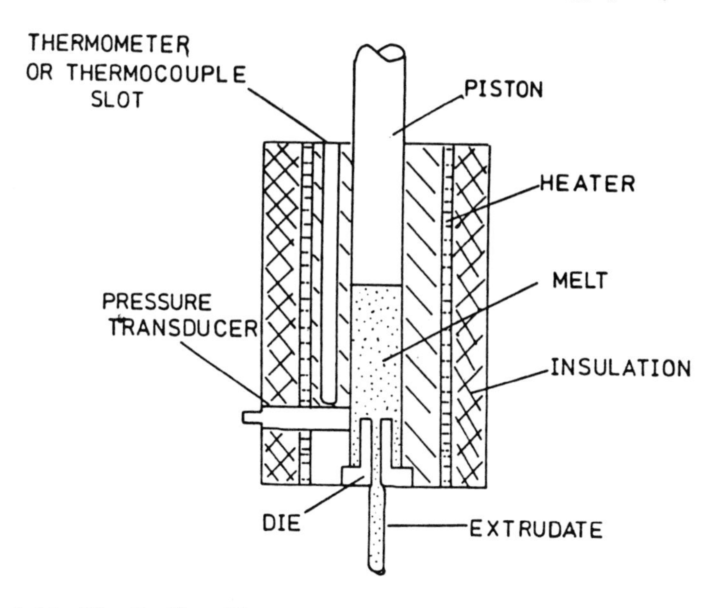

Fig. 6.24 The Capillary Rheometer.

The Capillary Rheometer or Extrusion Rheometer is shown in Fig. 6.24. Like the Melt Flow Indexer it, too, consists of a heated barrel with a changeable die fixed at the bottom. The polymer melt is pushed through the die by a piston that moves at a controllable rate. The piston speed S and the pressure P at the capillary entrance are measured for a numer of values of S. In order to eliminate the entrance pressure P_0 due to the extensional flow at the die inlet, the experiment is repeated with a zero length die or orifice plate of the same diameter for the same values of S. The long die should have an L/D ratio of at least 16:1 to ensure that the tensile stresses have relaxed before the outlet is reached.

The shear stress at the wall of the capillary is given by

$$\tau = \frac{(P - P_o)r}{2l} \qquad (6.20)$$

where r is the die radius and l is the length of the long die.

$$\dot{\gamma} = \frac{4Q}{\pi r^3} \qquad (6.21)$$

The flow rate Q can be related to the piston speed S

$$Q = \pi R^2 S$$

Therefore

$$\dot{\gamma} = \frac{4R^2 S}{r^3} \qquad (6.22)$$

The apparent shear viscosity η is given by

$$\eta = \tau/\dot{\gamma}$$

and the shear-thinning index n is given by

$$\eta = \frac{d(\log\tau)}{d(\log\dot{\gamma})} \qquad (6.23)$$

on the slope of the $\log \tau - \log \dot{\gamma}$ graph.

Sometimes the recoverable shear and tensile strains are related to the swelling ratio from a long die (shear) and for a zero length die (tensile). There are many discussions on the validity of the various equations used (7, 8) but one used for ICI data is

$$B^2{}_S = \frac{2}{3}\gamma_R \left[\left(1 + \frac{1}{\gamma_R{}^2} \right)^{3/2} - \frac{1}{\gamma_R{}^3} \right] \qquad (6.24)$$

where B_S is the swelling ratio due to shear flow and γ_R is the recoverable shear strain. This is the equation from which Fig. 6.11 is calculated for capillary dies.

The shear modulus is given by

$$G = \tau/\gamma_R$$

The rheological parameters η, η_E, G and E are sometimes given in ICI data but often only η is available in other manufacturers' data sheets. From the flow curves, n can be found.

The data sheets and brochures of different polymer manufacturers are extremely informative and should always be obtained when using a new material. Remember the old adage, 'When all else fails, read the instructions'.

REFERENCES

H.A. BARNES, J.F. HUTTON and K. WALTERS: *An Introduction to Rheology*, Elsevier Science Publishers, Amsterdam and New York, 1989.

A. BLUMSTEIN: *Polymeric Liquid Crystals*, Plenum Press, 1983.

J.A. BRYDSON: *The Flow Properties of Polymer Melts*, Iliffe, 1980.

L.L. CHAPOY: *Recent Advances In Liquid Crystal Polymers*, Elsevier Applied Science, London, 1985.

F.N. COGSWELL: *Polymer Melt Rheology*, George Godwin Ltd, 1981.

A.A. COLLYER: 'Thermotropic Liquid Crystal Polymers for Engineering Applications', *Materials Science Technology*, 1989, **5**, 309–322.

A.A. COLLYER and D.W. CLEGG: *Rheological Measurement*, Elsevier Applied Science Publishers, London, 1988.

R.J. CRAWFORD: *Plastics Engineering*, Pergamon Press, Oxford, 1981.

J.M. DEALY: *Rheometers for Molten Plastics*, Van Nostrand Reinhold, New York, 1982.

J.L. WHITE: in *The Mechanical Properties of Fibre Reinforced Thermoplastics*, D W Clegg and A A Collyer, eds, Elsevier Applied Science, London, 1985.

7 *The Effect of Process Variables and Molecular Architecture on the Rheological Properties*

INTRODUCTION

The mechanical and rheological properties of a polymer depend on its morphology, which in turn is controlled by the molecular behaviour during processing. The molecular behaviour will depend on the possible conformations and properties of the macromolecules and these are determined by the molecular architecture and modified by the process conditions.

It is the aim of the polymer scientist to be able to tailor the molecular architecture of a polymeric material to produce the desired morphology during processing, and as a result obtain the optimum mechanical properties in the final product. This approach is known as microstructural engineering, and the route by which the desired end-point characteristics are achieved gives the microstructural profile of the material.

Having discussed the basic rheological parameters, part of the way along the microstructural engineering route is to examine how these parameters are influenced by the process variables and the molecular architecture.

EFFECT OF TEMPERATURE ON SHEAR VISCOSITY

As the temperature of a fluid is increased, the molecules vibrate more rapidly and become more mobile. Fig. 7.1 shows a model describing the situation in fluids. A fluid can be regarded as a degenerate solid in which vacant sites exist into which fluid molecules can move if they have sufficient energy. Molecules a and b move into sites a' and b' leaving a vacancy where they had once been. As seen in Fig. 7.2, each molecule is situated in a potential well and can only move into the next potential well or site if it has sufficient energy to do so and if the site is unoccupied. If the fluid is stationary, the potential barriers

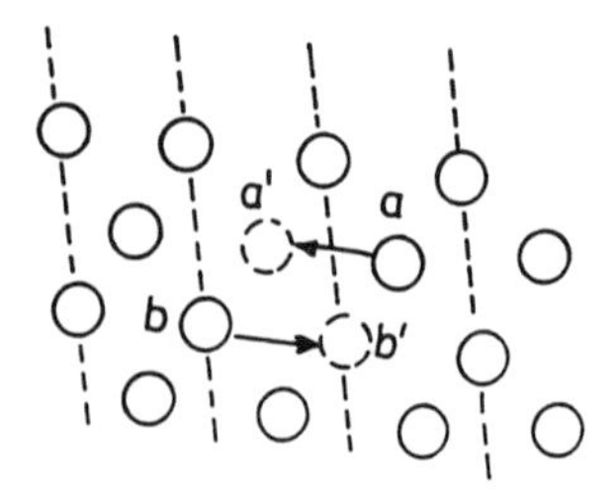

Fig. 7.1 A model to show vacant sites into which molecules can move in a fluid.

are symmetrical and of height ΔF_m, and the molecule is as likely to move to the left as to the right.

If a shearing force is applied from left to right, the potential barrier to the right becomes lower, $\Delta F_m'$ giving a greater probability of the molecule moving in the direction of the shearing force.

The ability of molecules to move from site to site, and hence provide flow, depends on the number of vacant sites n and the rate of moving from site to site, the jump frequency ϕ. If the temperature is increased, the free volume will increase giving an increase in the number of vacant sites, and the jump

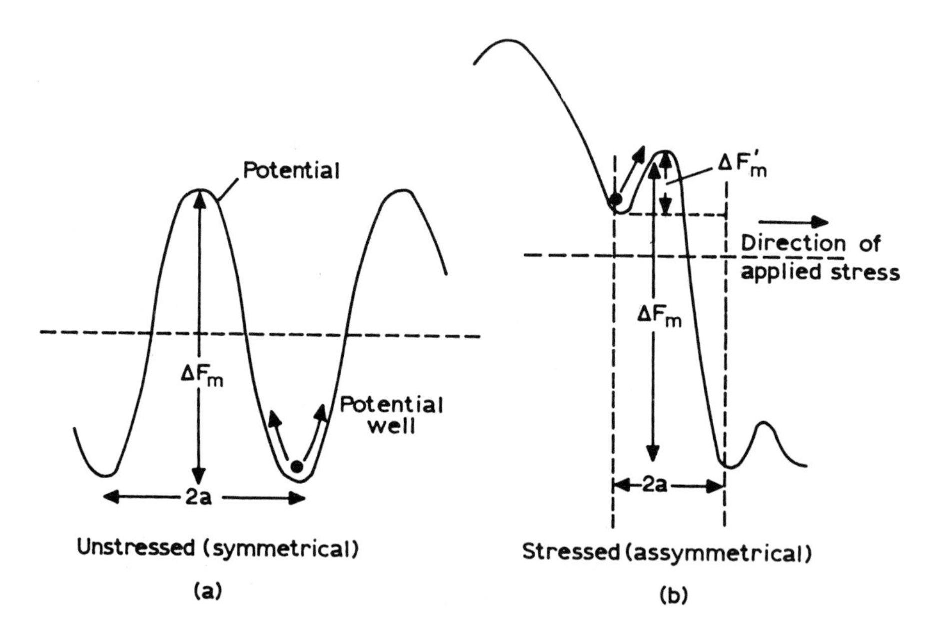

Fig. 7.2 The symmetrical and asymmetrical potential barriers in (a) a stationary fluid and (b) a fluid moving from left to right.

frequency will increase because each molecule has more thermal energy. Therefore the viscosity will be given by

$$\eta \, \alpha \, \frac{1}{n} \times \frac{1}{\phi} \tag{7.1}$$

It can be shown that the number of vacant sites at a given temperature T (in kelvin) depends on

$$n = A \exp \frac{-\Delta E_n}{RT} \tag{7.2}$$

where ΔE_n is the activation energy for hole creation, R is the gas constant and A is a pre-exponential constant.

Similarly it can be shown that ϕ is given by:

$$\phi = B \exp \frac{-\Delta E_B}{RT} \tag{7.3}$$

where ΔE_B is the energy of the potential barrier and B is a pre-exponential factor.

Combining equations (7.2) and (7.3) and including a constant of proportionality for equation (7.1) gives

$$\eta = K \exp \left[\frac{\Delta E_n + \Delta E_B}{\mathrm{RT}} \right]$$

$$= K \exp \left(\frac{\Delta E}{RT} \right) \tag{7.4}$$

where K is a pre-exponential constant, R = 8.31 J mol^{-1} and ΔE is the activation energy for flow. A graph between $\ln \eta$ ad l/T gives a straight line

$$l n \eta = \ln K + \frac{\Delta E}{RT} \tag{7.5}$$

where the slope = $\Delta E/R$. If the graph was drawn on log graph paper (log to base 10), giving $\log \eta$ versus l/T, the slope is ($0.4343\, \Delta E/R$).

This approach is based on the Eyring theory for rate processes and equation (7.4) describes Arrhenius type behaviour.

Newtonian fluids show Arrhenius type behaviour as do polymer melts at low shear stresses. However, when the polymer melt is behaving in a shear-thinning manner, the accuracy of equation (7.4) may be limited to a range of 50–60°C and the value of the activation energy for flow, ΔE, will depend on whether it was obtained at constant shear rate $\Delta E_{\dot{\gamma}}$ or at constant shear stress ΔE_{τ}. At zero shear rate $\Delta E_{\tau} = \Delta E_{\dot{\gamma}}$, whereas at other shear rates

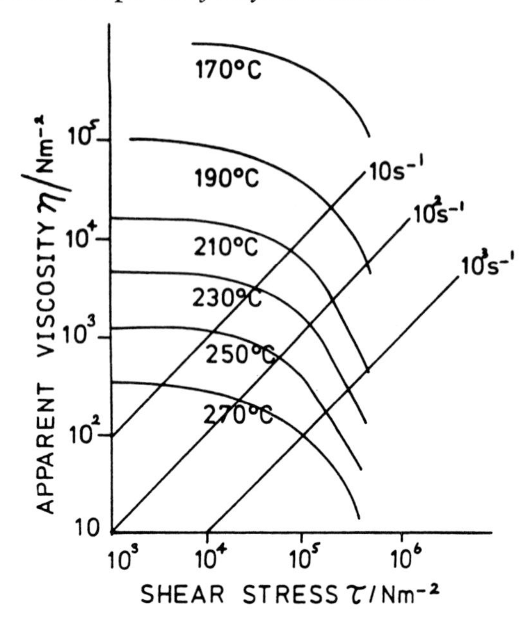

Fig. 7.3 The variation of shear viscosity with shear stress and temperature for poly methyl-methacrylate PMMA.

$$\Delta E_{\gamma} = n\Delta E_{\tau} \quad (7.6)$$

where n is the shear-thinning index (1).

The activation energy for flow at zero shear rate for polymer melts lies in the range 2×10^4 to 2×10^5 J mol^{-1}. The lowest value known is 16.7 J mol^{-1} for dimethyl silicone polymers, and this is attributed to the great flexibility of the silicone backbone chain. Stiffer backbone chains give higher values of ΔE. For a value of ΔE of 8×10^4 J mol^{-1} and a temperature change from 300°C to 310°C, the viscosity is reduced to one third.

Fig. 7.3 shows the variation of the shear viscosity of PMMA with shear stress and temperature. The effect of temperature is less at the higher shear stresses so that $(d\eta/dT)_{\tau}$ depends on shear stress. The increase in temperature reduces the tensile viscosity too, but the temperature effect on the shear and tensile moduli is less marked. Both E and G decrease as T increases.

Williams, Landel and Ferry (2) developed a theory based on free volume to account for the variation of viscosity with temperature. At a certain temperature below T_g, $(T_g - 52)$, they supposed that no free volume exists between polymer chains. As the temperature increases, the free volume increases, as in the Eyring theory mentioned earlier, giving easier molecular flow. The resulting equation called the WLF equation is

$$\log A_T = \log \frac{\eta}{\eta_R} = \frac{C_1 (T - T_g)}{C_2 (T - T_g)} \quad (7.7)$$

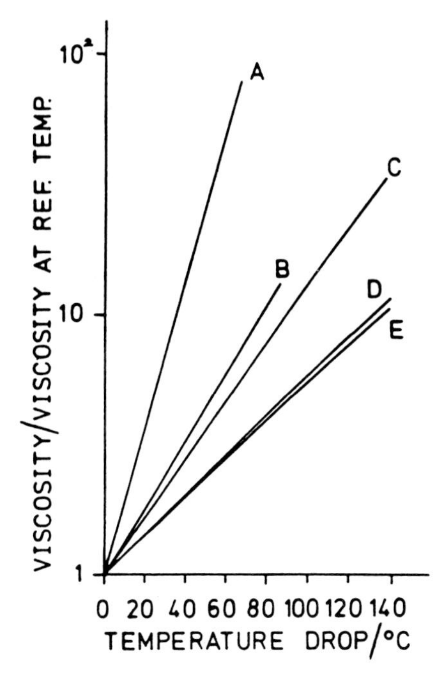

Fig. 7.4 Normalised shear viscosity as a function of $T - T_R$ for A) poly-methyl methacrylate, B) Nylon 66, C) low density polyethylene, D) acetal copolymer, and E) polypropylene.

where A_T is a shift factor necessary to move the graphs of η versus T vertically such that they superimpose; η_R is the zero shear rate viscosity at the reference temperature, which is generally chosen to be T_g and C_1 and C_2 are universal constants.

$$C_1 = 17.44 \text{ and } C_2 = 51.6 \text{ K}$$

In practice, these values are not universal and a greater accuracy can be obtained measuring the zero shear rate or low shear rate viscosity at low temperatures and substituting values in equation (7.7) to find C_1 and C_2 for the polymer under examination. Sometimes the question arises which model, the Eyring or the WLF, to use when evaluating temperature effects. It is best to use the WLF when working less than 50 to 100°C above T_g, but after that the Arrhenius equation (7.4) gives values close to the WLF equation.

Rather than make calculations, it is useful to have graphical data in the form of Fig. 7.4. This shows the variation of the ratio of viscosity at temperature T to that at a reference temperature T_R as a function of $(T - T_R)$. The viscosity ratio is often referred to as a normalised viscosity. From these graphs in Fig. 7.4, it is obvious that the polypropylene is less sensitive to temperature change than the polymethylmethacrylate.

Temperature can also affect the shear viscosity by modifying the polymer molecules. PVC and PMMA are both thermally sensitive polymers, which are rarely processed at temperatures much higher than $(T_g + 100)$°C. The type of thermal modification that can occur depends on the polymer and five possibilities can be identified:

1) Chain scission, giving a reduction in $\overline{M}_w$ and a decrease in viscosity.
2) Unzipper reactions that result in a complete depolymerisation.
3) Cleavage of sidegroups giving discolouration and embrittlement.
4) Crosslinking, which fixes the structure into a three-dimensional network. The effect is extensive in vulcanised rubbers and thermosets.
5) A combination of 1) and 3) may occur.

A similar chain scission may occur due to high shear rates.

EFFECT OF HYDROSTATIC PRESSURE ON SHEAR VISCOSITY

As mentioned earlier, viscosity is dependent on free volume, and an increase in hydrostatic pressure will compress the polymer and reduce free volume. This effect is far more noticeable in amorphous polymers than in crystalline ones. This effect is important in injection moulding where hydrostatic pressures involved may easily be as high as 100 MN m^{-2}.

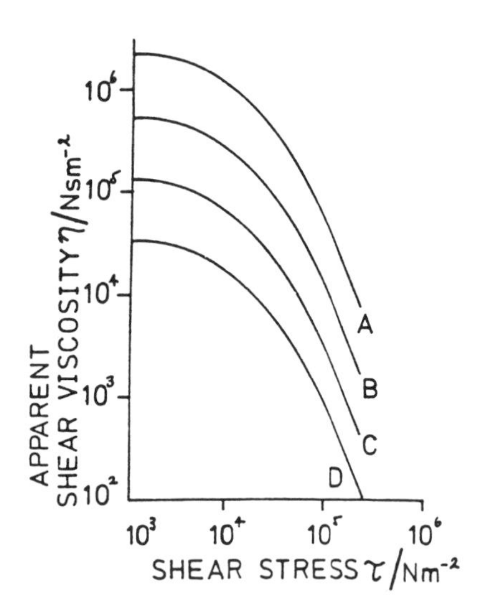

Fig. 7.5 The variation of shear viscosity with hydrostatic pressure for low-density polyethylene at *A)* 300 MN m^{-2}, *B)* 200 MN m^{-2}, *C)* 100 MN m^{-2}, *D)* atmospheric pressure.

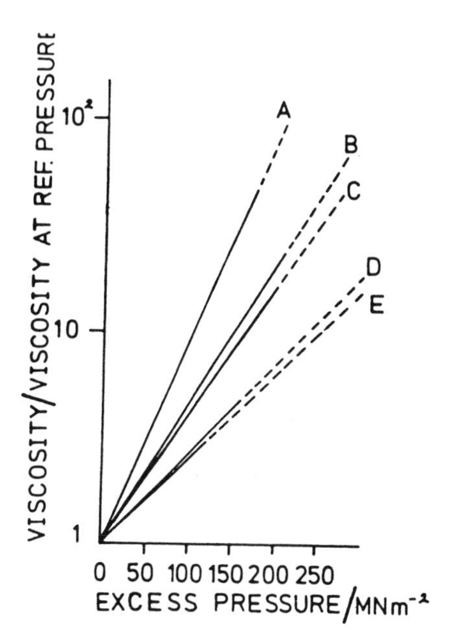

Fig. 7.6 The variation of normalised viscosity (η/η_A) with ($P - P_A$), where η_A is the shear viscosity at atmospheric pressure, P_A, for A polymethylmethacrylate, B polypropylene, C low density polyethylene D Nylon 66 and E acetal copolymer.

Fig. 7.5 shows the effect of hydrostatic pressures on shear viscosity and Fig. 7.6 shows the variation in the normalised shear viscosity with the pressure difference ($P - P_A$), where P_A is atmospheric pressure. This is a useful graph in that it shows which polymers are most sensitive to changes in pressure. The acrylic shows the greatest variation and acetal copolymer the least.

This pressure increase, which increases shear viscosity, can be expressed in terms of a decrease in melt temperature, as shown in Fig. 7.7. This expresses the decrease in melt temperature that has the same effect as the increase in hydrostatic pressure.

PROCESS VARIABLES

The process variables usually encountered are shear rate, melt temperature and hydrostatic pressure. In Chapter 6 and the early part of this chapter, the effect of these variables on the shear viscosity, in particular, has been described. One of the other important features, however, that has a great bearing on the viscosity and the elasticity of melts is the molecular architecture and the

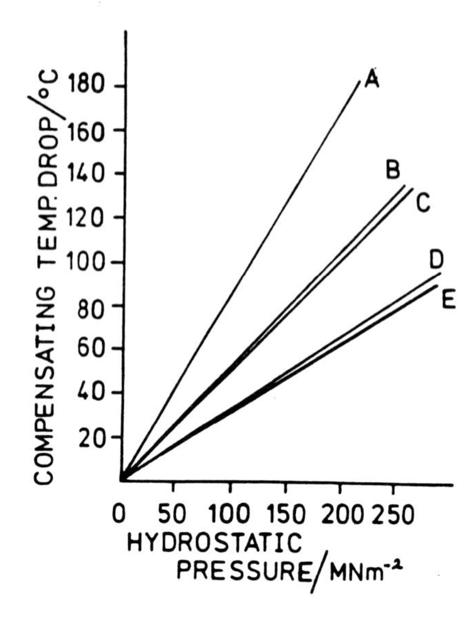

Fig. 7.7 The equivalence of pressure and temperature for a constant normalised viscosity for *A* polypropylene, *B* low density polyethylene, *C* acetal copolymer, *D* polymethyl methyacrylate and *E* Nylon 66.

resultant morphology. Often, an understanding of this is neglected, and this can lead to problems encountered in processing that are difficult to solve.

The different aspects of molecular architecture are described below.

Chain Stiffness

As already mentioned, chain stiffness has a great bearing on T_g, and, as such, will influence the activation energy for flow, ΔE, and the shear and tensile viscosities, η and η_E, which all increase as T_g increases.

The covalent bonds in a linear chain fix the distances between atoms, and the bond angles are fixed rigidly too. Rotation about the bonds is, therefore,

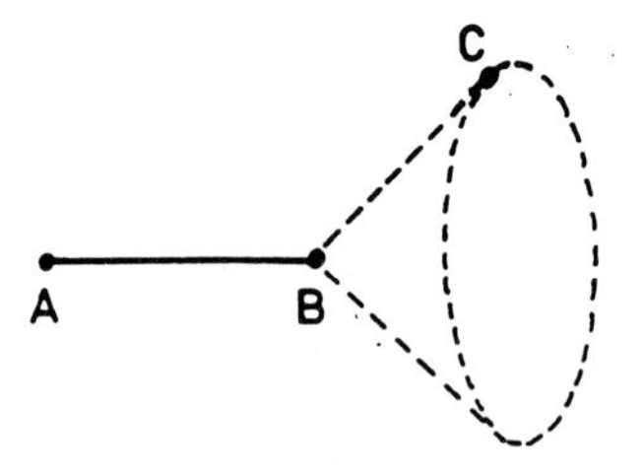

Fig. 7.8 The rotation about C–C bonds in a backbone chain such that the angle *ABC* remains constant.

the only simple way that a molecule can take up different conformations. This is shown in Fig. 7.8, in which during the rotation of the molecular segment of a long chain, the angle *ABC* remains constant. *B* can rotate about *A* too, so that many conformations are possible for long chains even though rotation is the only permitted mode.

Silicones and polymers with ether linkages are highly flexible and have low values of T_g and ΔE. These materials are processed at low temperatures. C–C links in a linear chain give rise to stiffer molecules and higher values of T_g, ΔE, η, η_E and tensile modulus, E.

More restrictions to movement occur when the sidegroups are aromatic, as in the case of polystyrene and polyphenyleneoxide (shown in Fig. 7.9). The T_g and the upper service temperature are increased, and in the case of

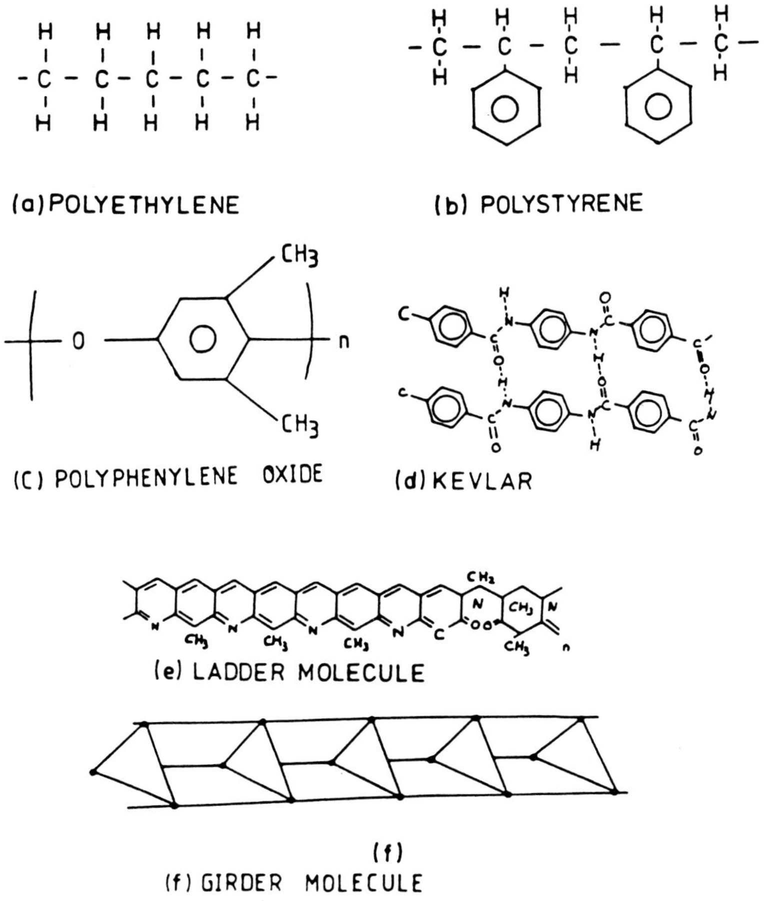

(a) POLYETHYLENE

(b) POLYSTYRENE

(C) POLYPHENYLENE OXIDE

(d) KEVLAR

(e) LADDER MOLECULE

(f) GIRDER MOLECULE

Fig. 7.9 The repeat units of polystyrene, polyphenyleneoxide, polyphenyleneterephthalamide (Kevlar), ladder structures and girder structures.

polyphenyleneoxide the restricted conformation leads to processing difficulties. This polymer is modified to ease its intractibility by a graft copolymerisation with polystyrene, with which it is compatible.

Aromatic rings may also be incorporated in the backbone chain, as in polyimide. This material is intractible, and like polytetrafluoroethylene is generally shaped by powder metallurgy techniques. Kevlar has an imidised structure, which is intractible, and this material is used as a reinforcing fibre. The ladder and girder structures shown in Fig. 7.9 represent the ultimate in chain stiffness and these materials are impossible to process by usual plastics methods. Ladder molecules exist but girder molecules are only a possibility.

From the preceding comments, it can be noted that even from cursory glances at the molecular structure of polymers, much information can be gleaned about the likely rheological behaviour. Sometimes, using models or computer aids to build up structures gives an insight into the flexibility of the molecule.

Molecular Weight Average $\overline{M}_w$ and Polydispersity Q

For a given polymeric material, by far the greatest effect on the zero shear rate viscosity η_0 is that due to the molecular weight average, $\overline{M}_w$, of the polymer molecules. This is demonstrated in Fig. 7.10. Provided that a certain critical molecular weight average, $\overline{M}_c$, has been exceeded, linear polymers obey equation (7.8).

$$\eta_0 = K\,\overline{M}_w^{\,3.5} \text{ for } \overline{M}_w > \overline{M}_c \tag{7.8}$$

For values of $\overline{M}_w < \overline{M}_c$ the relationship between η_0 and $\overline{M}_w$ is linear of index unity. $\overline{M}_c$ varies from polymer to polymer and is of the order of 15000.

The presence of highly polar groups along polymer chains increases the inter-molecular forces and these forces pull the chains closer together. This

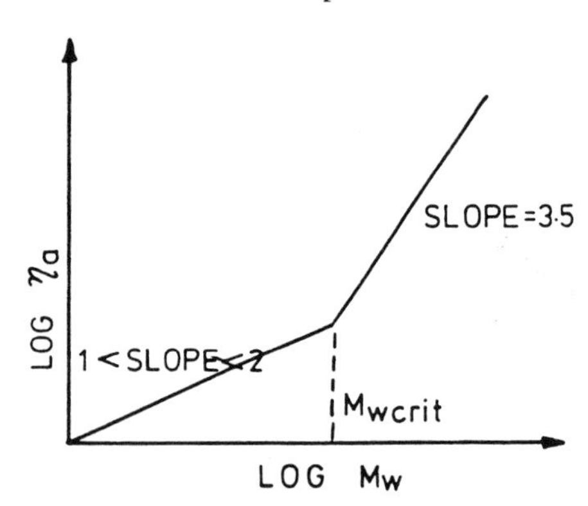

Fig. 7.10 The variation of zero shear rate viscosity η_0 with molecular weight average $\overline{M}_w$.

reduces the free volume and increases the rheological parameters T_g, ΔE, η, ηE and E.

Nylon 66 differs in its behaviour from other polymers because of the strong hydrogen bonding present, which increases T_g and non-Newtonian flow behaviour.

The addition of sidegroups can restrict the conformations of the backbone chain. If the sidegroups are stiff, as in polytetrafluoroethylene, or bulky, as in polymethylmethacrylate, the T_g will be increased. Flexible sidegroups as in polyisobutylene, hold the backbone chains apart and thereby increase free volume. This causes a decrease in T_g, ΔE, η, ηE. A more detailed and advanced discussion of this subject is given in (3, 4).

The explanation of the existence of $\overline{M}_c$ is that below a certain weight average there are insufficient long molecules to cause entanglement. It is these entanglements that increase inter-molecular interaction dramatically.

The value of K depends on temperature and on the flexibility of the polymer chain. The value of the index, however, is independent of polymer type or temperature for linear polymers. Plasticised PVC is an exception to this rule.

At higher shear rates, η is not as sensitive to $\overline{M}_w$, and the index decreases from 3.5 to between 1.5 and 2.0 at very high shear rates. It is found that under these conditions η depends more on $\overline{M}_n$ than $\overline{M}_w$. The reason for this decrease in sensitivity to $\overline{M}_w$ is that the resulting alignment of the polymer molecules lessens their entanglement.

Another approach to this situation is to consider free volume. Chain ends disrupt the packing of molecules and introduce extra free volume. Decreasing the density of chain ends by increasing the chain length, and hence $\overline{M}_w$, will reduce free volume. This will cause an increase in T_g, ΔE, η and E. One other observation is that as T_g and ΔE increase, the value of K in equation (7.8) becomes more sensitive to temperature.

The shear-thinning index, n, decreases as $\overline{M}_w$ increases, and highly shear-thinning behaviour often occurs. This is due to the increased entanglement, as $\overline{M}_w$ increases and the added difficulty for molecules to align along the streamlines. This gives rise to a greater probability of a region of intense entanglement such that large stresses sufficient to cause non-Newtonian behaviour are present at low shear rates. The fluid, therefore, gives an early departure from Newtonian behaviour.

Elastic effects are enhanced by increasing $\overline{M}_w$, owing to the increased entanglement. In processing, die swell will-increase and melt fracture will occur at lower shear rates. Sharkskin is not sensitive to $\overline{M}_w$.

Mechanical properties are generally improved by an increase in $\overline{M}_w$. The orientation of molecules in the flow direction will improve mechanical properties in that direction, at the expense of those in the transverse direction. There may be an optimum value of $\overline{M}_w$ for obtaining the best service properties.

In the above discussion, it was assumed that all the polymer molecules are identical — a monodisperse system. For commercial brands this is certainly not the case, and fortunately the lack of uniformity improves both the mechanical and rheological properties.

In a sample of broad molecular weight distribution quantified by a high value of Q, the polydispersity $Q = \frac{\overline{M}_w}{\overline{M}_n}$ the longer chains form more entanglements and give a protective network around the smaller chains. The longer chains, therefore, take up a disproportionate amount of stress. This results in these chains recoiling with a larger recoverable strain, which is retarded by the presence of the smaller molecules that have already conformed. The elastic response will be greater and slower than in a monodisperse system of the same $\overline{M}_w$. Thus in a polydisperse system, die swell is increased and melt fracture occurs at lower shear rates. Sharkskin is lessened however. Tensile viscosity is increased by polydispersity, but at low shear rates the shear viscosity is not much affected. However, the unequal distribution of stress will cause the longer molecules to disentangle at lower average shear stresses, giving an early onset of shear-thinning flow. At high shear rates the more polydisperse material will have a lower shear viscosity than a monodisperse system of the same $\overline{M}_w$. However, tensile viscosity will still remain higher than that of the monodisperse system.

This situation illustrates the limitation of the Melt Flow Index as a guide to reproducibility of product in that, if two batches of supposedly the same material have the same MFI, a greater polydispersity in the one will mean that at high processing shear rates it will have the lower shear viscosity.

The effect of temperature is less noticeable in polydisperse materials and the value of K in equation (7.8) is not as greatly changed by temperature as in monodisperse systems. The activation energy for flow, ΔE, is reduced by increasing polydispersity.

Earlier in this section, it was commented that polydispersity is desirable in a melt. An example of this is in the process of blow moulding. In this process a tube, or parison, of molten polymer has to hang vertically while a mould is clamped around it. The parison is then inflated by high pressure air until it conforms to the shape of the mould. A high tensile viscosity, and hence shear viscosity, is required during the hanging stage, whereas during inflation, a low viscosity is more desirable. The more polydisperse melt provides this compromise better.

Chain Branching

The effect of chain branching is to increase the density of chain ends and so increase the free volume. This will reduce T_g, ΔE and η. If the branches are very long indeed, there will be fewer of them, so that above a critical length for a polymer of a given $\overline{M}_w$ the density of chain ends will go through a turning point, and the free volume will start to decrease with increasing sidechain length.

Examining chain branching on a molecular level, it can be seen that a branched molecule is more compact than a linear one of the same $\overline{M}_w$. This will give rise to less entanglements in shear flows. If the sidechains are increased in length, the molecule will become less compact, giving an increase in entanglements. A further increase in sidechain length will eventually cause a greater compactness as the main chain shortens. Thus, by considering both free chain ends and molecular compactness, the free volume should go through a turning point, and this may explain some of the conflicting results obtained on the effects of chain branching.

Short sidechain branching reduces ΔE but LDPE with large sidechains has a higher value of ΔE than HDPE of the same $\overline{M}_w$. Branched polymers are more shear-thinning and more sensitive to shear degradation than linear ones of the same $\overline{M}_w$.

Equation (7.8) is not obeyed by branched polymers. For LDPE with long side branches

$$\log \eta_0 = A - B(\overline{M}_n)^{1/2} \qquad (7.9)$$

where A and B are constants.

Chain branching has a profound effect on the tensile viscosity of LDPE, as mentioned in Chapter 6. Only cis–1, 4–polyisoprene at low melt temperatures around 80°C shows similar tension stiffening behaviour. A model proposed suggests that, as tensile flows tend to uncoil molecules more than similar shear flows, it is in tensile flows that the sidechains play a more important role. It has been proposed before that the side branches act as 'hooks' and in this way increase the tensile viscosity at high tensile stresses.

Chain branching has an effect on the elastic properties of the melts. For the same $\overline{M}_w$ polymers with small side branches are less elastic so that swelling ratio and the elastic moduli are reduced and melt fracture occurs at higher shear rates.

Much work has been carried out on the effect of chain branching on rheological properties, but for a better understanding of the mechanisms it is imperative to obtain an estimate of the length of the sidechains. This is not easy. Much of the work has been carried out on polyethylene because of its three main forms: (1) LDPE (branched), (2) linear LDPE (small branches), and (3) HDPE (linear).

Crystallinity

At high hydrostatic pressures or at high shear rates when processing a melt close to its melting point, it is sometimes possible to induce crystallisation.

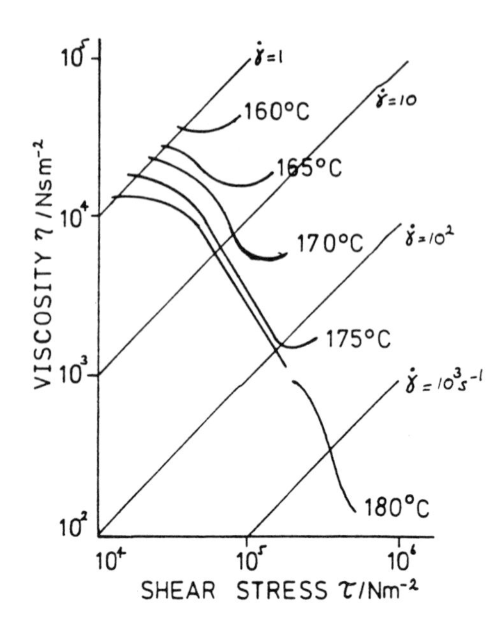

Fig. 7.11 The effect of shear-induced crystallisation in polypropylene.

The high hydrostatic pressure or the high shear rate cause the molecules to align and this is favourable for the onset of crystallisation.

The phenomenon shows itself as an upward sweep of the flow curve as shown in Fig. 7.11. Less crystalline materials are less susceptible to this effect and have far superior mould-filling capabilities because of their lower viscosities. Shear-induced crystallisation has been observed in polypropylene, cis–1, 4–polybutadiene and cis–1, 4–polyisoprene.

The degree of crystallinity can be partly controlled by the rate of cooling, as mentioned in Chapter 1. Crystallites readily form at low cooling rates, giving an opaque or translucent material, with increased mechanical strength and elastic moduli over the same material in amorphous form. Crystallinity, however, reduces toughness.

If the cooling rate is rapid, the melt will supercool, giving an amorphous material or one containing many small crystallites. The product will have improved clarity and impact resistance.

Often the cooling rate is uneven, resulting in a region of small crystallites surrounding material that has cooled more slowly, where there are large crystallites. This anisotropy will result in frozen-in stresses between the two regions, which may give rise to a warping of the product if it is raised to too high a temperature. Moreover, if crystallisation is incomplete after processing, it may take place gradually over a period of time while the product is in service. This will cause warping and shrinkage as the polymer chains align and reduce free volume. In both cases, incomplete crystallinity is undesirable and will result in a lower service temperature than the material is usually capable of. The choice of the optimum process temperatures to control crystallinity is, therefore, of great importance.

Frozen-in Orientation

When a polymer melt travels along a die or in the narrow channels (sprues, runners and gates) leading into an injection moulding cavity, the molecules tend to uncoil and align along the streamlines in the tensile and shear fields encountered. The degree of orientation will depend on both the molecular architecture and the process variables. When the melt cools, there may not be sufficient time for the polymer molecules to regain a random orientation before the material solidifies. The orientation is then said to be frozen in, and it gives rise to frozen-in stress and frozen-in strain. A certain amount of frozen-in stress is present in all products, and often the process variables are chosen to give a certain degree of orientation.

For given conditions of flow in a particular die, the degree of orientation will decrease with increasing $\overline{M}_w$ for a given polymer. This is because the longer molecules require more time or more stress to remove entanglements and to achieve the same degree of orientation as shorter ones. In polydisperse systems, the shorter molecules will be more aligned than the longer ones.

The amount of orientation that is frozen-in will depend on the initial amount and the rate of cooling. Clearly, the longer molecules take longer to orientate and to recover their random conformations. On rapid cooling, a more polydisperse sample will retain more orientation than a monodisperse one of the same $\overline{M}_w$.

Fig. 7.12 summarises the effects of frozen-in orientation on the mechanical properties of a product. In polyamides, orientation improves the mechanical strength at right angles to as well as along the direction of alignment.

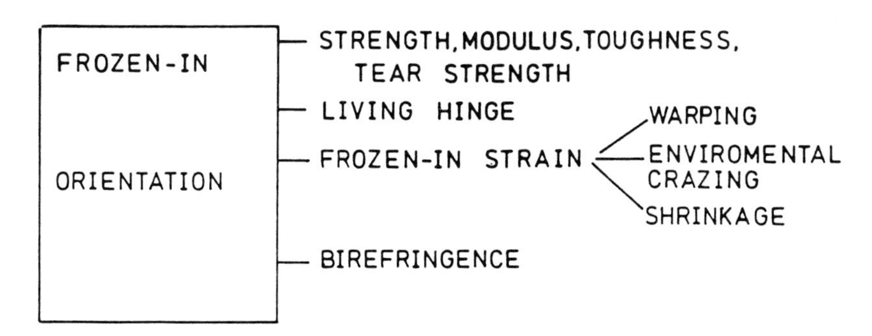

Fig. 7.12 The effects of frozen-in orientation on the mechanical properties of a product.

This is unusual, as although orientation promotes mechanical strength in the direction of alignment (called the machine direction), this is usually at the expense of mechanical strength in the two transvese directions. Orientation in films is desirable because the thickness direction is so small that low strength in that direction is unimportant, whereas the improved property in the machine direction is a boon. In the film blowing process biaxial orientation (that is orientation in two directions) is possible and this provides very strong films.

Uniaxial orientation reduces tear strength because it is so easy to cleave along the aligned molecules. A simple experiment can be carried out on thin-walled polystyrene drinking cups. These cups have considerable orientation up the side walls, and it is easy to tear down these walls. It is, however, very difficult to tear in the circumferential direction as the tear tends to propagate up or down the walls. In biaxially orientated films, the tear strength is improved over unorientated films, and packaging made from these films is often difficult to open.

Orientation in polyolefins has another unexpected beneficial property. A living hinge can be made from these materials. The hinge consists of a line of orientated molecules, which are very strong and resistant to cyclic fatigue. Such hinges are used on the covers of some pocket calculators and in some plastic lavatory seats and covers.

Very often frozen-in orientation causes more problems than it provides advantages. One of these problems is differential shrinkage, which may occur in service if the product is overheated, although still well below its upper service temperature. The greatest shrinkage occurs in the machine direction, when the polymer molecules become unaligned. This is accompanied by an increase in the transverse dimensions.

This phenomenon can also be demonstrated with thin-walled polystyrene drinking cups. If six of them are filled with boiling water for varying amounts of time, the one that receives the greatest heat will become the shortest and fattest. When heated in an oven at just over 100°C, a polystyrene cup can be reduced almost to a disc-like shape. Under these circumstances, all the molecules have relaxed and the orientation has disappeared. These materials, thus, seem to have a memory.

Orientation tends to promote environmental stress cracking or environmental crazing. The orientation causes frozen-in strain, which in combination with a mild solvent in the form of a liquid or a vapour causes surface crazing. This can lead to premature part failure. Polymers, like polystyrene, which are glass-like at room temperature with a low value of elongation to fracture, are very prone to environmental stress cracking.

The toughness of a product is adversely affected by orientation, because if the part receives a knock in a region where the frozen-in stress is high it is much more likely to break than a similar unstressed part.

In optical products, orientation gives rise to anisotropy which causes birefringence. This is totally unacceptable and optical lenses are carefully moulded to avoid frozen-in orientation. Optical lenses can be checked for birefringence by observing them between crossed polaroids.

In general, frozen-in orientation is desirable, and arises because the cooling cycle in the process is too short to permit the molecules to relax into random conformations. Injection moulding processes involve shear rates of around 10^5 – 10^6 s^{-1} in the gate regions, whereas in extrusion, shear rates of the order of 100 s^{-1} are encountered. Orientation is, therefore, much greater in the former process.

A designer tries to use orientation to advantage, as in the 'living hinge', and the complex inter-relationships between the material properties, process geometries and process variables make the optimisation of orientation very challenging.

REFERENCES

H.A. BARNES, J.F. HUTTON and K. WALTERS: *An Introduction to Rheology*, Elsevier Science Publishers, Amsterdam and New York, 1989.

J.A. BRYDSON: *Flow Properties of Polymer Melts*, Iliffe, 1980.

F.N. COGSWELL: *Polymer Melt Rheology*, George Godwin, 1981.

A.A. COLLYER: High Temperature Engineering Thermoplastics, *Progress in Rubber and Plast. Technol.*, **5**, (1), 1989, 36–87.

J.P. CRITCHLEY, G.J. KNIGHT and W.W. WRIGHT: *Heat-resistant Polymers*, Plenum Press, 1983.

R.W. DYSON: *Speciality Polymers*, Blackie, 1987.

J.W.S. HEARLE: *Polymers and their Properties*, Vol 1, Fundamentals of Structure and Mechanics, Elles Harwood, 1982.

S.L. ROSEN: Fundamental Principles of Polymeric Materials, Wiley, New York, 1982.

R.B. SEYMOUR and C.E. CARRAHER: *Structure Property Relationships In Polymers*, Plenum Press, 1984.

8 *Extrusion and Extruder-based Processes*

INTRODUCTION

A plastic material is seldom chosen for a particular application because its mechanical properties are ideal for that usage. The over-riding consideration is whether or not that material can be processed into the required article easily and economically. Consideration of the process is, therefore, an important part of the engineering design.

In this chapter, the main extrusion processing routes are briefly described, with a discussion on how the process variables influence the quality and output rate of the end product. For a full appreciation of this inter-relationship between process variables and rate of production of good product, the rheological behaviour of the melt must be understood.

The main processes by which plastic articles are made are by extrusion and by injection moulding. In the extrusion process, long, profiled articles such as pipes, sheet, film, covered wire or guttering are made by passing a melt through a shaping die. This is a continuous process. In injection moulding, the melt is forced into a mould, as in pressure die casting, and a large number of identical articles can be made at a rapid production rate. This is a repetitive process.

In both of these processes the rheological parameters of the melt have a profound influence on the success or failure of the product. Both the viscosity and the recoverable elastic strain can be controlled by the choice of operating variable (temperature and rate of processing). The melt shear viscosity is most important in determining the pressure required to drive the melt through the processing equipment, and the choice of operating temperature depends very much on the desired shear viscosity. The melt index indicates the ease of processing of the melt and some materials such as PTFE cannot be processed by either of the above processes.

The feedstock for both of these processes can be in one of the following forms, which are listed here in decreasing order of processability:

1) near spheres of 27 mm^3 volume (easiest);
2) lace-cut granules from dissecting extrudates;
3) cubes of side 3 mm;
4) irregular flakes; and
5) fine powder (most difficult).

In both processes, the machinery turns the feedstock into a homogeneous melt and performs the following functions:

a) pumping,
b) melting,
c) mixing, and
d) pressure (to drive the melt into the mould or die and to aid mixing).

In this respect the machinery used in extrusion and in injection moulding performs similar functions when judged on the basis of the chemical engineering unit operations, as shown in Fig. 8.1. The screw and barrel used in the two processes, though slightly different, act in the same way. Thereafter the two processes differ widely.

EXTRUSION	CHEMICAL ENGINEERING UNIT OPERATIONS	INJECTION MOULDING
Profiled Archimedian Screw and Barrel	Melting, Mixing and Homogenisation, Melt Transport	Reciprocating Screw and Barrel
Die	Primary Shaping	Mould
Often needed	Secondary Shaping	Not needed
Air or Cooling Bath	Shape Stabilisation	Mould
Cutting or Reeling	Finishing Operations	Not needed

Fig. 8.1 Chemical Engineering unit operation approach to polymer processes.

All the processes will be considered in the following way, where possible:

1) Aims of the process;
2) Advantages and disadvantages;
3) Basic principles involved;

4) Description of the equipment and suitable polymers for processing (intrinsic or design variables);
5) Description of the process variables and their effect on output and quality;
6) Precautions;
7) Diagnosing faults; and
8) Everyday applications.

This chapter cannot deal in detail with all the processes and further information can be obtained from the ICI Technical Literature.

EXTRUSION

The function of extrusion is to produce continuous products of good quality and dimensional uniformity at economic output rates from thermoplastic granules (nylon, polyacetal, polypropylene, polycarbonate, polyethersulphone, poly– ether– ether ketone, etc.), powders (PVC and acrylic) chips or melts (unusual).

The advantages of this process are:

a) Die costs are low in relation to output;
b) Long lengths of uniform cross-section are produced.

The disadvantages are:

a) The cost of extruders and auxiliary equipment is high;
b) Finishing or post-fabricating assembly operations are often required.

The process consists of the following stages:

1) the extruder;
2) the die;
3) the forming stage — involves the stabilisation of the melt, or some secondary shaping stage;
4) the post-forming or handling stage — the product is hauled off, collected and finished (trimming, cutting, reeling);
5) secondary processing — described later.

The Extruder Barrel and Screw

The simplest and most common way of achieving extrusion is to use a single, profiled Archimedian screw in a heated barrel, as shown in Fig. 8.2.

The barrel provides one of the surfaces for shear, and the heating of the three zones. It is actually made in one piece and coated with a hardened lining of nitrided steel, x-alloy or a similar material. The screw has a similar coating. The barrel must be able to withstand moderately high pressures and

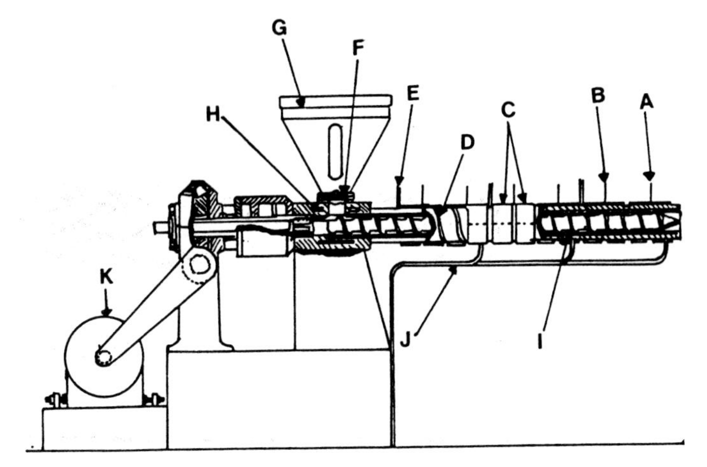

Fig. 8.2 Extruder barrel and screw. *A* deep-set thermocouples, *B* shallow-set thermocouples, *C* heater bands, *D* coiled tube, *E* water inlet for barrel cooling, *F* feed throat, *G* feed hopper, *H* water jacket for feed section, *I* barrel liner, *J* water outlet for barrel cooling *K* variable-speed motor.

provide temperatures of up 400°C. The size of the machine is classified in terms of the barrel internal diameter, e.g. 60 mm, 90 mm, etc.

Most extruder barrels have three heating and water or oil cooling regions, giving three independent temperature zones. The heating is provided by heater bands on the barrel or by an induction heater, either of which can be controlled by surface thermocouples. These give better temperature control than deep-set devices.

The heating is more necessary at the beginning of a production run, because, as the run continues, the friction heating caused by the movement of the granules and melt through the barrel is often so great that water or oil cooling is required to keep the zone temperature down to the selected level, until a thermal equilibrium is eventually reached.

Fig. 8.3 shows a typical general purpose profiled screw. The three regions (feed, compression and melt, or metering, region) are clearly illustrated. The feedstock is conveyed from the hopper into the feed zone by the rotation of the screw. The channel depth and screw width, as defined in Fig. 8.3, are kept constant so that no compression of the feedstock occurs. The design of this part of the screw is important so that neither too much or too little material is fed through to the melt zone at the far end of the extruder.

Melting occurs in the feed zone as follows. A thin film of molten material forms at the wall of the barrel, while the granules at the screw surface remain intact. As the screw rotates, it removes the film from the barrel wall, and this molten plastic flows down the front face of the screw flight. On reaching the surface of the screw it moves outwards again. This sets up a rotary motion

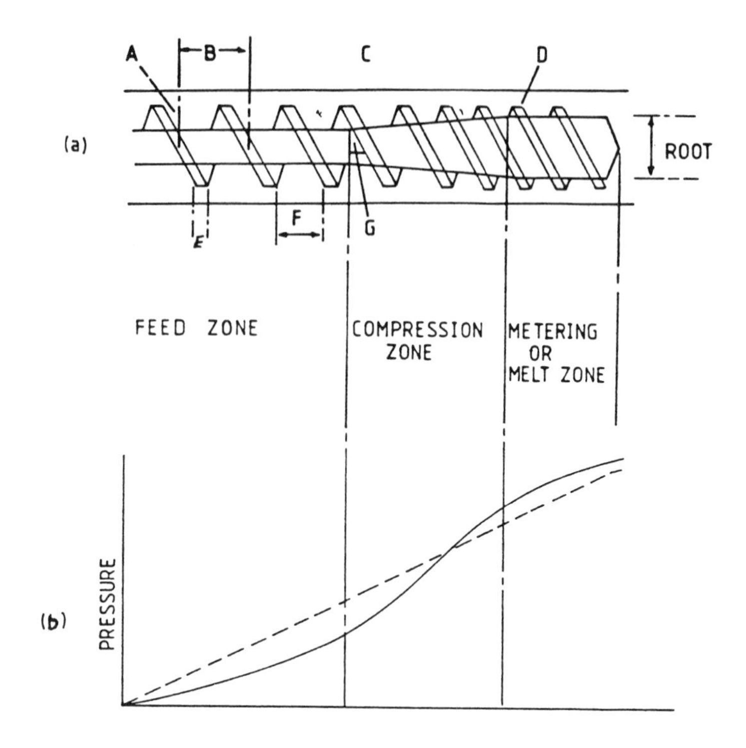

Fig. 8.3 a) Extruder barrel and screw showing *A* channel, *B* pitch, *C* barrel, *D* flight, *E* land, *F* lead and *G* helix angle.
b) Variation of pressure along the screw.

ahead of the leading edge of the screw flight. The granules are also subject to the motion and gradually all of them melt before entering the compression zone. This melting is achieved by heat from the barrel wall and by the viscous heat derived from the motion of the polymer.

In the compression zone the screw depth gradually decreases. This compression squeezes any trapped air back into the feed zone, and the decreasing screw depth improves the heat transfer from the barrel. In this region, the pressure of the melt increases as shown in Fig. 8.3 and the length of this zone depends, as discussed below, on the particular polymer being processed. At the end of this region the compression stops. The compression ratio of the screw is the ratio of the volume between flights in the feed zone to the volume between flights in the metering zone. Compression ratios of the order of 3:1 to 4:1 are usual.

The channel depth in the metering zone is kept constant because the melts are almost incompressible. In this zone the melt is homogenised and this is assisted by the secondary circulatory flow induced in the melt by the screw's

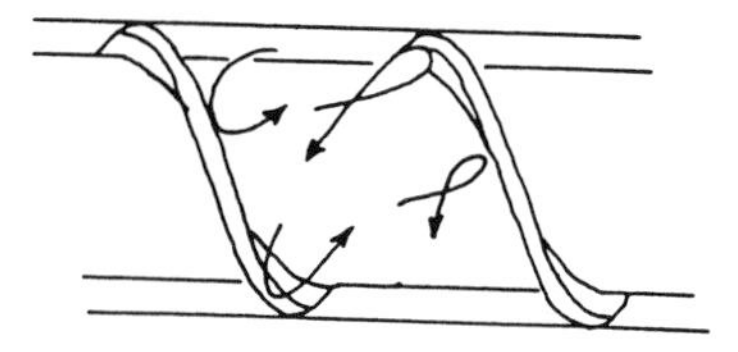

Fig. 8.4 Secondary circulatory motion induced by the rotation of the screw, which aids mixing.

rotation (Fig. 8.4). This transverse motion gives a circulatory flow that provides a good degree of mixing.

In the general purpose screw, the three regions are almost the same length, with the ratio of screw length L, to diameter D in the range 15 to 20, although more modern machines have L/D ratios of between 20–36, which are more suitable for extruding high temperature melts.

A special screw is used for nylons to accommodate their well-defined melting points (Fig. 8.5a). The compression zone may be only one-third to one-half of a screw flight in length, with a compression ratio of between 3.5 and 4.0 and an L/D ratio in the range 15 to 20. Clearly, using a general purpose screw for these materials would lead to a loss of efficiency.

For thermally sensitive materials, such as polyvinylchloride, a reduction in viscous heating (and therefore in the danger of localised overheating) is achieved by using a screw with a channel depth that decreases gradually all along the length of the screw (Fig. 8.5). This gives rise to a gentle compression and does not cause too much shear at an early stage.

In practice, it is neither possible nor economic to stock a different screw for every material, and the general purpose screw is often used. This results in the output rate being less than if a specially designed screw were used.

The last important intrinsic variable of the screw is the helix angle as defined in Fig. 8.3. As the helix angle increases the output increases, but the angle is never large — between 10 and 30°, with an optimum value of 17.6°, which is the most versatile.

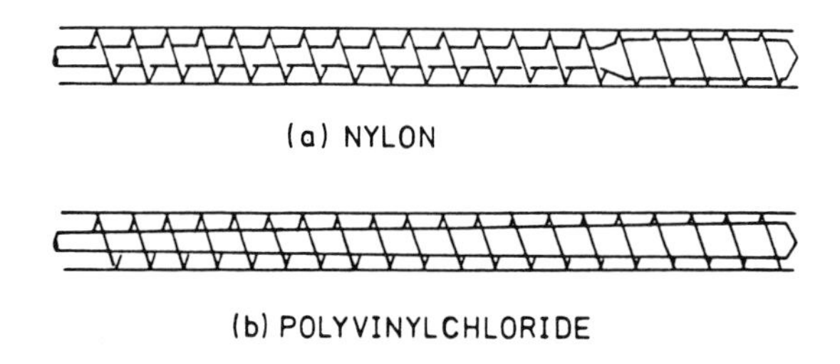

Fig. 8.5 Specialised screws for extruding nylon and polyvinylchloride.

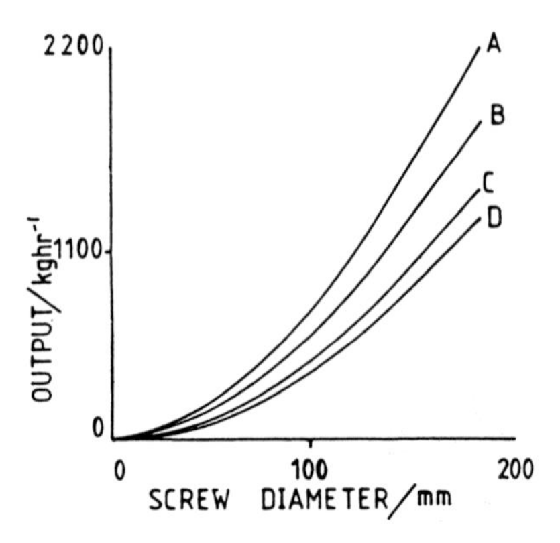

Fig. 8.6 Variation of output rate with barrel diameter for *A* plasticised PVC, *B* polystyrene, *C* low-density polyethylene and *D* polypropylene.

The output rate for a particular extruder barrel and screw depends on:

1) barrel diameter,
2) helix angle,
3) rotational velocity of the screw, and
4) the size of the die orifice

with the first variable having the greatest influence and the last the least. Fig. 8.6 shows the variation of output rate as a function of barrel diameter for

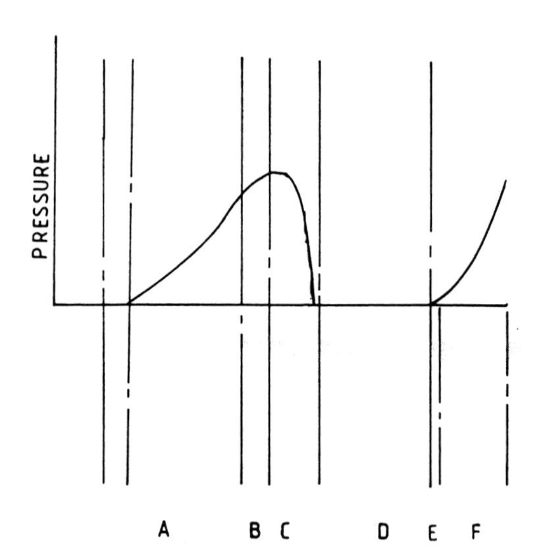

Fig. 8.7 Variation of pressure along the length of a vented screw. *A)* feed zone, *B)* first compression zone, *C)* first metering zone, *D)* decompression zone, *E)* second compression zone *F)* second metering zone.

several polymer melts. Plasticised polyvinylchloride gives the highest output rate, and high density polyethylene and polypropylene the lowest. Manufacturers usually quote values for low density polyethylene.

Some modern extruders are equipped with decompression zones and need special screws. These machines are particularly useful when extruding polycarbonate, polyethersulphone and other materials that absorb water, because they make pre-drying unnecessary. Manufacturers of these polymers give details of drying times and temperatures.

In these vented extruders, the materials are melted, compressed and mixed as before, but then the melt pressure is reduced, as shown in Fig. 8.7. The pressure in this decompression zone falls to atmospheric, and volatile substances, including water vapour, escape through a special port in the barrel. The melt is then compressed again in a second compression zone, the function of which is to discourage the formation of air pockets.

At extrusion temperatures of around 150°C, the water vapour in the polymer melt can develop a vapour pressure of about 4 MN m^{-2} (approx. 40 atmospheres). This is easily sufficient for the vapour to escape through the vent.

The Crosshead and Die

Before the melt enters the die, it passes through a filter pack and breaker plate, as shown in Fig. 8.8. The filter or screen pack is placed in front of the breaker plate or between two breaker plates, and its function is to filter out unmelted material and/or dirt. This is achieved by using screen meshes of between 40-200 mesh/linear inch.

The breaker plate consists of a number of holes through which the melt passes (Fig. 8.8). The effect of the holes is to turn the helical flow caused by the screw into a streamlined flow. In addition, the screen pack and breaker

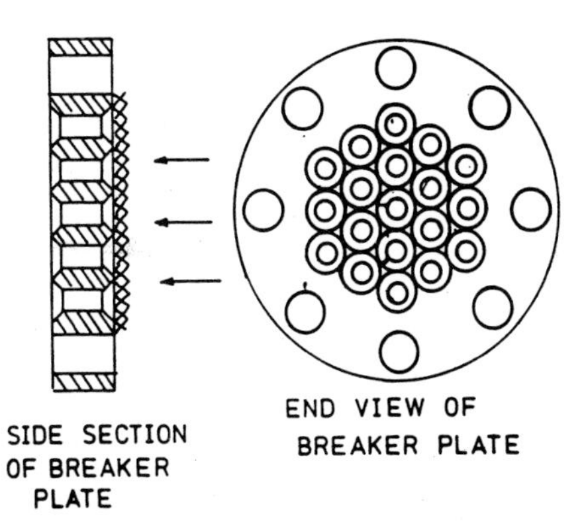

Fig. 8.8 Filter pack and breaker plate.

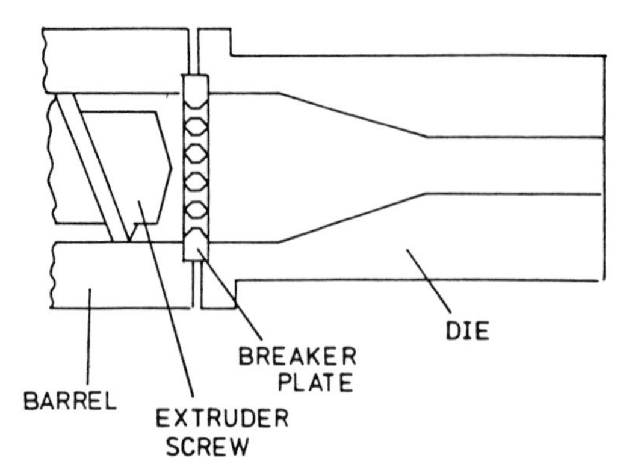

Fig. 8.9 Crosshead and die assembly.

plate present an obstruction to the melt flow and thus provide an additional means of controlling the back pressure in the barrel.

The barrel and screw and breaker plate assembly complete the four main functions, pumping, melting, mixing and developing pressure, such that the optimum melt conditions are provided in the die.

Fig. 8.9 shows a crosshead and die assembly, the purpose of which is to convey the melt to the tip of the die at the maximum rate with the minimum of overheating. These two regions have independent heater bands attached to them, giving five heating zones in all.

The length of the parallel parts of the die and crosshead and the degrees of taper will have a bearing on the onset of extrudate distortion and on the swelling ratio. Good design here enables a high output rate to be achieved; both melt fracture and die swell are reduced by small angles of taper and by ensuring that there are no dead spots. Not only do dead spots encourage extrudate distortion but they promote thermal degradation of melt trapped in them. The degradation can show itself as chain scission, unzipper reactions causing depolymerisation or crosslinking. All of these will give rise to inhomogeneity.

The reduction of the diameter of the die passages is necessary not only to give correct product dimensions, but also to develop sufficient pressure in the melt to recombine the numerous streams from the holes in the breaker plate into a homogeneous, streamlined flow.

As the polymer molecules move along the die, they will align along the streamlines as described in Chapter 6. It is this orientation and subsequent unrestricted emergence from the die that gives rise to die swell. This can be reduced by increasing the length of the die parallel. Other ways of reducing die swell are discussed later.

As a result of die swell, the design of dies is a very skilled task. Not only is the swelling ratio dependent on the process variables but also on the overall

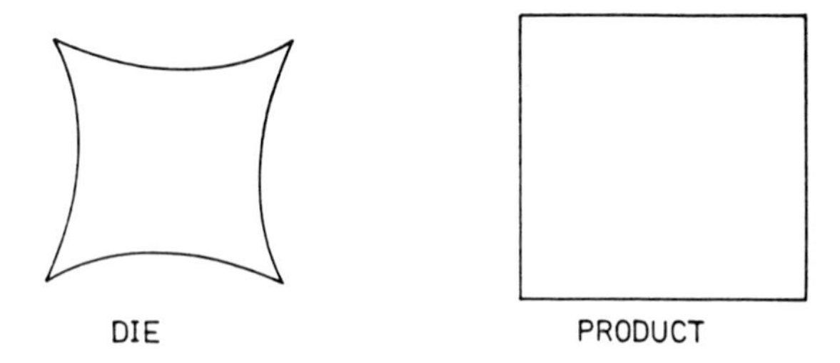

Fig. 8.10 Difference between the shape of the die and the final product formed by it.

size of the die. The die is, therefore, both smaller than and of a different shape from the final product (Fig. 8.10). The square shape is produced from a star-shaped die. Complex shapes such as those needed for gaskets, for refrigerators and windscreens need considerable thought. If these dies are used with different materials, the products may have different sizes and shapes. The only time that die swell is useful is in blow moulding.

In the extrusion line, however, a sizing die may be used. The extrudate is pulled from the die by the haul-off equipment and before the extrudate is cooled it is often pulled through a die that gives it its final shape and size. This gives the whole process greater versatility, and enables small batch to batch variation in feedstock to be compensated for more easily.

The Haul Off

The haul-off equipment is, as the name suggests, used for pulling the extrudate away from the die. It may simply grip the extrudate and pull it through air or a water bath, or additionally through a shaping die, or the equipment itself may give additional shaping — in which case the rollers are heated. To summarize, the main functions of the haul-off are:

a) to provide a means of removing the extrudate;
b) to modify die swell and extrudate distortion; and
c) to provide secondary shaping if necessary.

The kinds of equipment available are:

a) continuous caterpillar belt, with or without rubber foam blocks;
b) continuous caterpillar belt, either with or without pressure rollers;
c) capstans; and
d) nip rollers.

The type of haul-off used depends on the nature of the extrusion. The conditions that must be satisfied are:

a) an easily variable speed for the desired output rate;
b) a haul-off that is constant and free from pulsation; and
c) equipment that will not permit a tendency to slip back.

After the haul-off, a flexible extrudate may be wound on to a drum; otherwise it has to be cut to length.

The process variable associated with the haul-off is the draw ratio, which is the ratio of the haul-off velocity to the extrudate velocity at the die exit. The draw ratio will affect:

a) the die swell,
b) the extrudate distortion,
c) the draw resonance, and
d) the orientation in the product.

Materials for Extrusion

Most plastics materials can be extruded unless their shear viscosities are outside the range $10–10^4$ N sm^{-2} over the shear rate range $10–10^3$ s^{-1}, which is typically the range encountered in extrusion dies. If the shear viscosity is too low, the melt will migrate backwards between the screw flights and the wall of the barrel. A higher average molecular weight grade of the same polymer may have to be used instead. If the shear viscosity is too high, as is the case with polytetrafluoroethylene (PTFE) and polyimide, the pressure in the extruder will be too high or the material may decompose rather than melt, and would, therefore, be totally unsuitable for extrusion. Special extruders were needed to process the high-temperature softening polyethersulphone (ASTREL). This material is no longer manufactured.

Problems are also encountered with highly elastic melts, which give rise to extrudate distortions at low shear rates, and to low production rates. Raising the melt temperature will alleviate this at the risk of causing thermal degradation.

Some melts, such as those of polyvinylchloride, cause difficulties that are alleviated by the use of lubricants. In the case of the intractable polyphenylene oxide, processability was achieved by a graft copolymerisation with polystyrene. This blend is sold under the trade number 'Noryl'.

Quite apart from the consideration of polymer type, the ease of extrusion depends on the nature of the feedstock. In each case guidance should be sought from the manufacturers' literature, which provides a wealth of information.

PROCESS OR OPERATING VARIABLES

The discussions so far have been concerned with the design of the extruder system. This design is used in the belief that, or it is known that, it is the best design for the operation of the system. It is the design that the operator is left with, and as such, variables such as hopper design, screw design (pitch, screw depth, L/D and helix angle), die design, haul-off design and the shape and physical state of the polymer may be beyond the operator's control.

These variables are called intrinsic or design variables (how the machinery is made), and in this section are described the process or operating variables (how the machine is used). Unlike the intrinsic variables, the operating variables can be changed during a production run.

There is a large number of combinations of values of the operating variables. This often results in many extrusion lines being run well below their full capacity in order to avoid problems. In this respect, the choice of optimum design becomes of less importance.

In an extruder there are usually three independent process variables:

a) the temperature in the heating zones;
b) the screw rotation speed; and
c) the haul-off speed.

Choice of these variables affects the dependent process variables:

a) shear viscosity;
b) the recoverable elastic strain;
c) the melt pressure;
d) the melt zone position;
e) homogeneity;
f) output rate; and
g) orientation in the product.

The temperatures of each zone are chosen such that each part of the screw achieves its main objectives efficiently. The processing temperatures vary for

Table 8.1 Typical Processing Temperatures in °C for Plastics Extrusion.

Polymer	Feed zone	Compression zone	Melt zone	Cross-head	Die	Application
Nylon 11	170	190	210	205	185	Wire covering
Nylon 11	220	235	250	235	215–220	Blow moulding
HDPE	140–150	150–180	160–200	170–210	170–210	Blow moulding
PET	280–290	270–280	265–275	260–270	260–270	Blow moulding
PBT	260–270	255–265	250–260	245–255	245–255	Blow moulding

different grades of the same polymer and for different polymers, as shown in Table 8.1. Manufacturers and suppliers give details of these values, and operators gain experience of where in the ranges to set their process temperatures for their particular machines.

If low temperature values are chosen for the five zones, there will be a saving of power for heating and a reduction in the cooling necessary for the extrudate to harden; but the melt position may not be in the optimum place along the screw and more power will be necessary to drive the screw at the same rate. This may overtax the motor. Table 8.2 summarises the likely effects when the screw speed and haul-off speed are kept constant while the melt temperatures is lowered. In practice, the output rate may have to be reduced to eliminate the elastic effects of the melt and to reduce the power required to drive the screw.

Table 8.2 The likely effects on reducing melt temperature and maintaining the other process variables constant during extrusion.

Viscosity	Pressure	Power to the heaters	Power to the screw	Viscous heating	Mixing	Melt position	Recover- able strain	Die swell	Extru- date distortion
↑	↑	↓	↑	↑	↑	Nearer the die	↑	↑	↑

If high temperatures in the range are chosen, the arrows in Table 8.2 are reversed and the melt may suffer thermal degradation, involving a loss in molecular weight average or crosslinking. The former may result in a poor surface finish and a loss in mechanical properties, while the latter may improve the mechanical properties if the overheating is only modest.

Table 8.3 The likely effect of increasing the screw speed and haul-off speed to maintain a constant draw ratio, with temperature constant.

Viscosity	Pressure	Power to the heaters	Power to the screw	Viscous heating	Mixing	Melt position	Recover- able strain	Die swell	Extru- date distortion
↓	↑	↓	↑	↑	↓	Nearer the die	↑	↑	↑

As a further illustration, Table 8.3 shows the likely effects of increasing the screw speed and haul-off speed (to maintain a constant draw ratio) while keeping the melt temperature constant. In practice, this is what may be tried to work out the maximum output to achieve an acceptable product.

The last process variable that can be changed is the haul-off speed. If this is increased, the die swell and extrudate distortion are reduced and the orientation is increased.

Twin-Screw Extruders

Single screw extruders are not positive displacement pumps because their pumping action depends on the drag flow of the material between the rotating screw and the wall of the barrel. There is mixing due to the circulatory flow, but it is less effective mixing than in a twin-screw extruder with contra-rotating screws. These machines are truly positive displacement pumps and as such can more effectively process powders and can generate the higher pressures needed in some profile extrusions. Some twin-screw machines have co-rotating screws, and, although they are not positive displacement pumps, they still provide better mixing than in a single-screw machine.

Table 8.4 Advantages and disadvantages of twin-screw extruders.

Advantages	*Disadvantages*
Can process powders	More complex and expensive
Generate large pressures	Difficulty in heating melts
Better mixing	

The relative advantages and disadvantages of twin-screw over single-screw extruders are shown in Tale 8.4.

Precautions

Whenever materials are used for the first time, it is imperative to read all the manufacturers' literature to see if any special handling of material is necessary, and to see how best to purge the machine after use.

One special precaution that comes to mind involves the extrusion of polyacetal or acetal copolymer. This material should never be allowed to come into contact with polyvinylchloride at elevated temperatures because the mixture is explosive. Another point to note is that polyethersulphone left in a machine has an adverse effect on the nitride layer on screws and barrels. The authors discovered this to their own cost. There are many little instances like this which can be avoided by reading the brochures on each material used.

Variation in Feedstock

Sometimes there are variations in batch to batch properties of the same grade of a particular material. The kind of variation that might occur is a difference in $\bar{M}_w$ and in polydispersity. In this section, it is hoped that by examining this sort of problem it will be possible to illustrate the inter-relationships between molecular architecture, process variables, the output rate and the quality of the product.

Consider the following case. It is intended to use a higher grade of a certain polymer in the hope of producing better mechanical properties in the product. It is hoped to keep the same production rate. What are the likely problems that will be encountered?

The higher a grade of polymer means that it has a higher $\overline{M}_w$. This will affect the rheological parameters in the following way:

a) by increasing the shear and tensile viscosities, and
b) by increasing the recoverable elastic strain.

The effect of (a) is to increase the melt pressure and hence the power necessary to drive the screw.

Will this be possible?

The effect of (b) is to increase the die swell and cause extrudate distortion and draw resonance at lower output rates.

Clearly, unless the extruder is being used at its maximum rating, the increased viscosity can be accommodated. The increase in the recoverable elastic strain, however, could cause some problems, which can be alleviated by actions that will reduce it to its previous level. One method is to increase the melt temperature, and the relative merits of this are summarised in Table 8.5.

Table 8.5 The relative merits of increasing the melt temperature.

Advantages	*Disadvantages*
Reduces as well	More extrudate cooling needed
Reduces power to the screw	Thermal degradation
Reduces die swell, melt fracture and draw resonance	

This procedure may work provided that the melt is not already being processed at the maximum recommended temperature.

Another solution is to increase the haul-off rate or to decrease the screw speed; either way the draw ratio is increased. The effect of this is summarised in Tabe 8.6.

Table 8.6 Relative merits of increasing the draw ratio by increasing the haul-off rate or by decreasing the screw speed.

Advantages	*Disadvantages*
Reduces die swell	More likely to cause draw resonance
Reduces extrudate distortion	

This method may work provided that the melt is not already being drawn at its maximum rate. This seems a better method of attack and will produce a higher degree of orientation in the product in the machine direction. This will give a greater tensile strength (*TS*) in that direction, with a reduction of tensile strength (*TS*) in the two perpendicular directions. The product will be more likely to distort if overheated in service.

Redesigning the die to accommodate the increased die swell and possible earlier extrudate distortion is a very expensive solution. This illustrates how much more practical it is to manipulate the process rather than the design variables to solve problems.

Extrusion Faults

Many manufacturers and suppliers of polymers produce lists of typical extrusion faults and their probable causes. The faults covered in Table 8.7 are only intended as an introduction and fuller details are given in the manufacturers' brochures.

Table 8.7 Some common extrusion faults and their probable causes.

Fault	Probable Cause
Unfused particles	Fusion temperature too low, insufficient screening
Roughness	Insufficient screening, temperature too low, incomplete purging between runs
Uneven wall thickness	Faulty centering of the die
Bubbles	Feed zone temperature too high, moisture content too high, excessive movement between screen and barrel, compression ratio too low, screw speed too high
Tube strength low, weld lines	Material too cold in the head, melt passing through the head too quickly, head and die too short

PRODUCT APPLICATIONS

Suitable melts for extrusion must have sufficient melt strength to form a stable shape on leaving the die and prior to cooling to the finished product. If a given grade of a polymer is too runny, a higher grade will be needed. If the shear viscosity is too high, a lower grade of the material may suffice.

Polyvinylchloride is a polymer often processed by extrusion, making profiles such as rainwater gutters and pipes and uPVC sheet and curtain rails. In the unplasticised form (uPVC), it is difficult to extrude without the use of lubricants, and both forms being thermally sensitive have thermal stabilizers incorporated into them.

Nylon 66 is used in petrol pipes and hydraulic pipes in motor vehicles. Beer pipes are made from low density polyethylene and from ethylene–vinyl acetate (EVA), whose flexibility can be controlled by the vinyl acetate content of the copolymer. It is harmless in contact with foodstuffs and does not taint the beer. Other pipe materials are ABS and cellulose acetate. Pipe of outer and inner diameters 10 and 7.5 mm is usually drawn down to 8.5 and 6.2 mm. This gives a draw ratio of about 3:1. Tubes and pipes have draw ratios in the range 2 to 6 and 10 to 20% respectvely, where

$$\text{Draw ratio} = \frac{\text{Cross-sectional area of die}}{\text{Cross-sectional area of product}} \tag{8.1}$$

In the extrusion of pipes and tubes in which a close tolerance is needed on the inside and outside diameters, a vacuum sizing or calibration die is used. Use of this die precedes a cooling bath (often water) which is surrounded by a partial vacuum. When the melt enters the cool die, the air inside the tube, which is at atmospheric pressure, pushes the melt against the cold walls of the vacuum die, because of the reduced pressure outside the tube. This fixes the outside diameter of the pipe or tube. An adjustment of the screw speed alters the amount of material in the tube and so the wall thickness can be set.

A new invention made possible by plastic materials is flexible circuit boards. Polyethersulphone has a high enough T_g that it will not distort in the presence of liquid solder. This polymer is extruded into film and cut to the required size for the circuit board.

Extruder Based Processes

Extrusion usually leads to secondary processes as shown in Fig. 8.11. In the profile extrusion already described, the pipes, tubes, channel or sheets are cut to size. Those profiles that must be precise in dimensions, such as curtain track, use a water-cooled sizing die. This is generally about 2% larger than the final product to allow for the contraction of the melt, and possibly more for crystalline materials. Longer cooling bath times are needed for these as the crystallisation process is exothermic. Elastic effects limit the possible tolerances, and it is for this reason that polyethylene is not very suitable where great accuracy is required; nylon, however, is fairly inelastic.

A similar set-up is used in flat film and sheet extrusion. A film is regarded as less than 0.25 mm in thickness and sheet is greater than 1 mm. The middle range is termed foil. The film leaves a slit die and cools in a water bath or in a

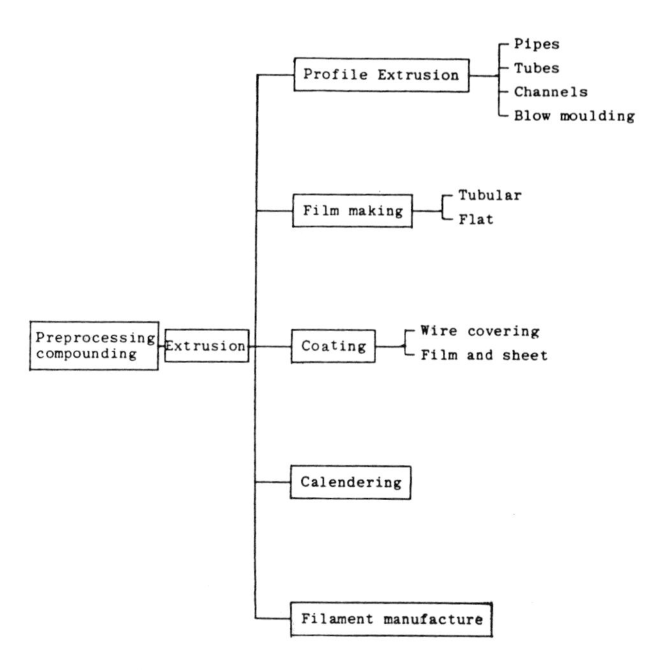

Fig. 8.11 Some extruder-based processes.

chill roll assembly. These systems produce uniaxial orientation in the machine direction. This gives strength in that direction but allows cleavage along the aligned molecules. Such films used as wrappings are easy to remove.

The remaining extruder-based processes involve more complexity and are considered individually, but not as in such detail as the primary process.

EXTRUSION BLOW MOULDING

The aim in this process is to use a pipe-shaped extrudate or parison and to inflate it with compressed air until it conforms to the shape of a mould that is clamped around it. This process makes hollow articles and is very similar to that used for making glass bottles. The containers made can vary in size from 2 ml phials to 9000 litre tanks.

The advantages of the process are:

a) The lower pressures in the moulds than those in the injection moulding process mean lower mould costs.
b) Lower machinery costs.
c) External threads can be moulded in.
d) Large open-ended parts can be made by splitting a cold moulding.

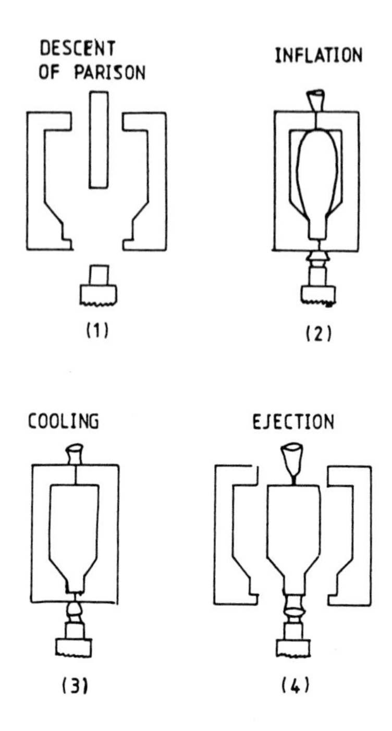

Fig. 8.12 Extrusion blow moulding process.

The disadvantages are:

a) Close tolerances are possible only on outer walls. The wall thickness varies, with thinning at the corners and at the largest dimensions of the moulding.
b) Holes can only be put in as a post-moulding operation.
c) Cooling times are longer than in injection moulding because the minimum wall thickness must be greater than the design wall thickness to accommodate the localised thinning. Therefore, the same product made by the injection moulding route would use less material.

The extrusion blow moulding process is shown in Fig. 8.12. A vertical parison produced from an annular die hangs freely until a mould is clamped around it. A hot knife separates the parison from the extruder, then the mould is moved away for inflation of the parison and ejection. Afterwards excess material is trimmed off.

Two different continuous extrusion processes are used to speed up the production rate, which is heavily dependent on the cooling time. In the first, shown in Fig. 8.13, multiple parisons are extruded, and moulds held in a manifold are clamped around the parisons. The moulds are removed for inflation, cooling and ejection, while another manifold brings empty moulds

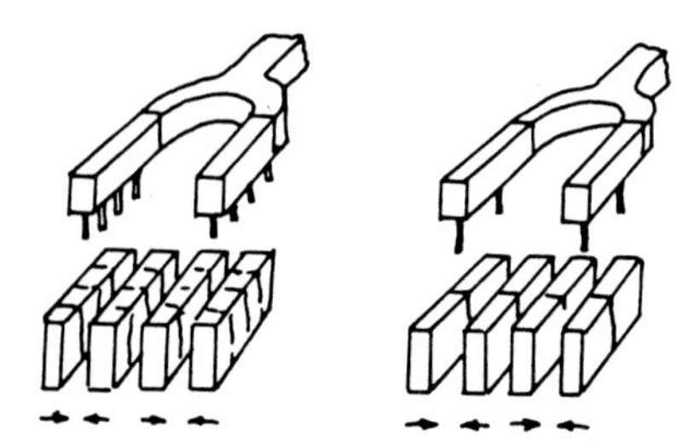

Fig. 8.13 Multiple parison rising and falling table machine.

to the new parisons that are being formed. These machines are called rising and falling table machines.

The main problem with this system is getting parisons of uniform length, which invariably means that more scrap is produced than with a single parison and mould. However, this method gives ample cooling time at an economic production rate.

In the second method, a rotating table of moulds is used, as shown in Fig. 8.14. A single parison is extruded and a mould is clamped around it. The table rotates and moves the mould to another position for blowing, cooling and eventually ejection. While this is going on, other moulds move in sequence to the injection position. Bottle ejection takes place at one of the ports. In Fig. 8.14, the blowing and cooling cycle lasts 75% of the process time because injection occurs at port 8 and ejection occurs at port 6, three-quarters of the way round the circle. Production rates of 6000 bottle/hour are achievable on these continuously rotating table machines, when a large number of moulds are used.

It is instructive to analyse the process more fully to understand the compromises necessarily imposed on the rheological properties. There are two main stages in the process, the hanging stage and the inflation stage.

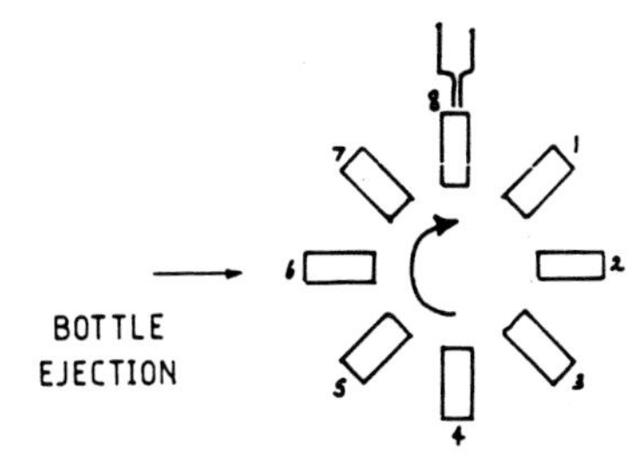

Fig. 8.14 Single parison rotary table machine.

During the hanging stage, there are three important influences:

1) The die swell that occurs causes the parison wall thickness to be larger than the die gap. The swelling ratio is, therefore, of great importance.
2) Extrudate distortions may give a curtaining effect in the parison, which may affect the quality of the product.
3) The wall thickness of the parison will decrease progressively from the upper end as the parison sags under its own weight. This sagging will affect the size of parison that can be successfully blow moulded.

The effect of this sagging on the wall thickness of the parison can be estimated. Suppose the wall thickness at the free end is h_0, and that at the suspended end after a time t is h, then the maximum Hencky strain ε is given by

$$\varepsilon = \ln h_0/h$$

This Hencky strain results from an elastic and a viscous component.

Therefore $$\varepsilon = \ln h_0/h = \frac{\sigma}{E} + \frac{\sigma t}{\eta_E} \qquad (8.2)$$

where σ is the maximum stress at the suspended end caused by the weight of the parison. If the density of the material of the parison is ρ and L is the length of the parison,

$$\sigma = L \rho g$$

Equation 8.2 can be used to estimate the degree of thinning as a function of time.

In practice, the combination of the die swell and sagging makes the prediction of the wall thickness in the final product very difficult, and this is compounded when articles of complex shape are to be blown. In these cases, the moulder programmes the output rate or the die gap to produce an uneven thickness in the parison such that the final article will have the correct distribution of material in it. This requires great experiences.

The blowing rate I of air during the inflation stage is generally rapid to give a stable inflation, during which the melt deforms elastically rather than in a viscous manner.

If the radius of the final product is R, the tensile strain rate $\dot{\varepsilon}$ is given by

$$\dot{\varepsilon} = \frac{I}{2\pi R^2 L} \qquad (8.3)$$

From the graphical data for the melt, the values of E and η_E, the tensile modulus and tensile viscosity, can be found for the tensile strain rate E. The natural time, λ, of the melt can be found ($= \eta_E/E$), and then the process time, t, is given by

$$t = \frac{\text{volume of bottle}}{I}$$

$$= \frac{\pi R^2 L}{I} \qquad (8.4)$$

Once the natural time of the melt and the process time are known, the Deborah number can be calculated

$$N_D = \lambda/t$$

It has been shown that if $N_D \geq 3/4$, the inflation will be elastic and should be stable. However, if the hoop stress developed during inflation is greater than the *TS* of the melt, the bottle will burst. A melt of higher tensile viscosity may be necessary if $N_D < 3/4$ and $\sigma > TS$.

If $N_D < 3/4$ the inflation will be controlled by the viscous component of the melt and will not be stable.

The combination of rheological parameters required for the ideal blow moulding melt is impossible to achieve because of the entirely different constraints necessary during drawdown (hanging) and inflation.

1) A high tensile modulus and a high tensile viscosity (giving a high shear viscosity) will give minimal drawdown during the hanging phase. These melts will have a high $\overline{M}_w$, which is likely to cause extrudate distortion.
2) A low value of tensile modulus and a high tensile viscosity are needed for a stable inflation. This gives conflict with (1).
3) A low shear modulus and a low shear viscosity will minimise sharkskin, which can give an unsightly appearance on the inside of clear bottles. It is not noticeable in opaque bottles.

The above combination is difficult to achieve but usually a material of large polydispersity is better than a monodisperse one because of the greater range in viscosity of the former over the latter in a large shear rate range. Polymers of low Melt Index are chosen for blow moulding. The problem of combining low shear viscosity with high tensile viscosity is eased in low-density polyethylene, in which the branching gives rise to tension stiffening behaviour.

Materials for blow moulding must have high tensile strength and good stretch properties at the extrusion temperature. For this reason, polyacetal is difficult to process by this route as its tensile strength in the melt is low. The melt should not have a sharp melting point. Nylon 66 is difficult to blow mould because of the high degree of orientation induced in it. It also has a sharp melting point. HDPE gives a higher degree of crystallinity than PP, which as a result gives glossier products. The increased amount of crystallinity in PE is due to the alignment of the linear chain. PP is much more likely to supercool and is best processed by injection blow moulding.

The product quality is judged by the distribution of the material in the bottle. The optimum distribution will:

a) produce adequate strength and rigidity; and
b) eliminate waste.

For this the moulder must allow for the fact that the parison will be thicker and cooler at the bottom than at the top and that the material that touches the cold mould wall first and solidifies will stretch less than other parts.

Apart from extrusion blow moulding, there are other related processes: blow moulding without a parison, injection blow moulding and use of preformed parisons.

The process variables associated with extrusion blow moulding are:

a) melt temperature;
b) extrusion rate;
c) blowing pressure;
d) blowing rate;
e) duration and timing of blowing; and
f) mould closing speed.

The melt temperature determines the extent to which the parison stretches under its own weight and so governs the thickness distribution in the moulding.

The extrusion rate determines the degree of die swell and extrudate distortion.

The blowing pressure does not have a great influence on the properties of the finished article, but the higher air pressures ensure that the fine detail of the mould cavity is faithfully reproduced by maintaining a good thermal contact between the melt and the cooling surfaces. The higher blowing pressures minimise mould shrinkage and the cooling time. Air pressures in the range 0.4–1.0 MN m^{-2} are commonly used.

The rate at which the blowing air is injected into the parison affects the stability of the process as mentioned earlier. Moreover, it affects the surface finish and the strength of the welds in the pinch-off region. As the parison is inflated inside the closed mould, residual air escapes from the cavity between the mould parting faces. If the blowing rate is too high, localised pockets of air may be trapped between the moulding and the surface of the cavity. This trapped air may cause dimpling and pockmarks on the surface of the moulding. In some circumstances, a high inflation rate may give rise to weak pinch-off welds.

One important point to note is that a reduction in blowing pressure does not necessarily have the same effect as a reduction in blowing rate.

The length of time that the air pressure is maintained inside the moulding should be sufficient to ensure maximum cooling and to minimise distortion in thick sections of the moulding, and yet give an economic production rate. The air must be removed from the moulding before the mould opens to prevent a rupture of the moulding or bottles.

The exact moment at which the air is injected into the parison relative to the timing of the closing of the mould affects the pinch-off weld strength. 'Early blowing', in which the injection of air into the open end of the parison commences 0.1–0.4 s before the mould finally closes, reduces the number of weak welds. This 'early blowing' is effective only when blowing is from the bottom of the parison, as in the Fig. 8.12, and the die effectively closes at the top of the parison. The rate of blowing often has to be reduced in this technique to reduce a sideways movement of the freely hanging parison.

Mould closing speed affects the strength of the pinch-off weld, as would be expected. A good weld is obtained when the pinch-off jaws exert a slow squeeze rather than a rapid one, which can completely shear the parison. Some machines employ a damping device that slows the moving platens over their last 25 mm or so of travel, as they close. The early part of the closure is usually rapid so that the moulding cycle is not extended. With troublesome pinch-off welds, the closing speed should be lowered when everything else fails.

The application of blow moulding to articles in everyday use has increased enormously since the advent of the use of polyethyleneteraphthalate for containment of beers, ciders and fizzy soft drinks. This polymer is ideal for this application because it is relatively impervious to pressurised carbon dioxide. The injection blow moulding of PET is an interesting variant of blow moulding and is described in the next chapter.

TUBULAR FILM OR BLOWN FILM PROCESSING

This process is similar to blow moulding in that an annular die is used to form a parison-shaped extrudate with very thin walls. The film is inflated into a wider tube, nipped at the far end from the die, and rolled up for either slitting down the side for sheet, or stamping and cutting for disposable loop. This sort of film is called layflat film.

The main advantage of this method over other methods of film manufacture is that the layflat film is biaxially orientated, giving strength along two axes. When used as a wrapping film, it is much harder to remove as there is no cleavage direction.

This process is illustrated in Fig. 8.15. A cylinder of molten polymer from an annular die is inflated to the correct diameter by air pressure, and is then cooled by an adjustable air ring, above which the melt freezes and the bubble

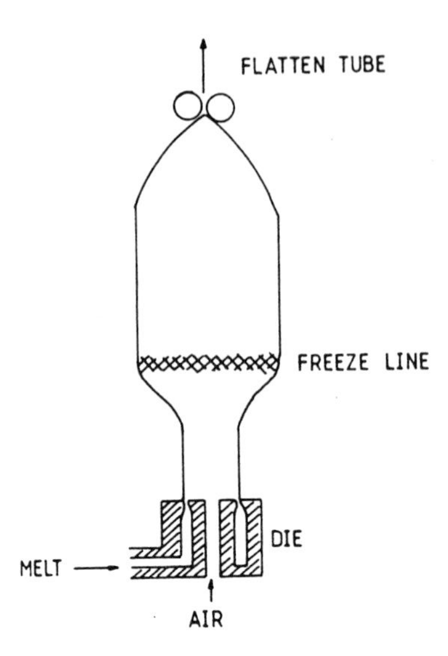

Fig. 8.15 The blown film process.

diameter remains stable. The bubble is finally collapsed between nip rollers and wound up.

In order to lessen the possibilities of weld lines caused where different flows meet, a special die called a spiral mandrel die is used. This gives a uniform flow rate, shear rate and thermal history to the different parts of the melts.

The nip rollers provide three main functions:

a) sealing the air;
b) collecting the product; and
c) controlling the orientation in the machine direction.

The maximum output rate in the manufacture of the thicker films is limited by surging of the melt in the barrel of the extruder. This phenomenon also occurs in other extruder processes at high output rates.

The process variables are:

1) The die gap — an intrinsic variable.
2) The melt temperature.
3) The blow up ratio which is $\dfrac{\text{the bubble diameter}}{\text{the extrudite diameter}}$
4) The freeze line height — dependent on the position of the cooling ring.
5) The haul-off speed.

These process variables will have a profound effect on the properties and quality of the product. The quality is jüdged by a consideration of:

a) film thickness,
b) layflat width,
c) mechanical strength,
d) surface and optical properties, and
e) biaxial orientation.

The thickness of the layflat film is initially determined by the die geometry and the temperature distribution in the area of the die. Once an extrusion line has reached equilibrium, the ultimate film thickness is controlled by the blow up ratio and the haul-off speed. A uniform film is produced when the die design is good and the cooling rate of the bubble is uniform. In practice, local thin spots occur as a result of inaccuracies in setting the die gap and the cooling device. This causes a lack of evenness in tension when the film is wound into a roll. For many applications, this is tolerable, but in the manufacture of good rolls the thicker spots must be dispersed. This is achieved by rotating or oscillating the die or the cooling ring or both.

The bubble width is strongly temperature dependent, and control of temperature can be achieved by:

a) surrounding the bubble with a thermostatically controlled environment; or
b) keeping the blowing air at a constant temperature.

As for the film thickness, the width depends on the haul-off speed and the blowing air pressure, and the latter can be altered by a feedback system that monitors changes in width. Typical blow-up ratios are in the range 1.5 to 4.5:1.

The durability of a film in service is difficult to assess, so that it is measured indirectly by measuring tensile, impact and tear strengths. The properties of the film are strongly influenced by conditions in the drawing zone situated between the die and the cooling ring. Table 8.8 shows the effect of the process variables on the durability. As the data in the table indicates, it is difficult to combine a high tear strength with a high impact strength and compromise is necessary in the choice of the process variables. High impact strengths are obtained from higher output rates and this gives rise to an interesting phenomenon. The shape of the bubble changes because at higher output rates the transverse orientation occurs later than that in the machine direction, whereas at lower output rates they occur simultaneously. The mechanical properties are influenced by this sequence of orientations.

The diffused light haze that occurs in films is caused by a combination of:

a) surface irregularities due to melt flow defects; and
b) crystallization behaviour.

During film forming, there will be a change in texture of the melt surface as a result of the reduction in depth of the extrusion defects. This reduction is

Table 8.8 The effect of an increase in the process variables on the tensile (machine and transverse direction), tear (machine and transversed directions) and impact strength of polyethylene films.

Process Variable	Tensile strength		Tear strength		Impact Strength
	Machine	transverse	Machine	transverse	
Melt temperature			↑		↑
Output from die	↑	↑	↓		↑
Blow-up ratio	↓	↑	↑	↓	↑
Freeze line height	↓	↑	↑	↓	↑

caused by an increase in the length and breadth of each defect region as the bubble is drawn lengthways and breadthways. The overall magnitude of the defects will decrease under the action of the surface tension in the melt. A high freeze line, giving a greater period during which the polymer is molten, will decrease the haze caused by melt flow defects.

The haze caused by the formation of crystallites at or near the surface of the film will be increased by raising the freeze line height because the melt will cool more slowly, encouraging crystallisation. Therefore, there is an optimum freeze line height at which the haze is minimised. The optimum freeze line height is also dependent on blow up ratio, haul-off speed, output from the die and extrusion temperature.

All blown tubular films are orientated depending on conditions. An increase in the degree of orientation gives improved clarity, strength and heat resistance. Films with balanced properties can be made by obtaining equal orientation of the molecules in the machine and transverse directions. This does not mean that the blow up ratio must equal the haul-off ratio as the two mechanisms of achieving the orientation are not the same.

The ratio of orientation in the machine direction O_M to that in the transverse direction O_T is given by

$$\frac{O_M}{O_T} = \frac{h_d}{h_b\, B_w{}^2} \tag{8.5}$$

where h_d is the film thickness coming from the die, h_b is the film thickness of the final product and B_w is the blow up ratio.

Orientation produces very strong films and is achieved with lower process temperatures. For the thicker films, the process variables have to be better controlled because any changes in orientation due to drifts in the process settings are greatly magnified.

This increased strength due to biaxial orientation allows thinner films to be made of equal strength to the extruded uniaxially orientated films.

Polypropylene is used in packaging, capacitors, carpet backing and sacks. Ethylene–vinyl acetate copolymer is used for tougher films, and control of the properties can be achieved by altering the ratios of the constituents of the copolymer. Polyethyleneteraphthalate is used as a substrate for magnetic tape and for boil-in-the-bag applications.

The polyethylenes illustrate an interesting development. Low density polyethylene (LDPE) is being replaced in many applications by the linear low density (LLDPE) form, which has better mechanical properties at a similar cost of manufacture. The linear low density polyethylene has many small branches and commercial grades have a higher $\overline{M}_w$ and lower polydispersity than low density polyethylene. This combination of parameters gives a higher tensile strength for the LLDPE and the mechanical properties are superior. This gain is at the expense of a higher melt viscosity and a higher shear thinning index, which means that higher temperatures, higher back pressures, higher motor loads and higher frictional heating are encountered when extruding this melt. The higher $\overline{M}_w$ and shorter chain branches gives a more elastic melt, which gives rise to extrudate distortion at low output rates and bubble instabilities are more common. There are three main difficulties in changing a production line from LDPE to LLDPE:

1) the need for machine modifications;
2) the development of expertise with the new materials; and
3) the higher energy utilization due to higher melt temperature and back pressures involved.

WIRE COVERING AND COATING PROCESSES

In wire covering and in other coating processes, a film of melt is extruded and brought rapidly into contact with the medium to be coated.

The die shown in Fig. 8.16 is a special one in which the wire to be coated travels at speeds of between 1 and 1800 m/min (depending on the diameter

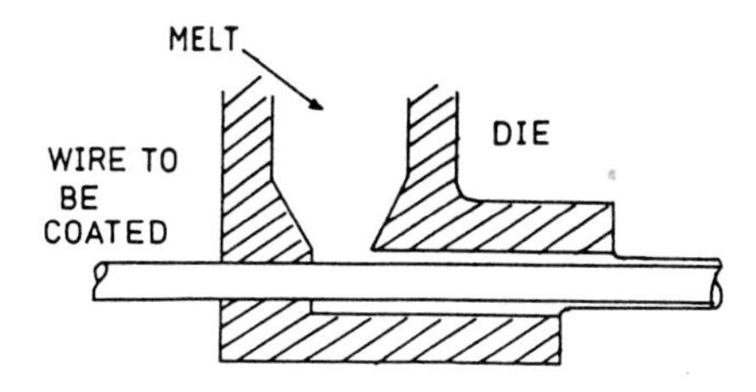

Fig. 8.16 A wire covering die.

of the wire) through a torpedo. When the wire emerges from the die, it has a coating of polymer on it. This coating arises from:

a) the drag flow due to the motion of the wire; and
b) the pressure flow due to the pressure difference between the extruder and the exit to the die.

By a simplified mathematical analysis of these two types of flow, the required pressure at the exit of the extruder can be calculated and the extrusion conditions set accordingly. The pressure at the exit of the extruder, P, is given by

$$P = \frac{6\eta l V}{H^3}(2h - H) \tag{8.6}$$

where l is the length of the die (Fig. 8.16), V is the velocity of the wire, H is the width of the annulus of the die and h is the thickness of the coating. Typical values of l are about 10–50 mm with draw ratios of 20:1 to 50:1 for coatings of greater than 0.25 mm and 50:1 to 100:1 for coatings of less than 0.25 mm. Here the draw ratio is defined as for tube extrusion.

After the wire is coated, it passes through a cooling trough, which may be several hundred metres long for high speed coating. The coated wire is then wound on to drums. In order to accommodate the high velocity of the wire, it is fed into and out of the die by dual pay-off and take-up systems, which are adjusted to prevent stretching of the wire.

In addition to wire covering with thermoplastics such as PVC and PEEK, rubbers and thermosets such as crosslinked polyethylene are often used to produce insulated and sheathed cables. These must be passed through a vulcanising tube to give the required crosslinking.

The polymer melts suitable for wire covering have lower viscosities than in other extrusion processes, certainly much lower than for blow moulding. Process temperatures are usually in the upper part of the range. Polypropylene causes some problems, which are associated with a surging of the melt at moderate draw ratios. At a critical value of draw ratio, the melt flow becomes unstable and a surge occurs giving rise to large thickness variations. This surging is not associated with the extruder screw. Melt surging limits the extrusion coating speed to about 35 m/min as compared with 200 m/min for LDPE.

The process variables associated with wire covering and coating include:

a) the usual operating and design variables for the extruder and die to provide melt at the correct temperature and pressure; and
b) the velocity of the wire.

The major applications include coating of paper, metal foil and fabrics. Polyethylene-coated paper for milk cartons represents the largest consumption of coated materials. The water barrier and heat sealing properties make PE ideal for this application.

For wire covering for primary electrical insulation or sheathing, polyethylenes in low or medium density or crosslinked form and vinyls are commonly used. Nylon is used for jacketing because of its strength, toughness, abrasion resistance and good heat resistance.

CALENDERING

In the calendering process, raw material in the form of continuous strip or rod from an extruder, or small strips from a mill, is fed between rollers to produce continuous sheet. This process was copied from those used in the paper, textile and metal industries.

The advantages of this process over sheet extrusion are:

1) Less chance of thermal degradation. PVC film and sheet is generally calendered, as is rubber sheeting. PP, PE and PS are usually extruded.
2) The calenders give greater output rates.

The disadvantages are:

1) Calenders are not as versatile — extruders can be used for something else.
2) They are less suitable for short production runs.
3) They take along time to reach the operating temperature and therefore longer runs are more economic.

In this process, the feedstock may be fed into the first set of heated contra-rotating rollers from an extruder with a low L/D ratio (for PE, PP and PS), which can be used to mix the polymer, plasticiser and colouring if required. Alternatively, the pre-compounding may be carried out in an internal mixer

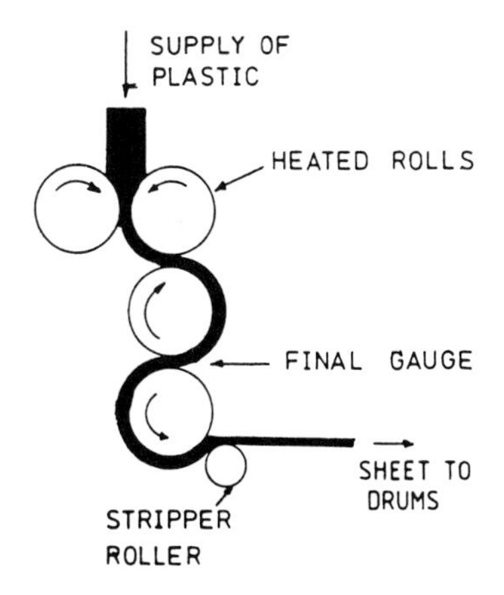

Fig. 8.17 A calendering process for sheet manufacture.

(for PVC) or on a mill (rubbers), which gives a dough-like consistency. In the extrusion mixing, solid matter is removed by the filter pack, but in the internal mixer pre-compounding a metal detector is used to find foreign matter.

A typical calender arrangement is shown in Fig. 8.17. The heated rollers squeeze the feedstock and the final rollers give the sheet its final width and thickness. The small stripper roller travels at high speed, peeling the hot plastic from the surface of the large roller. There is a tendency for the plastic materials to stick to the surface of the rollers. The stripper roller may be heated or cooled as needed. When the sheet leaves the calender, it passes between embossing rollers which give it its final surface finish, which may be glossy, matt or embossed. After this, the sheet goes to the cooling drums. It is then trimmed and stored on drums. The winding drum speed can be adjusted to give the correct drawdown in the sheets.

The production rate depends on:

a) The mixing and melting capacity at the input to the rolls.
b) The effect of the process variables on the material.
c) The surface finish.
d) Whether or not a post-calender stage is used to give the required surface quality.
e) The thickness of the sheet.

Production rates of greater than 60 m min^{-1} are attainable when making heavy gauge sheeting (> 0.25 mm), and even greater speeds are possible if a post-calendering stage such as embossing, laminating, coating or printing is used. Thin films often have lay-flat problems, but even with these, speeds of up to 100 m min^{-1} are possible. This may be reduced by a factor of 10 for high gloss, thin, rigid sheets.

The quality of the final product will depend on the following parameters:

a) the surface of the rollers (intrinsic variable);
b) the drive system (intrinsic);
c) the temperature; and
d) the accuracy of the nip between the rolls.

The calender rollers control the surface finish, stability of the film and the uniformity of heating in the product. For this reason they are made of chilled cast iron. Shock cooling suppresses the formation of graphite and encourages the production of very hard iron carbide. The rollers are driven by an infinitely variable drive system operating in the range 10–150 m min^{-1}. Together these represent very important intrinsic variables.

The quality of the product will depend on the close control and uniformity of the temperature throughout the polymer. The temperature control over the working face of the roller is about 1k in 200k, and is achieved by forcing high pressure hot water or steam through the rollers and by circulating hot oil through the bearings to minimise heat loss at the ends of the rollers.

During the process, the material being worked can generate heat, causing temperature rises of up to 20°C. Eventually the process will reach equilibrium, but it may be several hours before the quality is uniform. Only long runs are economical.

The control of the rollers to give a constant gauge product is most important. The calender rollers are subject to very high forces during processing, causing them to bend, and giving a product that is thicker at the centre. There are various ways in which manufacturers have counteracted this. Gauge thickness is constantly monitored in the machine and transverse directions and variations are fed back into the control system of the rollers.

Products may suffer from various faults, which are due to one or more of the factors below:

a) formulation errors;
b) inadequate compounding and mixing; and
c) calendering errors.

Whereas PE and PP are made into film and sheet by extrusion or blown film extrusion, plasticised PVC, because of its thermal sensitivity, is usually calendered into film. Calendered plasticised PVC, reinforced with fibres is commonly used as a flooring material and fabric reinforced thermoplastics are often calendered into sheet. Rubber sheet is processed in the same way.

EXTRUSION OF FILLED THERMOPLASTICS, FOAMS, RUBBERS AND THERMOSETS

The extrusion of filled thermoplastic is not difficult. The process parameters are chosen to give rise to melting in the correct part of the screw as for unfilled melts. The shear viscosities are higher but for commercial materials this does not usually present problems. It is, however, imperative that the walls of the barrel and the screw surfaces should be hardened, as the abrasive nature of fillers and fibres causes excessive wear.

Extrudates from cellular plastics may be produced by both chemical and physical methods of expansion. The free and controlled expansion techniques are used to make pipes, cables and profiles.

In physical expansion systems, n-pentane, fluorocarbons or other low boiling point liquids are used in the polymer melt to produce the cellular structure using conventional single screw extruders. Direct gas injection has also been used to produce cellular polyolefin, tubular blown films and slab stocks. The compression ratios involved are in the range 1.8–2.5:1.

In chemical expansion systems, both inorganic and organic materials are used as blowing agents to provide the cellular structure. The process variables are similar to those for the physical expansion systems, and in both types the object is to maintain the melt pressure above the gas solution pressure until the melt has passed through the die. Typical values of melt pressure to maximise the solubility of the gas in the melt are 12–14 MN m^{-2}.

When the melt leaves the die, it can either be allowed:

a) to expand freely (free expansion); or
b) to expand more gradually under constraint (controlled expansion).

In the free expansion technique, it is imperative that the melt strength and elasticity are large enough to prevent a release of gas, which would give at best a poor surface. The elasticity could be increased by crosslinking the melt. The success of this process depends on a good control over melt temperature and pressure.

In the controlled expansion technique as used in the Celuka process, the rate of expansion is controlled by inducing skin formation and by controlling the rate of decay of the melt pressure. The control over melt temperature and pressure is achieved by the use of an extended mandrel of decreasing diameter, which projects through the main die into a temperature controlled shaping die of increasing cross-section. This gives rise to a high density skin with a cellular core, which forms as the melt pressure slowly decays due to the increasing cross-section of the shaping die. The advantage of the controlled process is that larger products of more complex shape can be extruded.

In the extrusion of rubbers, there are two basic designs depending on whether the rubber is introduced into the extruder cold (cold feed) or hot (hot feed). In the latter system, one or more warm-up mills are used to feed the extruder, whereas in the former the extruder has to do this warming itself. As a result, cold feed extruders have L/D ratios in the range around 12 to 18:1, whereas cold feed extruders are shorter and in the range 4 to 8:1.

There have been many disagreements over the relative merits of cold feed and hot feed.

Advantages of cold feed extrusion include:

a) Warm-up mills, conveyor belts and the synchronisation between the equipment is not needed and investment costs are therefore lower.
b) The output and size of the product is more stable because of better control of the melt temperature and viscosity.
c) The extruder can easily be fed from different batches for blending purposes.

Disadvantages of cold feed extrusion include:

a) The extruder has to be more robust and the screw more complex, with a greater L/D ratio. This means greater expense on buying the extruder and in running it.

b) The cold feed machines usually have a lower output than hot feed machines of the same diameter.

In rubber extrusion, the die is used to give an approximate shape, as in profile extrusion of thermoplastics, and the final shape is given by a sizing die before vulcanisation is carried out.

There are several methods of achieving vulcanisation, and the more modern ones are the salt bath and the microwave unit. In the salt bath method the profile is passed into a stainless steel bath containing a molten salt. The bath is heated to 240°C by means of external electric strip heaters. The profile is conveyed through the bath on a driven stainless steel belt. At the far end the vulcanised profile is washed, dried and either reeled or cut to size.

In the microwave system, the profile passes through a resonator chamber if the profile is large, or a wave guide channel for small products. The shape of the microwave units is arranged so that the high frequency electromagnetic waves give a fast and even heating to the unvulcanised extrudate.

The above is intended only as a brief introduction to the extrusion of rubbers. Typical products include hoses, pipes, wire covering and sealing strip.

Screw extruders are not suitable for the extrusion of thermosets because it is difficult to obtain sufficient back pressure to get good product. A reciprocating hydraulic press is used. During the back stroke, powder falls into the barrel. This is pushed forward to the die and melting occurs. Dies are very long and are heated. The incoming powders are pressurised by being pushed against the material already in the die. This pressure is caused by the friction between the die wall and the curing material.

The pultrusion method of extrusion is used for producing profiles in continuous fibre reinforced thermosets. In this process, long glass fibres or other reinforcing material are pulled with a thermosetting resin through a heated steel die (Fig. 8.18). The reinforcement, in the form of continuous strand,

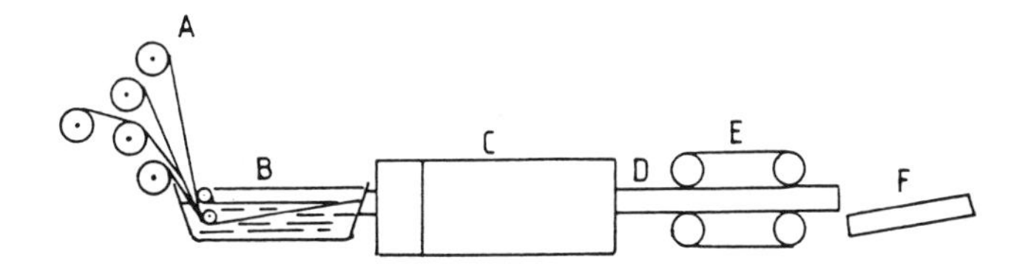

Fig. 8.18 The pultrusion process for continuous fibre-reinforced thermosets: *A)* roving and reinforcement, *B)* resin bath, *C)* die, *D)* finished product, *E)* pull rollers *F)* product cut to size.

woven fabric, woven roving, surfacing mat, reinforcing mat or a combination of these, is fed through a thermosetting resin bath and pulled over rollers to remove excess resin and thence to the shaping die. The die is water cooled at its entrance to prevent premature curing and binding. Rollers shaped to the product are used to pull the extrudate away from the die. Speeds of between 5–500 cm min^{-1} are common and depend on the resin, the shape and the die temperature and length.

Almost all pultrusions are made with glass reinforced polyesters because of the simplicity of use, economy and good electrical and chemical resistance.

REFERENCES

C. BLOW and C. HEPBURN, (eds): *Rubber Technology and Manufacture*, Butterworth, London, 1982.

R.L.E. BROWN: *Design and Manufacture of Plastic Parts*, Wiley, New York, 1980.

P.J. CRAWFORD: *Polymer Engineering*, Pergamon Press, Oxford, 1983.

R.J. CRAWFORD: *Plastics And Rubber*, Mechanical Engineering Publications, 1976.

Halar Fluoropolymer, Engineering Plastics, Morristown, NJ, USA.

W.A. HOLMES–WALKER: *Polymer Conversion*, Applied Science Publishers, 1975.

ICI Technical Literature, ICI, UK.

The Extrusion of Polypropylene – Processing Characteristics, ICI, Welwyn Garden City, UK.

R.W. MEYER: *Handbook of Pultrusion Technology*, Chapman Hall, London, 1985.

D.V. ROSATO and D.V. ROSATO: *Injection Moulding Handbook*, Van Nostrand Reinhold, New York, 1986.

S.S. SCHWARTZ and S.H. GOODMANN: *Plastics Materials and Processes*, Van Nostrand Reinhold, New York, 1982.

9 *Injection Moulding and Moulding Processes*

INJECTION MOULDING

The function of injection moulding is to make identical products at a very high repetition rate from granules or powders of thermoplastics, rubbers and thermosets. This process is the plastics' equivalent of pressure die-casting of metals. Machines vary in size from making a few grams of moulded product to those capable of a shot weight of over 30 kg. The machines are graded according to shot weight and mould clamping force, which may vary between 2 to 3500 tonne.

The advantages of injection moulding as a process are:

1) Mouldings are produced completely finished.
2) The process is accurately repeatable.
3) Metal inserts, threads and holes can be moulded in.
4) High output rates can be achieved with automatic operation and multicavity moulds.

The disadvantages are:

1) The capital cost of the injection moulding machine is high compared with that of other moulding machinery.
2) Tool costs are also similarly high.
3) Unless foaming the material, it is difficult to mould parts with large variations in wall thickness.

In injection moulding, as in extrusion, the raw material is pumped, melted, mixed and pressurised before being pushed into a cold mould for shaping, where the product solidifies and from which it is ejected in its finished form.

As in Chapter 8 on extrusion, the intrinsic or machine variables will be described first. Fig. 9.1 shows the most versatile form of injection moulding machine, the kind that uses a reciprocating screw plunger.

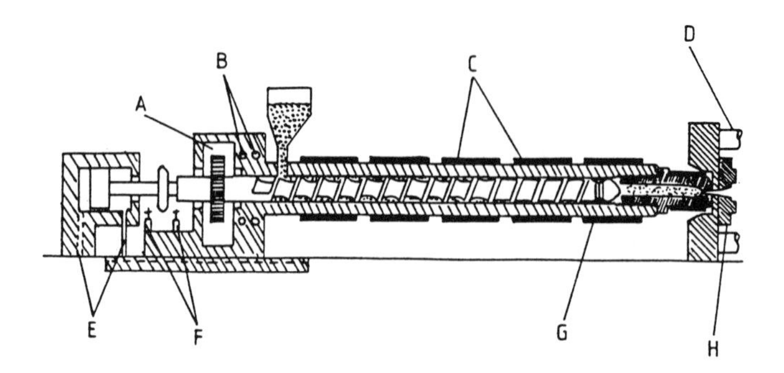

Fig. 9.1 Reciprocating screw and plunger injection moulding machine. *A* hydraulic motor, *B* water cooling channels, *C* heater bands, *D* tie bar, *E* hydraulic fluid pipes, *F* adjustable screw travel limit switches, *G* back-flow stop valve and *H* the mould.

The cylinder has to withstand very large pressures of the order of 100 MN m^{-2} and is made of high tensile steel with walls often 25 mm thick. This is the sort of specification for a 100 g machine (shot weight 100 g). The maximum barrel temperature is usually 400°C.

The profiled screw usually has three separate temperature zones provided by heater bands on the barrel. The compression ratios are in the range 2.5–4:1 with L/D ratios of 15–20. The maximum pressure that can be developed is 100 MN m^{-2}. The main difference in design from an extruder screw is the provision of a back-flow check valve, which prevents back flow across the flights when the screw is used as a plunger. This valve is a place where material can collect, and as such is not used with thermally sensitive polymers such as PVC. Some machines have decompression zones, but if they do not, many polymers must be pre-dried.

As the screw rotates, granules or powder from the hopper are melted by heat conducted from the barrel and by viscous heating from the screw. The check valve opens and melt is pushed into the space between the screw and the nozzle. As the screw continues to rotate, the pressure developed in the melt by the nozzle pushes the screw backwards towards the hopper. A limit switch is set such that the exact volume of material to fill the mould is melted ahead of the screw. When the backward movement of the rotating screw trips the limit switch, the back-flow check valve closes and the screw ceases to rotate. The mould closes and the injection cycle shown below (Fig. 9.2) commences as follows:

1) The screw stops rotating and moves along the barrel, acting as a plunger. It pushes plasticised material into the mould cavity.
2) The screw remains in the forward position maintaining pressure long enough for the melt in the gate region to solidify.

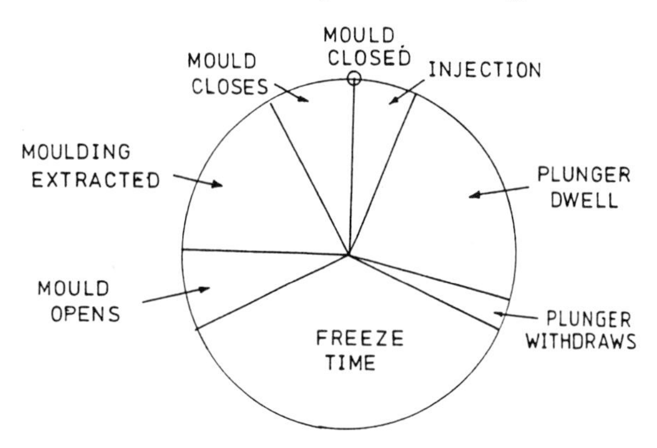

Fig. 9.2 The injection moulding cycle.

3) While the rest of the moulding cools in the mould, the screw starts to rotate. This draws in new melt ahead of the screw which causes the screw to move backwards until it trips a limit switch. The rotation stops.
4) The mould opens to eject the product and then closes to receive the next shot of melt.

Fig. 9.2 shows a typical injection moulding cycle. A good proportion of the time is taken in cooling the product so that it will not deform on being ejected from the mould. Cycle times depend on the rate of plasticisation of the material or the rate of cooling – whichever is the longer – and range from 15 seconds to two minutes.

When the material passes through the nozzle it receives heat from a heater band and through viscous heating. The melt then runs along the sprue and runner system into the mould (Fig. 8.3). The runner is narrowed by the entrance to the mould cavity so that the sprue and runner system can be separated easily from the product. This region is called the gate, and as it is small in cross-section, shear rates of the order of 10^5–10^6 s^{-1} are generated, causing orientation in the melt.

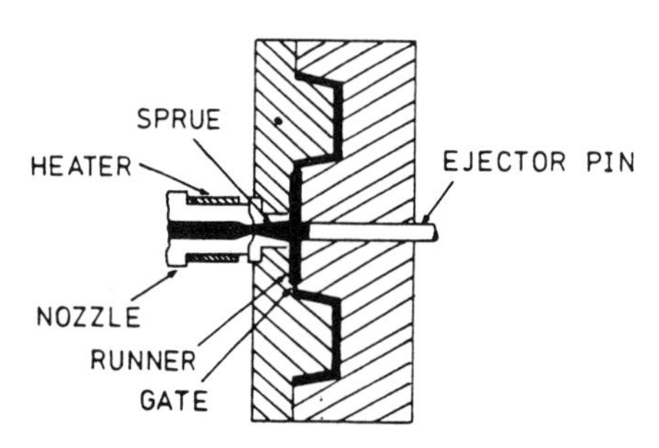

Fig. 9.3 A typical injection mould.

The mould is made in two parts that are held together by the mould locking force. This must be sufficient to keep the mould closed during the high pressure injection of the melt.

The mould facilities include:

1) Backing plates to enable the mould to be bolted on to the machine platens.
2) Channels to provide mould cooling or heating.
3) Ejector pins so that the moulding can be removed.
4) Venting to provide for the escape of the air from the mould. Manufacturers give details of gate, runner and vent sizes for their products.

Mould design is highly complex, especially when multi-cavity designs are necessary. Moreover, one of the main problems with the simple design shown is that the sprues and runners generate scrap, which ideally should be reground and recycled. Up to 25% of regrind is acceptable, except in optical components, which require uniformity of optical properties and high precision.

In order to minimise the creation of scrap three modern developments are often used:

1) hot runner systems,
2) hot nozzles injecting directly into the mould cavity, and
3) insulated runner systems.

Before analysing the process variables, mention should be made of the manner in which the melt enters the mould cavity. It is often assumed that the melt enters as a front, but usually a jet of fluid precedes the main front. This gives rise to weld lines where materials with different flow paths meet. This is particularly so when there are regions of varying thickness or the mould is fed from several gates. Weld lines are a source of weakness.

Control of the Operating Variables

The injection moulding process is far more complex than extrusion. As well as having more operating variables, the process must be properly sequenced so that the mould is opened and closed at the right time, and the screw is moved forward as required.

The operating variables include:

a) The setting of the limit switches to control the forward and reverse traverse of the screw to give the required shot into the mould.
b) The injection speed.
c) The plunger dwell time.
d) The freeze time.
e) The injection pressure and its variation throughout the cycle.

f) The mould locking force.
g) The barrel temperatures.
h) The mould temperature.

These variables lend themselves well to computer control and manufacturers have such systems available. The injection moulding cycle can be so accurately controlled by this means that during the course of a day the shot weight may vary by only 0.6% once an equilibrium has been reached. It is even possible to discern batch to batch variation in a given polymer grade. This sophistication of control enables:

1) high precision moulding, even of optical parts;
2) more economic moulding; and
3) the use of complex pressure control to mould parts previously found difficult.

Future developments, no doubt, will improve and extend injection moulding processes.

Choice of Operating Variables

a) The setting of the limit switches should be such that the mould cavity just fills.
b) The injection speed is usually chosen to be as rapid as possible, but when this causes the injection pressure to rise too much, the speed must be decreased. In addition, too high a speed may cause the melt to overheat at the gate, causing splash or mica marks and excessive jetting. The advantage of high speed is that high shear rates are evolved, which in these shear-thinning materials is an effective way of reducing melt viscosity. This method of reducing viscosity is often preferable to raising the barrel temperature because no additional cooling time is needed.
c) The plunger dwell time is chosen to be sufficient for the material in the gate region to freeze, and to keep the mould filled in order to lessen voiding and sink marks. If the pressure is kept too high, orientation and hence frozen-in-strain will be encouraged, and if the component is subject to heat in service it will have a tendency to warp. The frozen-in-strain enhances brittleness in the gate region.

   In practice, the dwell time is increased until the moulding reaches maximum weight, and this value is maintained for the process. This method will reduce shrinkage but never eliminate it, particularly when the shrinkage is partly due to crystallisation.
d) The freeze time is set such that the mould opens only when the moulding is sufficiently solid to retain its shape. The correct setting of (c) and (d) will give a good product at a good production rate. Overall cycle times will vary with the moulding thickness and the barrel and mould temperatures.

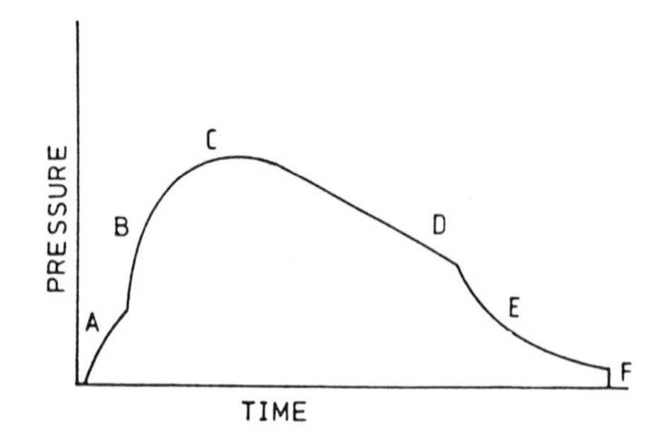

Fig. 9.4 Variation of pressure through the moulding cycle. *A* to *B* the cavity fills, *B* to *C* packing of the cavity, *C* peak pressure, *D* the discharge, *E* gate sealing, *F* residual pressure. The plunger moves forward until *D* and returns after *D*.

e) The injection pressure and its variation during the moulding cycle are shown in Fig. 9.4. The aim is to fill the mould sufficiently to avoid sink marks and voids. As the screw moves forward, melt is pushed through the nozzle into the mould cavity. The pressure increases to a maximum as melt is forced into the mould, and then decreases as a small additional amount of polymer is forced into the mould to counteract shrinkage due to the contracting of the melt as it cools or crystallises. When the material at the gate freezes, which occurs first because this is the narrowest part, the pressure is released. It can be noted that a small residual pressure exists that has been frozen into the moulding. This is caused by orientation.
f) The value of the mould-locking force is chosen to keep the mould closed. If this value is insufficient the parts of the mould will separate and flashing will occur, giving a moulding with at least a noticeable seam and often a web all around it.

Table 9.1 Injection moulding process temperatures.

Melt	Cylinder Temperature °C	Mould Temperature °C
High-density polyethylene	220–260	. . .
Nylon 6	230–260	60–90
Polybutyleneterephthalate	230–260	30–90
'Noryl'	230–260	50–80
Polycarbonate	270–320	80–120
'Vectra' Liquid crystal polymer	285–325	30–150
Polyetherimide	340–425	65–175
Polysulphone	375–395	160
Polyethersulphone	340–360	150
'Xydar' Liquid crystal polymer	400–425	205–280

g) The cylinder temperatures are chosen to melt and mix the polymer so that it enters the mould cavity at the optimum temperature. Different polymers require different moulding temperatures, as shown in Table 9.1. Manufacturers' literature provides this information.

The choice of cylinder temperature range will affect (a) the production rate, (b) the degree of orientation and (c) the surface quality of the product. If low cylinder temperatures are chosen, the production rate will be controlled by the rate at which the polymer can be melted (see Table 9.2). This will depend on:

i) The specific heat,
ii) the temperature rise, and
iii) the latent heat of fusion in partly crystalline polymers.

Table 9.2 Shows the heat required to plasticise various polymers.

Polymer	Total heat input at moulding temperature $MJ\,kg^{-1}$
Polystyrene	0.28
Polyacetal	0.42
Polypropylene	0.58
LDPE	0.6–0.7
HDPE	0.7–0.8

If cylinder temperatures are high the production rate will be controlled by the rate of cooling. Fig. 9.5 shows how the production rate for a given product goes through a maximum value with cylinder temperature. If conditions are chosen to give the maximum production rate, it is found

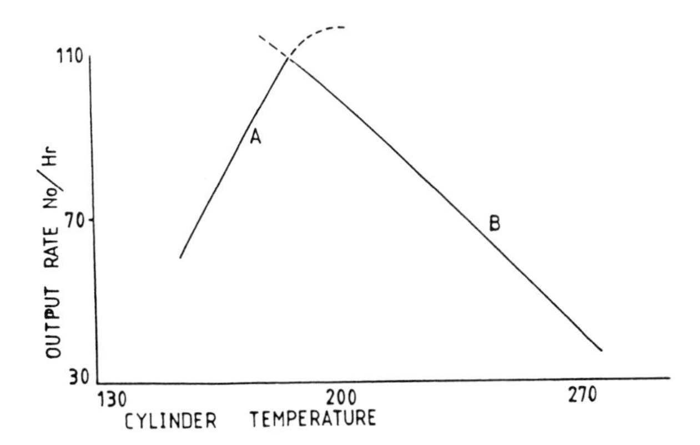

Fig. 9.5 The effect of cylinder temperature on production rate for a given moulding, showing how the output rate is controlled by *A* the rate of plasticisation and *B* the rate of cooling.

that the surface finish is inferior to that obtained at higher cylinder temperatures. The melts most faithfully reproduce the surface of the mould if they are less viscous. Another reason that this maximum rate may not be obtainable is that the melt may be too viscous to fill narrow mould cavities.

The melt temperature has little effect on shrinkage, which is slightly larger for the higher temperatures.

h) The mould temperature is chosen to cool the moulding at the optimum rate to give a good product. Needless to say, this is found with experience. Table 9.1 gives some typical values of mould temperature.
The rate of cooling of the moulding will control:

i) the degree of orientation,
ii) the surface quality, and
iii) the degree of crystallinity.

If the mould temperature is high, better surface gloss is achieved and voiding reduced. However, flashing and sink marks are increased, and crystallinity is encouraged with its attendant greater shrinkage.

Too low a mould temperature may lead to short mouldings. In most cases manufacturers recommend temperatures for moulds in their data sheets.

Fig. 9.6 gives a summary of the effect of the operating variables on the degree of orientation. Orientation and shrinkage are very important and are dealt with more fully.

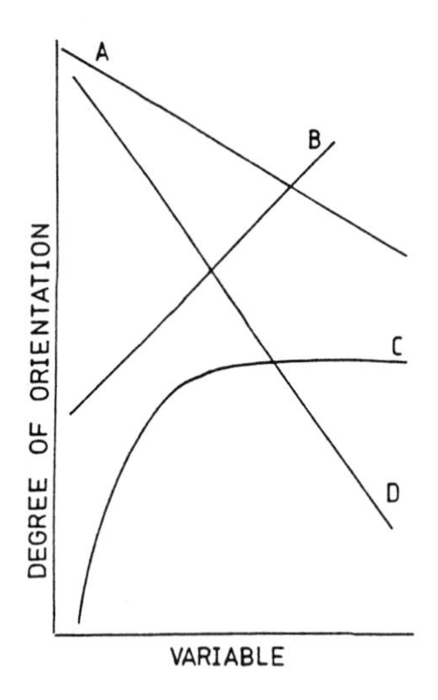

Fig. 9.6 Variation of the degree of orientation with *A* mould temperature, *B* injection pressure, *C* packing time, and *D* cavity thickness.

ORIENTATION

Orientation is very important in injection moulded articles. The degree of orientation will depend on the molecular architecture such as $\overline{M}_w$, polydispensity, degree of branching and additives, as well as on the process variables. Longer molecules will orientate less readily than shorter ones because of the greater entanglement. In a material of wide molecular weight distribution, the shorter chains will align while the longer ones remain closer to a random orientation. The effect of orientation is much greater in injection moulding than in extrusion because of the higher shear rates in the gate region of the mould.

Such orientation can be demonstrated using the thin-walled polystyrene drinking cups. In these the molecules are oriented radially across the bottom of the cup and up the sides. It is easy to tear up and down the sides but almost impossible to tear circumferentially.

Sometimes a designer may be able to use this orientation to advantage as in the 'living hinge', in which the continual bending of polypropylene creates a hinge that is very strong. In general however, the complexity of the interrelationships between the material properties, the operating variables and the mould geometries make orientation hard to control. Orientation varies across and along a moulding.

Orientation gives rise to frozen-in-strain, which again varies across and along a moulding, and causes warping if the product encounters high service temperatures. This can again be demonstrated using the thin-walled polystyrene drinking cups. Half a dozen of these cups when filled with boiling water and left for varying amounts of time will suffer varying amounts of distortion. The ones in contact with the boiling water for the longest time will be shorter and wider than the others. If the heat treatment is intensified the cup becomes a distorted disc. The shrinkage is much greater in the direction of orientation.

This differential moulded-in-strain often gives rise to an unacceptable product, because although the orientation that causes it enhances the strength in one direction, it also makes the product more brittle. A sharp blow to a product with moulded-in-strain will cause a breakage that an unstrained product may have withstood. For this reason sharp corners are designed out of plastic products as they act as stress concentrators.

Moulded-in-strain also increases the susceptibility to environmental stress cracking. Some polymers when in contact with mild solvents suffer surface cracks which can propagate more readily if the product has moulded-in-strain. Polystyrene has a low elongation to failure and is in a glass-like state at room temperature. It is, therefore, highly susceptible to the formation of small cracks (crazes) in its surface when in contact with mild solvents or vapours at room temperature. Polystyrene dipped into kerosene crazes readily. Crazing leads to premature part failure.

When moulded-in-strain is undesirable or excessive, it can be removed by annealing. This consists of heating the moulding uniformly to just below the softening point of the material, and holding it in shape until the frozen-in strain has relaxed out. The moulding is then slowly cooled. Annealing improves the overall mechanical properties but is an expensive and uneconomic secondary process. Operating variables are therefore chosen such that orientation is minimised.

This is particularly important when designing optical products such as plastic lenses, because orientation gives rise to birefringence. Indeed, the orientation in a transparent product can be examined by placing the product between crossed polaroids. The moulding times are generally longer for optical products than for other products of similar size.

Orientation is reduced by altering the operating variables such that shear rates are lower and ample time is given for the relaxation of the molecules to a random configuration. This involves minimum cylinder pressure and hold-on pressure and maximum cylinder and mould temperatures for the polymer. The variation of the degree of orientation with the operating variables is shown in Fig. 9.6.

It has already been mentioned that the degree of orientation varies across a moulding. This is due to the complex way in which a melt enters a mould cavity. Often jetting occurs, in which a stream of melt from the hot centre of the incoming front shoots ahead of the front. By the time it strikes the cold mould wall the material is unrestrained and not oriented. Thus, the material that forms the skin of the moulding is relatively unoriented (Fig. 9.7).

The next material entering the mould cavity is more oriented because it is more restrained in a narrower channel than the initial melt, and is in a higher shear rate regime. It cools more slowly than the outer skin material because of the poor thermal conductivity of the skin. Nevertheless, it retains more orientation. The material that enters the mould last suffers lower shear rates and occupies the centre of the moulding. It is less oriented, and cools far more

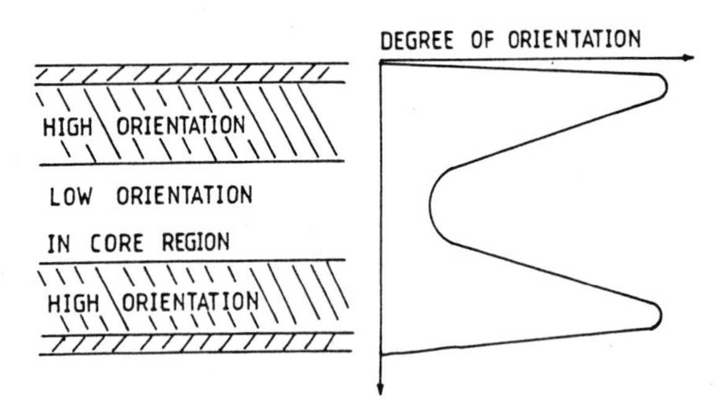

Fig. 9.7 Variation of degree of orientation across the moulding.

slowly, giving time for the molecules to adopt a random conformation. This gives the variation in orientation shown in Fig. 9.7. The orientation will vary along the moulding as well, and in multi-gated systems the orientation profile will be complicated.

MOULD SHRINKAGE

All polymeric materials shrink when cooled in a mould, so that designers always make mould cavities larger than the finished article. There are three causes of shrinkage:

1) fall in temperature,
2) crystallisation, and
3) crosslinking.

Cause 1 affects all polymeric materials, giving mould shrinkages in the range 0.4–0.7%. This may vary along and across the orientation directions with highly oriented molecules. Additives such as glass spheres or fibres reduce mould shrinkage. For glass fibre-reinforced thermoplastic polyesters the shrinkage is in the range 0.5–0.35% in the flow direction and 0.5–0.75% in the transverse direction.

Crystalline polymers show greater amounts of shrinkage (1.5–2.5%) than amorphous polymers because the alignment of chains during crystallisation gives closer packing and hence a greater contraction.

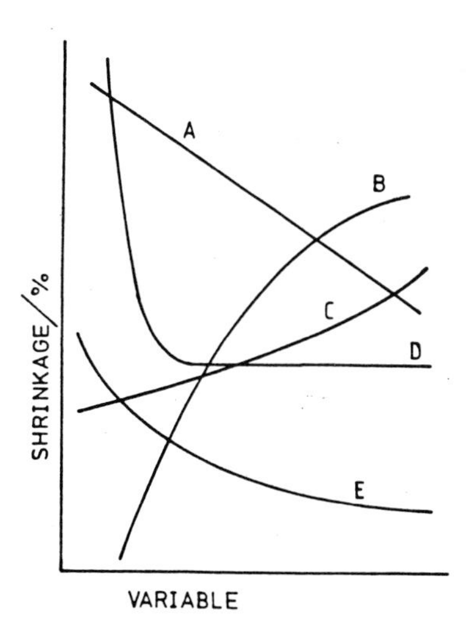

Fig. 9.8 Variation of mould shrinkage with *A* pressure, *B* thickness of moulding, *C* mould temperature, *D* hold-on time *E* gate size.

Thermosetting materials show even greater mould shrinkage, because the materials contract considerably during the crosslinking reaction. Some contraction may be as high as 14%, which makes high product precision difficult to achieve.

The product designer must know beforehand the likely degree of mould shrinkage, and where high precision is required materials with low shrinkage must be used. Details of mould shrinkage are given in manufacturers' literature and in BS2026 (1953), BS4042 (1966) and DIN 16901 (1973). Fig. 9.8 shows how the operating variables affect mould shrinkage.

CRYSTALLINITY

Crystallinity and its effect on rheology and processing has already been mentioned. Crystalline injection mouldings have faster production rates than with similar amorphous material because the change in viscosity around the melting point is marked. This allows an earlier ejection from the mould than for the amorphous product.

The crystallinity will vary throughout the moulding due to the differing rates of cooling at different layers. This gives rise to a source of mechanical weakness between the zones of fine and coarse crystallites. An increase in crystallinity gives increased strength and modulus at the expense of decreased impact properties. For crystallinity to be retained the mould temperature must be sufficiently high to prevent supercooling of the product.

QUALITY OF MOULDING AND THE USE OF COMPUTER AIDED DESIGN (CAD)

The quality of a moulding can be judged by its mechanical properties, which are partly dependent upon the choice of operating conditions and on the design of the mould.

In the designing of the mould it is necessary to consider the positioning of potential weld lines to regions that are not load bearing. Weld lines are formed where two melt fronts meet and join. They occur wherever there is a change in the cross-section of the mould, where a melt divides around an obstacle and rejoins later and when multi-gated moulds are used.

Weld lines represent regions of weakness and the need to position them away from sensitive regions of the product has been an important designing art — which now, with the aid of the computer, is less of an art than it was.

The strength of a weld line can be improved by:

a) having sufficient temperature and pressure at the weld line to give good welding. (There are two possibilities here: higher *cylinder* temperatures and pressures may increase orientation so that it may be better to increase the *mould* temperature to improve welding, and suffer the increased cycle time);

b) providing venting at the position of a weld line to prevent the entrapment of air, which could give rise to voids.

Recent developments in computer aided design (CAD) have led to the use of flow simulation programmes, which make mould design easier. One of the uses of these programmes is to assist in the design of multi-gated moulds and in 'family' moulds, in which several different mould cavities are to be filled simultaneously. The latter, however, introduce problems associated with the optimum filling of each cavity. Flow simulation programmes allow the engineer to design his mould on the computer instead of by trial and error.

The engineer inputs the information on the mould size and shape into the computer, or he builds up a picture of the mould cavity in the computer by using a digitizer on the technical drawing. The computer then divides up the mould cavity into a mesh, the coarseness of which can be determined by the designer. The flow programme is based on finite element analysis, and from a bank of rheological and thermal data and with details of the operating parameters, it produces isobars in the mould cavity. From knowledge of the positions of these isobars, it is possible to analyse the mould filling in family moulds and to manipulate the positions of weld lines in multi-gated systems. Computer aided design allows the engineer to position gates and to dimension runners and gates to give the best design before cutting a single cavity.

SPIRAL FLOW MOULDING

Satisfactory mouldings depend on the ability of the polymer to fill the mould and to replicate the surface features of the mould. This ability will be mainly determined by the choice of operating temperatures, pressures and length of the injection stroke. ICI has developed a spiral flow moulding test that manufacturers can use to find the optimum moulding conditions for their polymer products.

The mould has a long calibrated channel in which flow lengths can be measured. The channel is half-round in section and is in the form of an archimedian spiral. The radius of the spiral increases by 12.5 mm per revolution. The polymer is injected into the centre of the spiral and the less viscous melts flow for great lengths outwards as shown in Fig. 9.9. The overall length of the channel is 1.96 m. This is accommodated in a mould platen measuring 300 mm × 230 mm and a means of controlling the mould temperature is provided.

The mould can be used on a small injection moulding machine provided that the accurate setting of melt temperature, injection pressure, feed time

Fig. 9.9 One half of a spiral mould and a moulding (By courtesy of ICI plc).

and cycle time are possible. The shot capacity of the machine must be adequate to plasticise the right amount of material to fill the mould in the cycle time.

The temperature, times and pressure controls are set at the desired values and moulding is commenced. The feed is adjusted until the reciprocating screw just fails to complete its full forward stroke. When the conditions are steady the average of ten spiral lengths is taken as a function of cylinder temperature for a constant injection pressure.

Graphs are drawn to show spiral length versus melt temperature for a constant injection pressure, as shown in Fig. 9.10. Graphs of spiral length versus injection pressure at constant temperature may also be useful.

Curves such as those in Fig. 9.10 provide a means of comparing the flow behaviour of materials under real processing conditions. Interpretations from the graphs are as follows:

1) A material with a flow/temperature curve that gives a shallow slope over a wide range of temperature is easily moulded, since a comparatively large range of temperatures has a small affect on the spiral length (Polyethylene).
2) A material having a flow/temperature curve with a shallow slope over a narrow range of temperature at the lower end of the flow scale is difficult to mould and requires maximum available pressure (Acrylic heat-resistant grade).
3) A material having a steep flow/temperature curve requires very accurate control of both temperature and pressure, since there is very little latitude

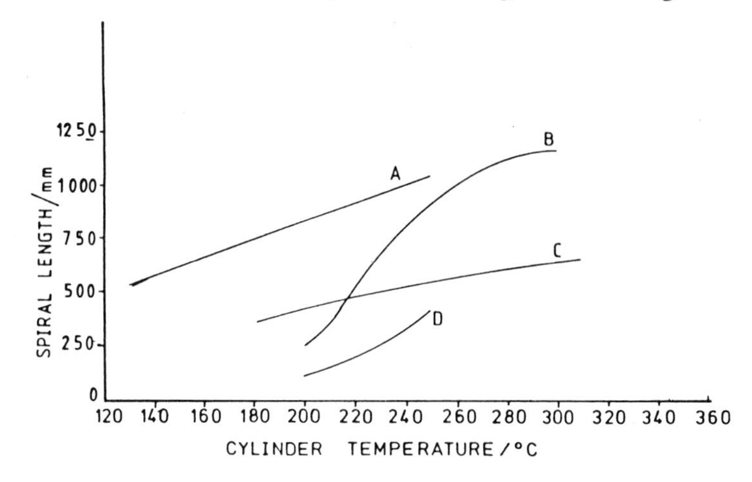

Fig. 9.10 Spiral flow moulding curves for *A* a low density polyethylene, *B* nylon 11, *C* high density polyethylene *D* heat resistant acrylic.

between being unable to fill the mould (viscosity too high) and causing the mould to partially open and allow a web of melt to surround the moulding (viscosity too low). This is called flashing (e.g. Nylon 11).

4) A material with a long shallow flow curve that cuts across the curves of several other materials can be used to purge a machine before use for another melt. Purging must be carried out using a more viscous melt than the one in the cylinder. A melt of the above type is, therefore, highly suitable because at some temperatures it will be of higher viscosity than other melts, whereas at higher temperatures it can purge out materials and then be purged out itself by the new material to be used (e.g. High density polyethylene).

One word about hazards when purging melts. Melts based on polyformaldehyde such as polyacetal and acetal copolymer should never be mixed with PVC otherwise an explosion will occur.

INJECTION MOULDING OF FIBRE REINFORCED THERMOPLASTICS

The shear and tensile viscosities of fibre reinforced thermoplastics are higher than for the matrix material. Higher cylinder pressures are, therefore, needed to mould these melts. This is rarely a problem.

The abrasiveness of the fibres and fillers causes wear of the screws and barrels of injection moulding machines, which must therefore be specially hardened. The high shear rates generated in the gate regions if they are too narrow, may cause some breakage of the fibres so that some of their benefits may be reduced.

The presence of the fibres causes a considerable improvement in mechanical properties along the direction of orientation, at the expense of the properties at right angles. Usually turbulence disturbs the orientation of the fibres that arrive first at the mould cavity, but as the mould fills the last material retains the fibre alignment around the gate region, which has a weakness at right angles to the flow direction.

INJECTION MOULDING OF RUBBERS AND THERMOSETS

Rubber granules quite often cause feeding problems in injection moulding machines, which as a consequence are sometimes modified to deal with strip feedstock. Thermoset feedstock is in the form of granules, which consist of partly polymerised resin, fillers and additives.

In both cases the materials and the barrel temperatures are such that the melts will flow easily into the mould with minimal crosslinking and polymerisation. The curing of the rubber and the crosslinking of the thermoset takes place in a hot mould and must not occur in the barrel.

For thermosets a special barrel and screw design is used in which there is no check value (as also for PVC) and hence less likelihood of accumulating material that may crosslink and cause a blockage. The screw is practically the same depth all the way along. The pressure (220 MN m^{-2}) is generally higher than that used for thermoplastics because the barrel temperatures are kept lower to prevent crosslinking.

The moulds are generally made from harder steels than for unreinforced thermoplastics, because the thermosets and rubbers contain abrasive fillers. The other main difference is that the sprue and runners are kept cool to minimise crosslinking. They can then be retained in the mould during the ejection of the hot crosslinked moulding. The runner and sprue section are then injected into the mould cavity during the next moulding cycle. This reduces scrap, which in these materials cannot be recycled.

The advantages of injection-moulding rubbers and thermosets over the more traditional compression-moulding are:

a) fast cure rates,
b) efficient metering of material,
c) efficient pre-heating of the melt,
d) thinner flash, and
e) lower mould costs.

The disadvantage is that there is more waste in the runners and sprues.

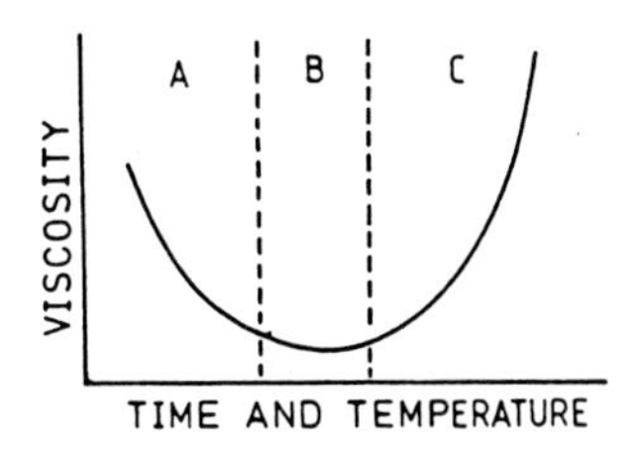

Fig. 9.11 Flow curve of a thermoset moulding compound during processing, showing *A* plasticisation period, *B* injection stage *C* curing stage.

The setting of the process variables in the moulding of thermosets must be far more precise than for thermoplastics. Fig. 9.10 shows the variation in the viscosity of a thermoset with time during the moulding cycle. The temperature is low during the plasticisation stage and the viscosity falls as the thermoset heats up. In the injection stage the viscosity starts to rise as crosslinking occurs. If crosslinking is slightly premature the mould may not fill adequately, but adjustments to injection speed and barrel temperatures may be made to give a satisfactory moulding. The crosslinking increases in the mould until the product is sufficiently form-stable to be ejected. In practice, the curing time is minimised to lessen the chance of blistering and distortion. In injection moulding, the material is more crosslinked during the injection stage than in compression moulding, giving a shorter curing time and higher output rate.

Typical applications for injection moulded rubber products include sealing rings, bushes, shoe soles, stoppers, floor mats, tyres and solid wheels. (8)

INJECTION MOULDED STRUCTURAL FOAM

Foamed products have a cellular core with a relatively dense skin, the thickness of which can be controlled.

The advantages of foamed structures are:

1) for a given weight, they are up to five times more rigid than a solid moulding;
2) they are free from orientation effects, with uniform shrinkage; and
3) very thick sections can be moulded with narrow parts without voiding and sinking.

The range of applications is large and the uses can be categorised as follows:

1) anti-sink applications — polystyrene hulls;

2) density reductions — weight reduction in large items, structural foams and upholstery;
3) thermal insulation.

There are two processes by which structural foams are injection moulded:

1) a) low pressure chemical process b) low pressure nitrogen process.
2) high pressure process.

All these processes can be carried out on slightly modified injection moulding machines.

In the chemical process a blowing agent is incorporated with the feedstock and plasticised with the raw material in the usual way. The nozzle shut-off valve is closed, and once the required shot size has been plasticised, the nozzle valve opens and the mixture is rapidly forced into the mould. The shot size is chosen to be less than the volume of the mould cavity so that under the reduced pressure in the cavity the foaming agent expands and pushes the melt against the mould walls.

The bubbles on the surface of the melt contact the walls and burst. This gives a solid skin around a cellular interior. The cold mould walls inhibit bubble formation, and in order to obtain a thinner skin more blowing agent has to be incorporated. The overall density may be as low as 75% that of the solid resin and the mould pressures are low, about 1.4–2.8 MN m^{-2}.

In the low pressure nitrogen process, gaseous nitrogen is mixed with the engineering plastics materials in the heating cylinder. The mixture is transferred to an accumulator — if the moulding is too large — and then injected into the mould as above.

Both of these processes are slow and a carousel type of arrangement for the moulds is used, as described in blow moulding. The nitrogen process is marginally cheaper than the chemical process.

In the high pressure process, the blowing agent is incorporated, as in the low pressure chemical process. Then the mixture is injected into the mould until it is full. No foaming takes place. The mould is then partially opened until the product dimensions are correct. The foaming agent then produces the cellular structure. The mould pressures are about 100–135 MN m^{-2}.

The advantages of the low pressure process are:

a) lower density mouldings,
b) less moulded in stress,
c) lower mould and injection pressures (1.4–2.8 MN m^{-2})
d) lower mould locking force,
e) standard mould design, and
f) lower machine and mould cost.

The advantages of the high pressure process are:

a) improved uniformity of cell structure, and
b) no swirl marks on the surface.

These foaming processes can be used with any engineering thermoplastic but nylons and polyacetals are not foamed as much as other thermoplastic materials. Fibres and fillers can be incorporated into the polymers and give products of very high stiffness to weight ratios.

Foams can also be made from thermosets, although this is usually carried out by reaction injection moulding (RIM). These processes give a spectrum of behaviour from flexible to rigid foams. The advantage of the RIM method over the conventional injection moulding is that it is cheaper. RIM is described later.

SANDWICH FOAM MOULDING

In sandwich foam moulding (sometimes called co-injection moulding) separate skin and foam melts are simultaneously injected into a mould. The nozzle allows the skin material to be injected from an annular ring while the foam material is injected through the central portion of the nozzle. This kind of technique has been much used in coextrusion.

The cycle of events proceeds as follows:

1) The skin material is injected. The tensile viscosity of the melt keeps the sheath of material intact.
2) Shortly afterwards the core melt is injected and again the tensile viscosity ensures that the two streams do not intermix. Both melts are injected together.
3) After partially filling the mould the core melt is shut off.
4) The skin material is shut off shortly afterwards.
5) The core melt foams giving a two-layered product.

This method permits a product to be made from two different polymers simultaneously and the sequence of starting and ending with skin material flow ensures that the product is completely covered with skin material and that the nozzle has been purged of core melt in preparation for the next cycle.

MATERIALS AND APPLICATIONS

Injection moulding is an extensively used process and applications are too numerous to list. Examples include styrene–acrylonitrile (record player covers), polymethyl methacrylate (motor vehicle rear lenses), acrylonitrile–

butadience-styrene ABS (cover for the 'Flymo'), 'Noryl' (car fascia panels), cellulose acetate (translucent screwdriver handles), glass reinforced nylon (Black and Decker drill cover) and polyacetal (kettles).

Blends of polycarbonate and ABS — 'Bayblend' — are used for car bumpers in Europe but in the United States RIM polyurethane foams are preferred because of the large absorption of energy from the microcellular structure.

Polyethersulphone (PES), because of its high temperature resistance, can be used in circuit boards. It can withstand contact with hot solder without distorting, whereas polysulphone and polyetherimide just fail this test. The use of PES has led to the invention of three-dimensional circuit boards made possible by injection moulding.

Polycarbonate is often moulded, but if metal inserts are to be used this material must be reinforced otherwise cracks, which start from the metal/polymer interface, occur in service.

COMPRESSION MOULDING

In this process feedstock, such as fine powder, a sheet moulding compound, tablets, a roughly-shaped moulding or a melt from an extruder, is fed into an open mould, which is used to form the final shape. This technique can be used for thermoplastics and is usually used for thermosets and rubbers.

The advantages of this process over other moulding processes are:

1) Cheapness of the mould and the tooling.
2) Flow rates are small and mouldings are free from orientation. This is particularly useful for the moulding of optical components.

The disadvantages include:

1) The production rate is low even with multi-cavity moulds.
2) It is more labour intensive.
3) A lack of homogeneity and temperature uniformity may give rise to warping.

Fig. 9.12 shows the basic compression moulding process. The correct amount of feedstock is fed into the mould, which is then closed and the moulding is achieved by pressure and temperature (~ 150–200°C) for thermosets and rubbers.

The temperature softens the thermosetting material or rubber, which melts and flows under pressure to take up the shape of the mould. The heat crosslinks the thermoset or rubber, which becomes form-stable and can be released from the mould. The temperature, pressure and cure time will depend on the material and on the size and the shape of the moulding. For phenol formaldehyde, temperatures and pressures used are 150–200°C and 14–23 MN m^{-2} respectively, with cure times from a few seconds to several minutes.

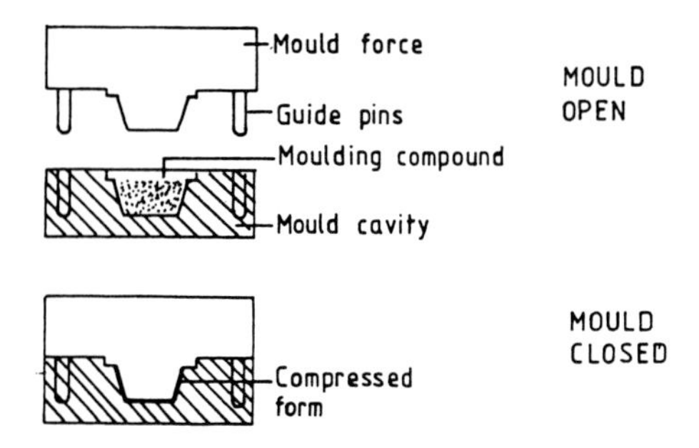

Fig. 9.12 The compression moulding process.

Poor heat transfer can lead to differing thermal histories in different parts of the melt. This can cause warpage in the product. Moreover, thick-walled items may not receive the correct amount of heat at the centre and consequently may not cure properly. One solution to this is to heat the material in an extruder and fill the mould from that directly.

Compression moulding can be used with thermoplastics but the need for cooling in the mould lengthens the cycle time and complicates the operation. This method, however, is used for making optical parts from pre-forms of roughly the correct shape (contact lenses from polymethylmethacrylate).

The mould itself has the following facilities: mould location, ejector pins and vents. One trick used to ensure the moulding adheres to the top half of the mould is to use differential heating. This can also be used in deterring moulded-in-stress and warpage. One problem is that the mould components wear readily because they bear the moulding pressures.

Transfer moulding (Fig. 9.13) is an improvement on the basic compression moulding process. The material is heated in a separate chamber and then injected into the mould. The advantages are:

1) Better homogenisation and temperature distribution caused by flow through the nozzle into the mould.
2) Larger items can be made.
3) The mould components do not transmit the pressure directly on the charge of melt and less wear results.

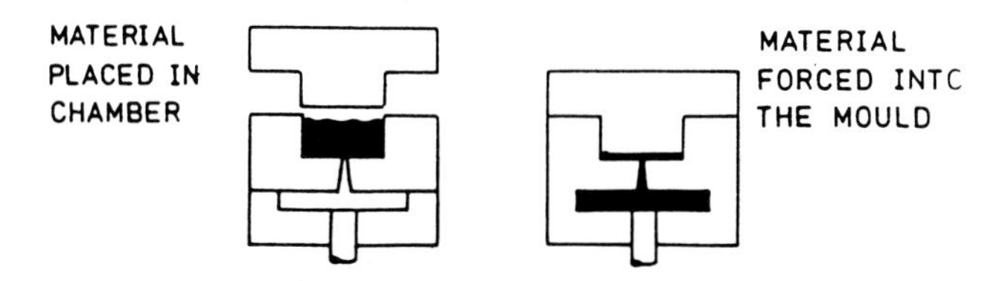

Fig. 9.13 Transfer moulding process.

4) Intricate shapes can be moulded and inserts can be used.
5) The surface finish tends to be better.
6) The cycle times are shorter.

The process variables involved in compression and transfer moulding are pressure and temperature. The latter is most important because it controls the curing rate. If the material cures prematurely the narrower parts of a mould may not fill. Material suppliers try to optimise the cure characteristics to give minimum cycle times.

The main precautions are that mould shrinkage is large with thermosets and the designer must allow for this. In addition, in compression moulding, care must be taken to ensure that the material does not cure before transfer.

Both the above moulding techniques are used for a whole host of products, although if a thermoplastic can be used instead it is preferred because injection moulding produces articles at a faster rate and scrap can be recycled. This is not possible with thermosets and rubbers.

INJECTION OR STRETCH BLOW MOULDING

This process differs from extrusion blow moulding in that the injection moulding process is used to mould a pre-form in the shape of a smaller bottle than the final article. This pre-form is then transferred to another mould where it is stretched by a mandrel and blown in a radial direction to produce the bottle. This process is used widely for making carbonated drinks bottles from PET. Its main advantages over extrusion blow moulding are:

1) the biaxial orientation that results from the process gives greater strength and hence thinner wall sections may be used;
2) there is no waste;
3) the wall thicknesses are more uniform so that overdesign is unnecessary, and
4) seamless neck and base regions and high surface quality are possible.

There are several ways in which injection or stretch blow moulding can be achieved. Injection moulded pre-forms are made as many as 12 to 16 at a time, using a hot runner system. The pre-forms are rapidly cooled to prevent any crystallization of the material, which would give the bottle an unwanted cloudy appearance. The above process is carried out in the plastic state and the orientation is not large.

The pre-forms are then placed on a rotating heating wheel which passes them through static infra-red heaters. These soften the material, which remains in an elastic rather than a plastic state.

The pre-forms are transferred from the heating wheel to a rotating blowing wheel. Each blowing station receives a transfer mandrel and a heated pre-form. The mould closes around the pre-form, which is stretched by the mandrel to the required length and blown at a low pressure to conform to the sides of the mould cavity. The blowing pressure is then increased to push the material against the mould walls, which are water cooled. The low blowing pressure is applied for a longer time than that of the high pressure blow. A similar time is allowed to exhaust the air.

The process variables associated with the post injection phase are:

a) the mould temperature (2–5°C);
b) the temperature for stretching (90–95°C);
c) the blowing pressures and times (low pressure 1.4–1.6 MN m^{-2} for 0.7 s and high pressure 2.7 MN m^{-2} for 0.45 s);
d) the exhaust time (0.3 s).

The figures shown in brackets are for polyethyleneterephthalate.

Biaxially Orientated PET Bottles

As mentioned earlier, the process temperatures and cooling rates are important in the control of the crystallinity. Fig. 9.14 shows the variation of the rate of crystallisation as a function of temperature for PET. The enclosed area of the graph represents the range in which crystallites will form, with the maximum rate of crystallisation being between the glass transition temperature T_g and the melt temperature T_m. In order to produce clear bottles the crystallisation of PET must be prevented, and the melt is supercooled from its injection temperature of about 280°C to room temperature. The stretching temperature is just above the T_g of PET (T_g = 85°C). This stretching occurs when the material is in a thermoelastic state and the process gives rise to an orientation of the molecules in the longitudinal and in the circumferential

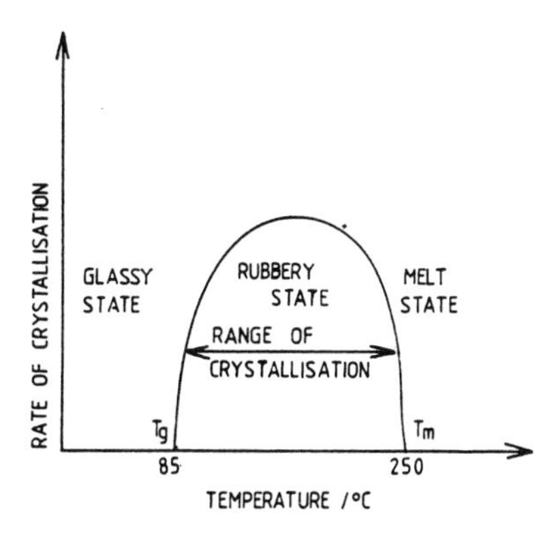

Fig. 9.14 Rate of crystallisation versus temperature for PET.

directions, giving biaxial orientation. Any crystallinity would hinder this process and the bottle would have inferior qualities.

The advantages of biaxially oriented products over similar but unoriented ones are:

1) the tensile and impact strengths are increased;
2) gas permeation through the container wall is reduced because a tighter network results from the molecular biaxial orientation;
3) creep is reduced and
4) clarity is improved.

The mechanical properties of PET bottles are very good. A filled 2 litre PET bottle at a working pressure of 0.4 MN m^{-2} (a typical pressure encountered in containing carbonated drinks) will survive a drop of 3 m on to a concrete floor. The bottles will withstand an internal pressure of 1.2 MN m^{-2}.

The barrier properties of PET are excellent and for this reason it is chosen in preference to PP, PVC polyacrylonitrite and styrene–acrylonitrite (SAN) for the containment of carbonated drinks. PVC is used for soft drinks. A coating of polyvinylidenechloride (PVdC) on PET bottles improves their barrier qualities and doubles shelf life.

PET is regarded as a safe material for use in contact with foodstuffs. Its main problem is the production of acetaldehyde during processing, which can taint foodstuffs. Acetaldehyde production is minimised by limiting the residence time of PET melts to less than four minutes. This gives acceptable products.

Reaction Injection Moulding (RIM)

In the RIM process two reactive, low viscosity fluids are mixed rapidly together and injected into a mould. The reaction is completed there and gives rise to a microcellular structure. Moulding times vary from 1–10 minutes depending on the reaction rate, the thickness of the part and the processing equipment.

This process was originally developed for the manufacture of fascias and bumpers in motor vehicles but has since diversified its application. Three types of parts are manufactured:

a) Rigid, structural polyurethane (PU) foams, which have a dense integral skin surrounding a lower density core.
b) Solid PU parts with thin wall sections, which are used in sports footwear.
c) High performance microcellular parts similar to a) but of smaller size.

The advantages of the process over injection moulding are:

1) Low energy consumption.
2) Larger parts can be made on smaller machinery.
3) Complex shapes can be made.
4) Swirl marks are absent because of the initial low viscosity of the material.
5) The parts are very light in weight.
6) The capital investment is lower due to lower injection pressures.
7) The chemistry is more variable giving a wider range of mechanical properties.

The disadvantages are:

1) The large thermal coefficient of expansion makes it difficult to join RIM materials to metals.
2) The cycle times are longer.
3) Difficulties are encountered in releasing the moulding, which prolongs the cycle time.

The polyurethane is made by reacting a polyol pre-mix which contains OH groups and an isocyanate with some additives in a mould. This gives rise to the evolution of gas, which provides the cellular or microcellular structure and the urethane links to form the polymer. A polymerisation and a crosslinking reaction in the mould make the product form stable.

A whole spectrum of grades, from rigid plastic to elastomeric, can be made depending on the concentrations of the polyol and the isocyanate. The polyurethanes are segmented block copolymers in which soft blocks of the long chain polyether are separated into individual domains from hard blocks arising from the reaction product of the di-isocyanate with the low molecular weight diol.

The soft segments have a glass transition temperature well below room temperature, and the hard segments are usually glassy or highly crystalline.

Catalysts control the rate and direction of the polymerisation and the chain extenders increase the chain size of the isocyanate-ended prepolymers. This gives rise to polymerisation and crosslinking.

The formulations usually contain a polyol (polyester or polyether), isocyanate, catalyst, chain extender, a surfactant to improve cell structure, a blowing agent, colorant and an evaluator.

The mechanical properties are dependent on the following:

a) the microphase separation,
b) the crosslinking and branching, and
c) the length of the hard and soft blocks.

For automotive fascias, the important property is a relatively unchanging modulus over the temperature range –30–120°C, the latter temperature being the paint curing temperature.

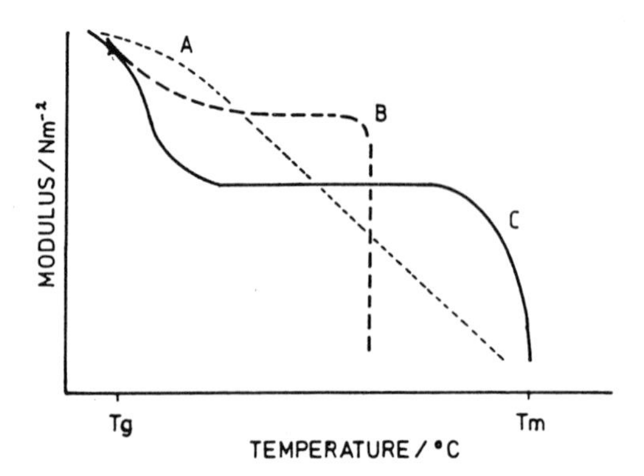

Fig. 9.15 Theoretical modulus versus temperature curve for segmented copolymers. *A* typical curve for polyurethanes, *B* curve for compatible hard and soft segments *C* curve for incompatible hard and soft segments.

Fig. 9.15 shows the theoretical variation of elastic modulus with temperature for segmented block copolymers. The modulus behaviour depends on the compatibility between the hard and soft blocks. The greater the insolubility of the hard blocks in the soft, the flatter the modulus. The glass transition temperature is determined by the nature of the soft or flexible segments in the backbone chain (polyether or polyesters), and the melting temperature is determined by the melting of softening point of the hard segments. The modulus plateau can be raised by increasing the concentration of the hard blocks.

Fig. 9.16 illustrates the RIM process. The polyol is a primary-capped polypropylene ether glycol to give a high reactivity in the RIM process. The other major ingredient is the isocyanate. This constituent is held above room temperature to keep it molten and both constituents are kept free of water vapour.

The high pressure metering unit must take the highly reactive polyol premix and the isocyanate and within a few seconds inject them into the mould prior to the onset of the reaction. The synchronisation of the delivery of the two streams will affect the curing of the material in the mould, and hence influence the final mechanical properties of the article. The throughput must be high to give turbulent mixing of the ingredients.

The main part of the metering system is the pump. The two types used are the axial pump or the radial piston pump. They provide precise metering, infinitely variable output, and a reduced tendency toward 'lead-lag' problems, which cause incorrect synchronisation of the metered streams.

The two main disadvantages of the above two types of pump are:

1) The upper viscosity limit for each of the streams at the operating temperature is 5 N s m^{-2}.

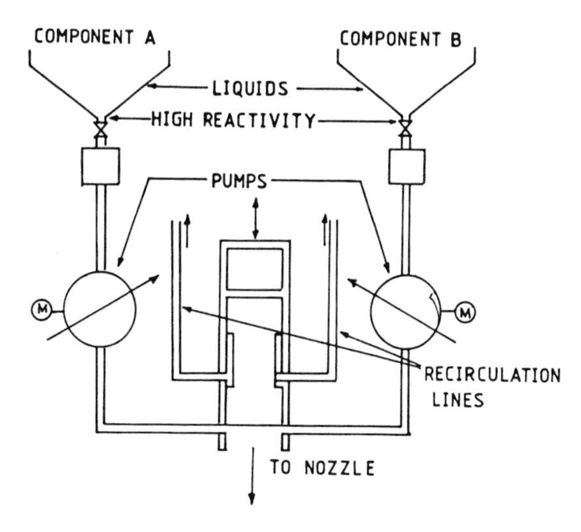

Fig. 9.16 Reaction injection moulding (RIM) process.

2) The close pump tolerances will not permit the metering of fibre reinforced or filled materials. These materials are abrasive and cause a severe reduction in the pump life and are difficult to control in streams.

Sometimes dosing displacement pumps are used. These are satisfactory for use with fillers provided that a recirculating scheme for controlling temperature and flow is employed.

The metering system takes the materials from the storage tanks to the mixhead. The metering units have two modes of operation; the recirculation and delivery modes. The former mode is of long duration at low pressure to give each stream a uniform temperature and composition. The delivery mode is of relatively short duration at high pressure to deliver a precise shot of material to the mixhead under controlled high pressure conditions to give the required degree of momentum for mixing. This produces a shear rate in the injector nozzle of around 10^5 s^{-1}. The metering equipment must switch between modes instantaneously and in a synchronised manner. The maintenance of the stoichiometric ratio or isocyanate index in the RIM formulation is imperative. The pressures of the two constituents must be balanced to keep both the flow rate and isocyanate index constant during injection into the mixhead and during recirculation.

Most commercial equipment measures ratios by measuring the delivery into the return lines. Microprocessors can be used to monitor and regulate the metering functions and subsequent mixing and pressing operations.

Shot weights of between 0.5–25 kg can be achieved with a maximum delivery time of 5 s.

The high pressure, impingement, self-clearing mixhead is important to RIM. At high flow rates and short residence times it must:

a) Develop high velocity during the delivery of each stream to the mix chamber.
b) Carry out the delivery under synchronised conditions.
c) Develop turbulence in the mix chamber to give good mixing.
d) Clean the mix chamber to prevent the build up of polymer from previous cycles.

An aftermixer placed between the mixhead and the mould (in the transition zone) is sometimes used. The transition zone includes the sprue, runner and gate system and the aftermixer. One function of the transition zone is to convert the turbulent flow into laminar flow before entering the mould. This prevents entrapped air, and the mould filling is most efficient if the rapidly gelling mixture enters as a stable, solid wavefront.

The mixed ingredients flow out and take up the shape of the mould. The melt gels within 5–10 s of the start of the injection and it polymerises and crosslinks *in situ*, eventually attaining its final mechanical properties. The cycle times are usually in the range 2–4 min, depending on the formulation and the product size. Some rigid foam formulations may be cycled in about one minute.

The success of the moulding step depends on:

a) The distribution and polymerisation of the material in the mould.
b) The mould cavity itself.
c) The mould carrier.

The RIM formulations currently used cure in the mould in 30–90 s, the balance of the time (90–150 s) being devoted to open mould operations. It is not difficult to obtain a cure in 30 s, which would reduce cycle times to 1–2 min if the open mould operations could be carried out more quickly.

The heat of the exothermic reactions of the process are retained in the moulding because of its microcellular structure. The retained heat effectively cures the foam core. The urethane material at the skin, however, can lose or gain heat from the mould surface. For this reason moulds are usually heated so that the skin material will cure evenly. Under these conditions the part can be de-moulded in the shortest possible time.

The mould surface must be very smooth because the low viscosity reactants readily reproduce surface flaws. This can be used to advantage if for aesthetic reasons a grained or textured surface is required.

All moving parts and mould surface must fit well to reduce flash, but if the fit is too tight the 5 s required for injection may be too short to allow all the air to escape. Mould release agent must be applied to the mould cavity and regular cleaning is necessary to prevent the build up of flash material. These open mould operations prevent fast cycle times.

The moulding pressures in the RIM process are about 1/80 that for injection moulding ($3.5–7 \times 10^5$ N m^{-2}). Therefore more materials are suitable for RIM moulds than for injection moulds. Machined steel, machined aluminium or formed nickel liners in aluminium or epoxy bases may be used, giving both lighter and cheaper moulds.

There are many options available for mould carriers in the RIM process. These include:

a) Stationary presses or clamps, where the mould and the delivery system are fixed.
b) Conveyor belt mould operations, where the delivery system is fixed and the mould moves.
c) Carousels, where the delivery system is fixed and the mould is attached to a revolving carrier (as in blow moulding).
d) Turreted systems, in which the delivery system moves from mould to mould.

The fixed mould approach is used in the manufacture of larger parts, such as motor vehicle fascias. Smaller products are made from moving mould or moving carrier systems.

In less than 150 s after the injection of the reactants the cured RIM urethane part can be ejected from the mould. It must have sufficient integrity to withstand removal from the mould by the automatic slides, flexible parts can be easily removed by compression.

As in injection moulding, there will be some degree of mould shrinkage. This is a complex function of:

a) chemical system,
b) density of the product,
c) geometry of the product, and
d) the entire process history.

A part held in the mould for a long time will have a greater crosslink density, which will have been achieved under constraint in the mould. It will, therefore, have a greater memory effect. Shrinkages are in the range 0.5–1% for rigid foams and 1–1.5% for flexible foams.

After demoulding, the part possesses a certain strength referred to as the green strength. The full mechanical properties are developed after post-curing. The way in which curing is carried out affects the final properties, and the post-curing temperature and the duration of cure are determined to give the optimum properties in the final product. For automotive fascias the product is placed in an oven at 120°C, since this is the temperature to which the product is subjected during the paint curing process.

Post-curing gives the following advantages:

a) It flattens the tensile modulus plateau.
b) It increases the tensile strength.
c) It increases the flexural modulus.
d) It increases the heat deflection temperature.

Recent research has resulted in RIM formulations that do not need post-curing. The post-moulding operations may be summarised as follows:

1) demoulding,
2) trimming operation in which flash is removed,
3) cleaning, in which the mould release agent is cleaned from the moulding,
4) post-curing, and
5) priming and painting, as unpainted urethanes yellow with age.

The process variables and their effect on the properties of the mouldings will now be described. The important variables are:

a) the formulation of the chemical systems,
b) the mixing temperature,
c) the mixing efficiency,
d) the mould temperature, and
e) the post-curing temperature and duration.

The molecular weights of the polyol affects the flexural modulus. Increases in molecular weight average $\overline{M}_w$ causes a greater incompatibility between the hard and soft blocks. Experiments show that the flexural modulus reaches a maximum, after which increases in $\overline{M}_w$ cause the modulus to decrease.

The chain extender has a large influence on the soft block/hard block compatibility. Flexural modulus is increased by the hard block concentration, but the slope of the modulus-temperature curve is not affected. Different chain extenders give rise to differing degrees of compatibility between hard and soft blocks. In the case of ethylene glycol, as extender, the compatibility is reduced because of the increased polarity of the polymer when extended by this chemical. A mathematical model developed by Scientific Time-Sharing Corporation has been used to predict the change in elastomeric urethane properties as a function of extender concentration.

There are several isocyanates available for RIM formulations. Some isocyanates are unsuitable because of their low reactivity. The ratio of isocyanate to hydroxyl is chosen to give the best balance of elastomeric properties, processing ease and economical production. An index of 100 represents the stoichiometric $NCO/(OH + NH_2 + H_2O)$ equivalents, and in general indices of greater than 100 are used to avoid under-indexing. Indices greater than 110 give rise to low green strength, less dimensional stability on ejection from the mould, post moulding expansion and poor mould release. In the region of 100–105, the tensile and tear strengths are improved as the index is

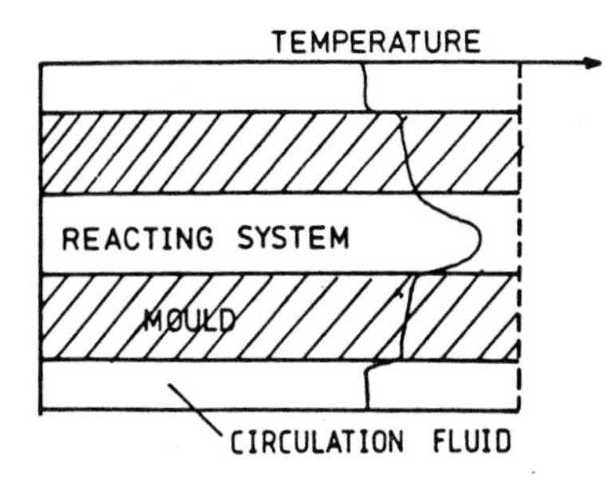

Fig. 9.17 Variation of temperature with depth in a moulding.

increased, but the deflection temperature is lowered. The most appropriate index has to be found for each RIM formulation and application.

The modulus/temperature behaviour can be improved by introducing branching into the isocyanate, but this is sometimes at the expense of lower green strength, and excessive branching reduces elongation.

The mixing and moulding temperatures affect the rate of reaction of the polymer, which will influence the sequencing of the chain's makeup. During moulding the mould temperature and the temperature of the reacting fluids will affect the temperature profile and the maximum temperature in the mould, because the polymerisation reaction is exothermic. A typical variation of temperature through the moulding is shown in Fig. 9.17. The microcellular structure gives rise to good thermal insulation and the temperature gradients can be large. This gives rise to greater $\overline{M}_w$ and affects the growth of the soft block and hard block domains and crystallinity. The mould temperature affects the surface finish and quality.

The density of the final product depends on the mould filling, and has an important influence on the tensile strength, tensile modulus, elongation and flexural modulus. All parameters increase linearly with density.

Post-curing increases the heat distortion temperature and causes a kind of annealing, in which domain structure is reorganised, and some crosslinking (post-curing or conditioning). Post-curing temperatures are usually 80–120°C for durations of 0.25–2.0 hr.

The skin quality is important and poor quality can occur for a combination of reasons, such as:

a) poor mould surface,
b) incorrectly applied paint, and/or
c) surface pitting caused by a poor choice of process conditions.

RIM was originally developed to make polyurethane automotive fascias and bumpers. This is particularly true in the United States, but in Europe the cheapness of polypropylene and composites based on it have meant that polyurethane has not been much used for this application. Polyurethane's resistance to sea water makes it ideal as a fender to cushion off-shore drilling platforms, for damping the impact of boat collisions or in more everyday applications such as cabinets for electrical goods and as a general replacement for metals.

The RIM technology has been extended to other materials in which two reacting fluids are brought together to form the polymer. These include epoxies, polyamides, unsaturated polyesters and polystyrene.

REINFORCED RIM (RRIM)

The three main drawbacks to RIM are:

1) the large coefficient of thermal expansion (15 times that of steel) makes it difficult to affix RIM parts to metal;
2) the low stiffness, and
3) sag at elevated temperatures.

For this reason milled glass fibre reinforcement is used to improve performance. However, the existing process had to be modified for the following reasons:

1) the glass fibre/liquid mixtures are much more viscous,
2) the fluids are more abrasive, and
3) the fibres are broken at the nozzle opening of the impingement mixhead.

The high pressure positive displacement pump cannot be used and is replaced by single-acting piston dosing units as add-on devices to existing RIM equipment. The fibres are added to one or both reacting fluids.

The microstructure of the reinforced material is different from the same RIM material because the milled glass fibres or chopped strands act as a heat sink and are enclosed in matrix material. As a consequence no fibre penetrates the cell walls. This gives a much finer cell structure and improved mechanical properties. This is true of high density foams in the region 900–1200 kg m^{-3}.

Short chopped fibres of 3, 6 or 12 mm length are used to reinforce low density foams. The glass fibres are coated with the matrix material and form composite struts between several cells. There may be local variation in cell size where bundles of fibres have formed. These bundles act as a heat sink and this causes a reduction in cell size.

The reinforcement is usually glass fibre but other materials can be used, such as mica flakes, carbon fibres, calcium metasilicate and polymeric fibres. Each have their relative merits and drawbacks.

DMC, SMC AND ZMC

DMC or dough moulding compound (called bulk moulding compound in the USA) is an unsaturated polyester resin mixed with Norwegian talc

(Dolomite) and glass fibre of lengths in the region 3–12 mm. This thermosetting blend has been developed for injection moulding.

The resin is blended with a catalyst to harden it for about 20 minutes in a dip mixer. The blend is then mixed with the filler and the glass fibre in a Z blade mixer with contra-rotating blades. This takes about an hour. Once mixed, this compound has a shelf life of up to seven days.

The injection moulder barrel is kept cooled by water to around 30°C. The screw has a large uniform flight depth and constant pitch and so gives no compression. This minimises fibre degradation while maintaining a large output.

The hydraulic pressure is relatively low and low mould-locking forces are necessary (1.5–2.0MN). The mould is heated to 150–200°C for rapid curing, with cycle times of the order of 30 s. The product is ejected from the mould while hot and post-moulding operations may be necessary.

This particular material and process is used in the motor vehicle industry and is used to manufacture headlamps, which must withstand working temperatures of 200°C and have complex shapes.

SMC or sheet moulding compound is usually an unsaturated resin, as above, with a filler, chopped glass strands and a metal oxide or hydroxide (often $Mg(OH)_2$). This gives a sheet of leathery consistency, which is usually supplied between protective sheets of polyethylene. The protective sheets are removed and the SMC cut to size to be compression moulded into the finished product. This process has been developed to make body panels for the automotive industry.

The two main problems associated with compression moulding are:

1) poor surface finish, and
2) low production rates.

For this reason the ZMC process has been developed using SMC. This process uses a modified injection moulding machine with a positive pressured feed that is in synchronisation with that of the screw. This gives no slip between the screw and the compound and thus minimises fibre degradation. The material is conveyed by the screw to a reservoir, after which it is injected through a wide nozzle into the mould. The mould is hot and induces cross-linking in the product.

This is a new and developing process, which has been successfully used to make the tailgate of the Citroen BX car.

ROTATIONAL MOULDING

In this process thermoplastic or thermosetting powders are put into a heated mould, which is rotated such that these powders fuse and the polymer adheres to the wall of the mould, forming a moulding. Very large mouldings of 10 m^3 capacity are possible.

The advantages of this process are:

a) Low pressures are involved with the attendant low mould costs.
b) Bosses, lugs and metal inserts can be moulded-in.
c) Open-ended parts can be made by splitting a closed symmetrical moulding.
d) There is no waste as all the material is used in the moulding.
e) Stress-free mouldings are made.
f) The wall thickness is more uniform than in extrusion blow moulding, and the wall is often thicker at corners, which is an advantage.
g) It is possible to obtain laminated mouldings by co-roto moulding.

The disadvantages are:

a) Cycle times are longer because both heating and cooling are necessary and can only be done outside the mould.
b) Raw material costs for powders are higher than for granules.
c) Only outside dimensions can be accurately reproduced.
d) The number of suitable polymers is small.
e) Holes cannot be moulded-in and have to be drilled afterwards.

The feedstock for this process is a powder. The average particle size of the powder is important, particularly when there are intricate surface features on the mould wall. Under these conditions the powder should be very fine. The finer powders are more expensive, but the particles fuse together more readily than the coarser powders and the cycle time will be slightly shorter. The coarser powders are quite adequate for simple mouldings. Particle shape also affects the ease of moulding. Spherical particles are more difficult to mould.

The powders are placed into the mould, which is then closed and heated while being rotated. Two of the more common methods of achieving this are shown in Figs. 9.18 and 9.19; the 'rock and roll' machine and the hot air machine.

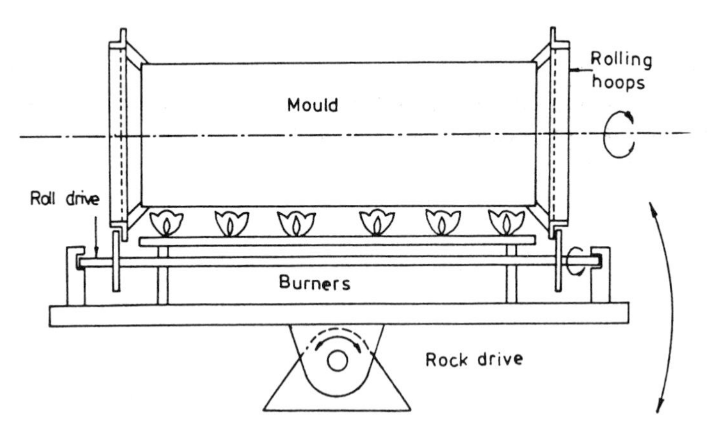

Fig. 9.18 'Rock and roll' rotational moulding machine.

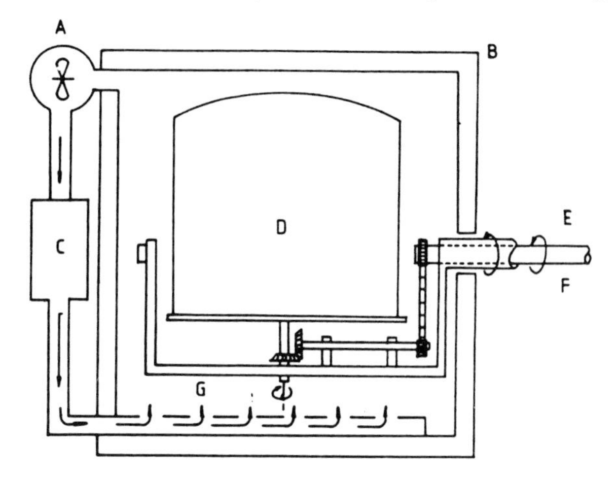

Fig. 9.19 The hot air rotational moulding machine *A* air circulating fan, *B* oven, *C* heat source, *D* mould, *E* inner first drive, *F* outer second drive, *G* hot air.

In the 'rock and roll' machine, which employs direct heating, the mould continuously rotates about a horizontal axis, while oscillating back and forth some 20–25° about a vertical axis. Heating is usually provided by gas jets playing directly on to the mould.

Fig. 9.19 shows the indirect heating method or hot air machine. The mould is welded on to a table that can be fully rotated about two perpendicular axes in a gas-fired oven. When the heating cycle is complete the mould is removed to a cooling bay, where it is cooled by air and water.

'Rock and roll' machines are considerably cheaper than the hot air machines and permit shorter heating times. However, more safety precautions are necessary and only one mould can be heated at a time.

The moulds do not suffer pressures and are usually made from sheet steel or cast aluminium. For large simple shapes, steel is the preferred material, but for complex shapes aluminium castings are used. The surface finish of the moulding will depend on the surface of the mould. For cosmetic reasons the number of welds should be minimised.

The kind of melt suitable for rotational moulding must be of low molecular weight average and of relatively high melt index (MFI of between 2 and 50 for low density polyethylene). This type of material gives a much better spreading quality, a better surface gloss, a smoother internal surface and requires less heating than melts of lower melt index. Unfortunately, this means that the mechanical properties of the finished article will be inferior to those of a similar article made from the same polymer of lower melt index by the blow moulding process. The high melt index gives rise to a reduction in toughness and a lower resistance to environmental stress cracking.

Chain branching, as in low density polyethylene, is desirable for a roto-moulding polymer because this increases the entanglement, the recoverable strain and the cohesiveness of the melt.

Crystallinity will give a higher tensile modulus and higher tensile strength but a lower elongation to break and reduced impact strength. Crystallinity can be controlled by the cooling rate. The density of the material will be related to the degree of crystallinity. For the higher degrees of crystallinity there is greater shrinkage and consequent distortion. The distortion can be reduced by lowering the mould temperature and by increasing the cooling time, at the expense of longer cycle times. In practice, polymers have a critical density above which the impact strength decreases rapidly.

The suitable materials for rotational moulding are polyethylene, cross-linked polyethylene, ethylene–chlorotrifluoro–ethylene, ethylene–vinyl acetate copolymer, polyvinylchloride, nylon 6, 11 and 12, polycarbonate, ionomers, plasticised PVC and cellulosics.

The process variables are:

a) the mould temperature,
b) the heating time,
c) the cooling rate,
d) the cooling time, and
e) the wall thickness.

The mould temperature and heating time are considered together as an enhanced mould temperature can be offset by a shorter heating time or vice versa.

Fig. 9.20 shows the variation of impact strength with heating time for a low density polyethylene. At a critical heat input the impact strength reaches

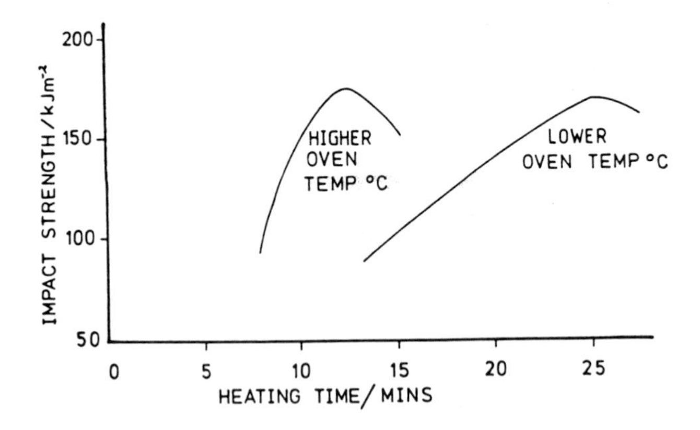

Fig. 9.20 The effect of heating time on impact strength for two oven temperatures.

a maximum value, and it seems reasonable to design for this. It should be noted that although the oven temperature is known, the temperature of the inside surface of the moulding and of the interface between the mould and the melt will be vastly lower.

The build up of impact strength corresponds to the completion of fusing of the particles, as seen from examination of the surface of the mouldings. This fusing will reduce irregularities that may act as stress concentrators. The rapid decline in impact strength of low-density polyethylene is due to oxidative degradation, causing crosslinking.

The problem for the moulder is to determine the best oven temperatures and heating time to give the best mechanical properties. One method is the hole count. The progress of a powder towards a complete fusion can be monitored in the completed moulding by counting the small bubbles present in the cross-section of the wall of the moulding. A small piece is cut from the wall (20 mm × 10 mm) and examined under a microscope. The sample is scanned on the 20 mm edge and the holes present in a number of representative areas are counted. The sample is scanned in three places, corresponding to the outside, centre and inside of the moulding. The hole number is expressed per cm^2. The diameters of the holes are in the range 0.1 to 0.3 mm.

As fusion proceeds the number of holes decreases to a minimum. Further heating will not remove the remaining holes, if any are left. The moulder should mould a number of products at different heating times for a constant oven temperature and after counting the holes choose the conditions that give the minimum hole count.

Minimum hole counts are higher in black mouldings, and in mouldings of wall thickness greater than 6 mm. The lower viscosity polymers give lower minimum hole counts. The minimum hole count also corresponds with the best surface finish. In low density polyethylene, if fusion is incomplete the inner surface of the moulding is lumpy, and if overheated the outer surface is yellowed and becomes slightly sticky.

Cooling rate and cooling times have a profound effect on crystallinity and warping. If the cooling rate is too rapid, the increase in impact strength due to reduced crystallinity is offset by the stresses moulded-in and the resultant warping.

In practice, the cooling time is reduced as far as possible without serious loss of mechanical properties. It may be about half the length of the heating time.

The wall thickness is determined by the amount of powder put into the mould. The majority of mouldings have wall thickness in the range 3–15 mm, with 25 mm as the upper limit of practicability.

By using foaming agents it is possible to obtain microcellular structures in roto-moulded parts. Density reductions from 920 kg m^{-3} to ~ 200 kg m^{-3} are readily achievable, giving light, rigid mouldings with excellent heat insulating qualities.

In these microcellular mouldings, a little more care is needed with the moulding cycle in that a five minute slow air cooling of the mould should be carried out between the normal heating and cooling cycles. This ensures that all the blowing agent has decomposed and a full expansion of the polymer has occurred.

Typical products made by rotational moulding include liquid fertiliser tanks, dustbins (which are made two at a time and cut in half) and containers for storage and shipping. Nylon 6 is used as a replacement for metals in air intake ducts for lorries and in the 155 gallon fuel tank for the M2 infantry fighting vehicle. The latter application has enabled a very complex shape to be used so that the tank can take up all available space for fuel. Moreover, the tank is weld free. Ethylene–chlorotrifluoro–ethylene is used in containers for etching acids, double-walled heating tanks and for the roto-lining of metal tanks.

CO-ROTATIONAL MOULDING

The rotational moulding process lends itself well to multi-layer mouldings. These are particularly useful when a strong material needs to be protected by a weaker material that is resistant to the fluid to be contained. Under these conditions a third layer may be sandwiched between the two outer layers to give a good binding to both. This is necessary if the two polymers are grossly incompatible. Combinations of a special formulation of nylon 6 with PE or ethylene–chlorotrifluoro–ethylene (E–CTFE) have been successful. The process by which these mouldings are made is called co-rotational moulding or co-rotomoulding.

This process can be carried out in two ways. One method involves filling the mould with the powder of the outer coat and processing it, and when the moulding has formed, the mould is stopped and the powder for the second layer is admitted into the mould by an access port. The rotation is continued until this layer has flowed out, and so on.

The main problems with this method are:

a) The adhesive compatibility between the resins may be poor.
b) The rate of shrinkage of the resins may be different.
c) The viscosity of the moulded resin may be insufficient such that sagging may occur when the mould is stopped to add the second polymer.

Sagging in nylon 6 resins of wall thickness in excess of 3 mm occurs when the rotation ceases and PE resin is added. Sufficient heat remains in the nylon 6 to melt the PE while the mould is being cooled. There is an upper limit to the ratio of the thickness of PE to nylon 6 for which co-rotomoulding is possible.

$$\frac{t_{PE}}{t_N} = 0.0065\ [(T - 204.4) \times 1.08 + 40]$$

where t_{PE}/t_N is the ratio of thickness of the PE coat to that of the nylon and T is the bulk temperature of nylon in °C on adding the PE.

The second method involves the use of moulds equipped with insulated internal hoppers. First the exterior material is placed in the mould and the inner material is placed in an internal hopper fitted with a pneumatically operated door. In this process the mould rotation is never stopped. Once the exterior material has flowed out, the interior material is released from the hopper. If other layers are required more internal hoppers are used.

The main problem of this method is that it is restricted to large mouldings, but it provides the shorter process times.

THERMOFORMING

This process turns feedstock in the form of sheet into hollow or shaped articles. The advantage of the method are:

a) Inexpensive equipment is required.
b) Moulds do not have to withstand high pressures and temperatures and so can be made very cheaply.
c) The method is suited to producing large thin-walled parts that could not be produced as cheaply by injection moulding.

The disadvantages are:

a) Thinning occurs, particularly at the corners, and it depends on the draw-down.
b) Thinning produces orientation and frozen-in-strain.
c) The feedstock is in sheet form and is, therefore, more expensive.
d) Post-shaping and trimming operations are needed. These produce scrap.
e) Holes cannot be formed during shaping, and lugs and bosses cannot usually be incorporated.
f) The process is labour intensive with long cycle times.

The basic equipment is simple and includes:

1) a mould or tool,
2) a means of heating the sheet, and
3) a means of making the sheet conform to the shape of the mould.

The basic process is shown in Fig. 9.21. A thermoplastic sheet in the thickness range 0.025 to 6.35 mm is clamped in a frame and the heaters are

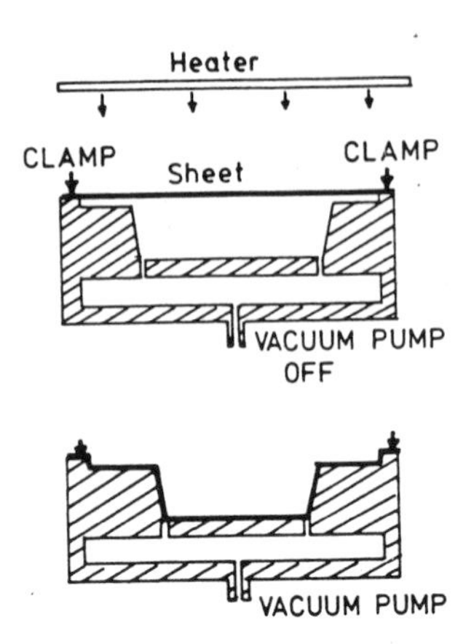

Fig. 9.21 Common form of thermoforming process using a female mould.

drawn across above (and for thick sheets below as well) the sheet. The sheet is heated until it is soft. It is important that the sheet should be heated uniformly to get the best results. The best method of heating is by controlling the heaters in zones, using energy regulators. By close control of the individual areas, heat losses at the edges can be reduced. The addition of reflectors at the sides will cut down peripheral heat losses, particularly where zone control is not available. The best reflectors are polished aluminium or plated steel. The positioning of them is by trial and error. It is also a good idea to cover the inner faces of the clamping frame with aluminium foil.

One method of testing the uniformity of the heating is to use a 'scorch test'. A piece of Bristol board is clamped in the place usually occupied by the plastic sheet. The heaters are put in position and kept on until the board scorches. This will reveal the local hot spots.

Thermoplastic sheets of up to 2.5 mm thick need only be heated from one side but the use of both sets of heaters reduces by half the time required to soften the sheet. The top heater should be between 10 and 30 kW m^{-2} power, with a value of half this for the lower heater. When the sheet is ready, the heaters are withdrawn and the mould table elevated until the mould touches the sheet. The process then divides into three main options:

1) A vacuum produced by a large rotary pump is used to draw the sheet over the mould (vacuum forming),
2) Pressure is supplied above the sheet to push it to the shape of the mould (pressure forming),
3) A matched male–female pair of moulds is used so that the sheet is pushed into the required shape (matched moulds forming).

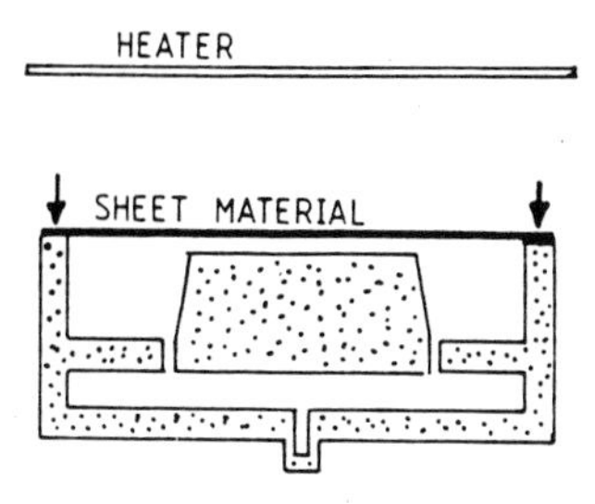

Fig. 9.22 Male mould for thermoforming.

The advantage of (2) and (1) is that heavier, tougher materials and more complex shapes can be moulded. Method (3) is used with laminated and foamed materials.

Whichever method is used, the moulding is allowed to cool in contact with the mould, and when cool the excess material is trimmed from the moulding. This process often generates a lot of scrap.

The best moulds are made of aluminium. Water cooling can be arranged with this material and the surface finish is good. Venting is necessary to permit the air to be drawn out from the gap between the sheet and the mould. These vents should be located at the bottom of the mould, and have diameters in the surface of the mould of no greater than half the sheet thickness. This will prevent the vents from marking the surface. Vent channels to these holes should be of 6 mm diameter and be 3 mm below the mould surface.

Practice moulds can be made of plaster and master moulds can be made from close-grained wood. Other suitable mould materials include laminates of phenolic resins, cast epoxy resins or magnesium oxychloride.

Fig. 9.22 shows a male mould. This type is preferred because a more uniform wall thickness is possible than with female moulds (Fig. 9.22). The male moulding will have a better outside surface finish, but is prone to webbing on sharp corners. When the authors were moulding a plastic rocket nose cone by this method, four webs appeared roughly at right angles to each other.

Female mouldings are less prone to webbing and a good surface finish is obtained on the inner surface of the moulding. Fig. 9.23 shows the variation in the wall thickness in female moulding. This can be alleviated by using a plug (plug-assisted moulding), as shown in Fig. 9.24.

Fig. 9.23 The variation of wall thickness during female moulding.

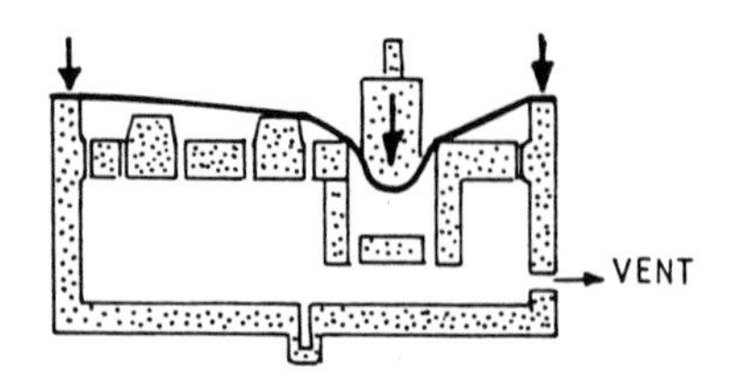

Fig. 9.24 Plug-assisted moulding.

In plug-assisted moulding, a male plug driven by a hydraulic ram pushes the sheet into the deeper parts of a female mould. The plug is made some 10 to 15% smaller than the cavity. This method is used when a fairly constant wall thickness is required in complex mouldings with deep draws.

There are a number of other variations such as box forming, drape forming and air slip or bubble forming.

Most thermoplastics can be thermoformed but the ideal candidates have the following qualities:

a) A low specific heat for rapid heating and cooling — hence preferably an amorphous polymer.
b) High thermal conductivity to enable the easy handling of sheets. Foamed materials present a difficulty here.
c) A formable temperature range below the softening point in which sagging does not occur.
d) A high molecular weight average and hence high tensile viscosity to reduce excessive thinning or tearing.

Comments on various thermoplastics for thermoforming are given in Table 9.3.

Table 9.3 Thermoplastics for thermoforming.

Thermoplastic	Comment on suitability
Polystyrene	Biaxial orientation of general purpose grades mechanical and optical properties. High impact grades are easily formed but have poor gloss and weathering characteristics. Expanded grades are suitable for matched moulding.
Acrylonitrile-butadience-styrene	Good properties and tougher than high impact polystyrene.
Polyvinylchloride	Easily formed.
Polymethylmethacrylate	Easily formed with excellent clarity and weathering performance.
Polyethylenes	Difficult to form because of high specific heat and crystallinity and low thermal conductivity.

The process variables involved here are the heating times, and the cooling times and rates, and the pressure available. For vacuum forming the maximum pressure is atmospheric.

The strength of a material decreases with increasing temperature, and, as sagging is undesirable, the heating time is chosen to be a minimum. If the sheet is not soft enough, however, there may be insufficient pressure to form the article. The heating time would then have to be increased.

A low sheet temperature will also give rise to increased orientation and internal strain, but the cycle time will be less.

The cooling rate can cause surface defects if it is too rapid. A heated mould is sometimes used. The heating and cooling rates will also affect the properties of the mouldings such as impact strength, tensile strength and tensile modulus. The parameters are affected in the same way as in other moulding techniques.

The main precautions in this process involve the operation of the moving parts, such that the mould table for instance does not rise while the heaters are in position, safety guards should be fitted at all times.

This process is ideal for the manufacture of large mouldings with fairly thin wall sections, such as polymethylmethacrylate baths, acrylonitrile–butadiene–styrene freezer compartments; and boat hulls are often made by this method.

CHOICE OF MOULDING PROCESS

It is found often that a given product can be moulded by several different routes. The cheapest route depends on the capital cost of the equipment and the number of mouldings required. Table 9.4 gives an approximate indication of the minimum numbers for economic production for each moulding process.

Table 9.4 Comparison of fabrication methods based on the minimum number of mouldings for economic output.

Production method	Number required
Thermoforming	10^2–10^3
Rotational moulding	10^2–10^3
Blow moulding	10^3–10^4
Injection moulding	10^4–10^5

REFERENCES

Capron for Rotational Moulding, Rotec-Chemicals Ltd., Bedford, UK.

A.A. COLLYER: *A Practical Guide to the Selection of High-temperature Engineering Thermoplastics.* Elsevier Advanced Technology, Oxford, 1990.

R.J. CRAWFORD: *Plastics Engineering*, Pergamon Press, Oxford, 1983.
R.J. CRAWFORD: *Plastics and Rubber*, Mechanical Engineering Publications, London, 1986.
J.B. DYM: *Injection Moulds and Moulding*, Van Nostrand Reinhold, New York, 1979.
A. WHELAN: *Injection Moulding Materials*, Elsevier Applied Science, London, 1982.
The Principles of Injection Moulding, Technical Service Note G103, ICI Welwyn Garden City, UK.
Rotational Moulding, Technical Services Note A122, ICI, UK, 1980.
The Principles of Vacuum Forming, Technical Service Note G109. ICI, UK.
C.W. MACOSCO: *RIM Fundamentals of Reaction Injection Moulding*, Hanser Publishers, Munich, 1989.
J. METHVEN and J.R. DAWSON in N.C. Hilyard (Ed.), *Mechanics of Cellular Plastics*, Ch. 10, Elsevier, *Applied Science*, London, 1982.
D.V. ROSATO and D.V. ROSATO: *Injection Moulding Handbook*, Van Nostrand Reinhold, New York, 1986.
J.L. THRONE: *Thermoforming*, Hanser Publishers, Munich, 1987.
B.C. WENDLE: *Engineering Guide to Structural Foams*, Technomic, 1976.
A. WHELAN: *Injection Moulding Materials*, Elsevier Applied Science, London, 1982.

10 *Joining with Polymers*

INTRODUCTION

There are several aspects to joining with polymers. Plastics may be moulded into complicated single components, integrating a number of separate sub-components which would probably be required if the equivalent component were to be made with a metal. As a result it could be thought that there is little need for joining plastics. However, this is not so. For instance it may be necessary to join two sub-components made from similar or dissimilar plastics. Sometimes joining of plastics is necessary for practical or economic reasons. Lengths of pipe must be joined, as pipes can only be transported to the installation site in finite lengths. In many instances only a few product items may be required, as in equipment for the chemical process industry. In these latter cases fabrication of the product from sheet or extruded semi-finished sections may be the only economic solution. All of these problems can be solved by welding or mechanical fastening of sub-components or preforms.

However, there is another effective joining technique and that is by the use of adhesives. These are all based on polymers of a wide range of types and joining with adhesives can be applied to nearly every material — plastic, rubber, metal, ceramic or composites. Dissimilar materials can also be joined effectively.

All of the above techniques of joining involving polymers will be considered here. Sections on the welding of thermoplastics, adhesive bonding and the mechanical fastening of plastics are included.

THE WELDING OF PLASTICS

The process of welding involves the melting or softening and flow of the appropriate boundaries of the pieces being joined. Clearly this process will only be applicable to thermoplastics: thermosets must normally be joined by other methods.

In the welding process heat must be generated and this can be done in two ways. One approach is to apply external heating to the workpiece and the other is to generate heat by mechanical displacement. The commonly used techniques employing these two approaches are described and some associated problems indicated.

Processes involving external heating

Heated Tool Welding

There are several variations of this technique depending on the geometries of the parts to be joined. The basic principal is straightforward. The surfaces of the two pieces to be joined are heated by a flat plate which is sandwiched between them. The plate or tool is nearly always electrically heated and is coated with PTFE to prevent sticking. The heating time and temperature is important and varies from plastic to plastic. Temperatures of between 180–230°C are used. After a predetermined time the plate is withdrawn and the surfaces are forced together under controlled pressure for an appropriate time to make the joint. The process is relatively slow, with times of from a few seconds for small components to 30 minutes for items such as large pipes being common.

Very strong joints can be made with typical strengths being at least 90% of the strength of the parent material. This makes the process attractive for

Fig. 10.1 Heated tool weld in a car radiator header tank.

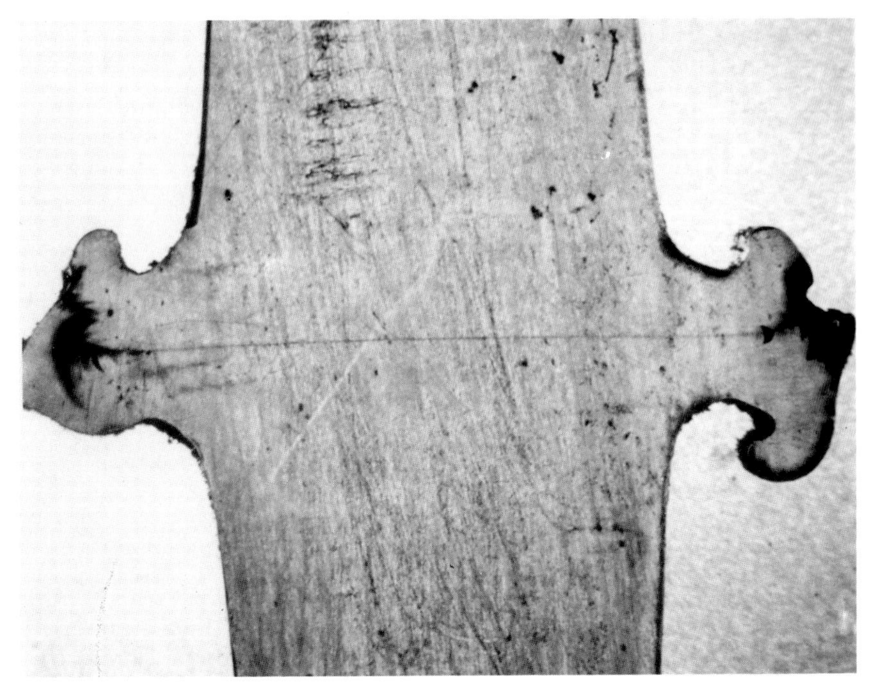

Fig. 10.2 Section through a heated tool weld in uPVC.

components containing pressurised or aggressive fluids such as header tanks for vehicle cooling systems (Fig. 10.1), batteries and large pipes. Gas, water and sewage distribution pipes up to 1600mm diameter are all welded in this way.

A section through a hot tool weld is shown in Fig. 10.2. The plastic (uPVC) has flowed considerably. However, the joint line can still be seen.

It is essential that the surfaces to be joined are clean, if maximum strength is to be achieved. This may not be easy to achieve in some circumstances such as the joining of previously buried pipework.

A variation of this method is the welding of films and tapes. In this case an electrically heated tool is applied to the outer surface of one or both of the workpieces while the two are held clamped together. The whole of the joint thickness is softened by a heat impulse and fuses under the applied pressure. Maximum film thicknesses using double sided heating are 1 mm for polyethylene but less for other plastics.

Hot Gas Welding

This technique is similar in some respects to the welding of metals with an oxyacetylene flame. The edges of the thermoplastic are preferably prepared by machining or scraping to a chamfered profile. A weld rod and the prepared

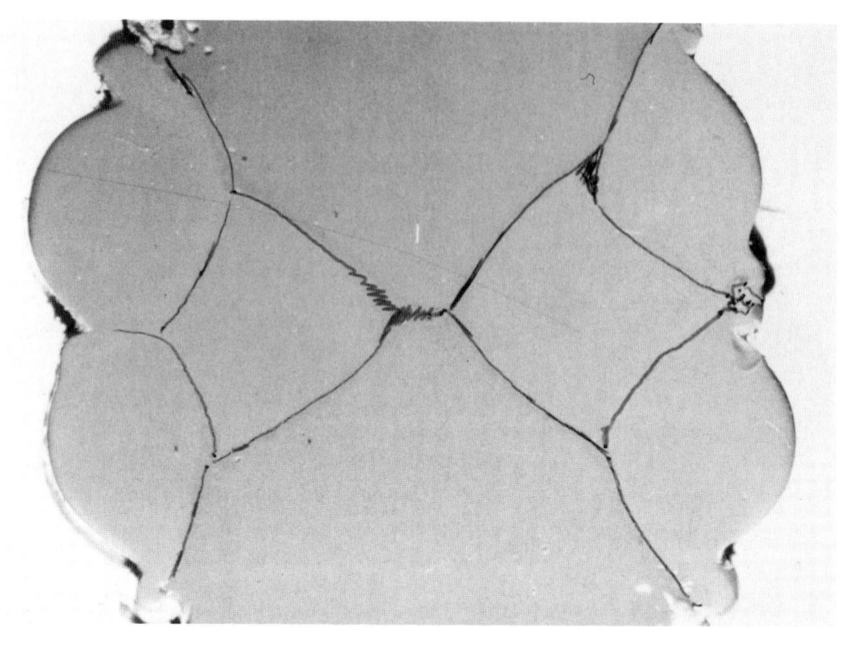

Fig. 10.3 Hot gas weld in uPVC.

edges are heated simultaneously by a hot gas stream from a welding gun and the rod is laid down in the joint. Depending on the plastic the gas temperature is between 200 and 300°C and the gas flow rate between 15 and 60 l min^{-1}. Various gases are used but usually compressed air is employed and gives satisfactory results in most cases. Nitrogen is used for plastics prone to oxidation. There are many variations of the technique involving different edge preparations and weld rod configurations. A section through a double vee weld containing three circular section rods in each vee is shown in Fig. 10.3. Sometimes triangular section rods are used. The thinking behind this is that better filling of the vee should result with less air entrapment. Generally, the thicker the sheets to be joined the more filler rods are necessary. Many plastics can be joined by this method including PVC, polyethylene, polypropylene, PMMA, polycarbonate and nylon. Applications include making chemical-resistant fabrications, pipework and ducting. Large, complex constructions can be produced. It is a slow process and the quality of the welds is mainly dependent on the skill of the welder.

It is difficult not to trap defects such as air pockets in the weld and this can have serious consequences for notch sensitive plastics such as uPVC. A fractured hot gas weld in transparent uPVC is shown in Fig. 10.4. In this case it was deduced by fracture mechanics that defects present were above the critical size for this material, resulting in brittle fracture at low stress.

Fig. 10.4 Fractured weld in transparent UPVC.

Table 10.1 compares weld strengths for a grey PVC with the transparent plastic. The hot gas welds were considerably stronger for the grey material because it was less notch sensitive, having a different formulation. Consequently the severest defects were below the critical size for brittle fracture. Higher strengths were thus obtained. Defects included those at the surface and those within the welds. The best strengths were obtained by a heated tool weld, where defects were virtually eliminated. It should be noted that PVC is a difficult material to process because of the difficulty of obtaining well fused material. This causes problems in hot gas welding where pressures employed are necessarily low. Other plastics can be welded quite satisfactorily.

Table 10.1 Mechanical properties of welds in uPVC.

Material	Weld Type	Tensile Strength of Weld (MPa)	Efficiency (%) (compared with unwelded material)
Grey uPVC	Hot gas	44.9	85
Transparent uPVC	Hot gas	18.5	28
Transparent uPVC	Heated tool	61	94

Processes involving internal heating

Friction Welding

Friction welding or spin welding involves the relative movement of two contacting thermoplastic surfaces such that sufficient heat is generated by friction to produce fusion. Where simple rotation is employed relative speeds of up to 20 ms^{-1} are employed with pressures of between 80 and 150 kPa. Weld times from 2–10s are used. High quality welds can be produced using simple equipment with little preparation of the workpieces. Friction welding is most suitable for joining circular components, although orbital motions can be used to join other geometries. The outer edges of components experience the most frictional heating and so residual stresses can be generated. Thin hollow sections or tubes can therefore be joined very satisfactorily. Welding can be carried out in liquids and so it is useful for the encapsulation of liquids in plastic containers with no entrapment of gases.

Vibration Welding

This is similar to friction welding except that the workpieces are moved relative to each other by linear oscillations thus producing frictional heat. When molten polymer has been produced the vibrations are stopped and the parts aligned and allowed to cool. Weld times of 1–5 s are attainable. Frequencies of 100–240 Hz with amplitudes of 2–5 mm are typical. Pressures are 80–150 kPa, as for friction welding. Large, complex joints can be made at high rates. For example joints of up to 1.5 m length can be made. Equipment is simple and several components can be produced at the same time. Nearly all plastics can be joined by this method and the technique is widely used in the automotive and domestic product manufacturing industries.

Ultrasonic welding

This method uses ultrasonic frequencies to generate mechanical vibrations at right angles to the area of contact between the workpieces while the parts to be joined are held together under pressure. Frequencies of 20–40 kHz are used. The high frequencies generate heat in the material and produce melting. The ultrasonic vibrations are produced by a lead zirconium titanate piezo-electric crystal and directed by a horn onto the workpieces. The parts to be joined must include an energy director which concentrates the energy in order to achieve rapid melting. This can take the form of a simple, small triangular ridge moulded onto the surface of one of the parts. Welding times are 0.5–1.5 s and during this time the softened plastic spreads through the joint and when solidified fixes the parts together. From a design point of view it is essential that the director contains enough material to spread over the joint surfaces when fluid. A recommended design is shown in Fig. 10.5. As the energy output of the machines is limited, only small welds can be made. Multihead machines, however, enable larger components to be joined.

The parts to be joined must be a loose fit so as not to inhibit vibration. In addition the plastics should have low damping capacity so as not to absorb

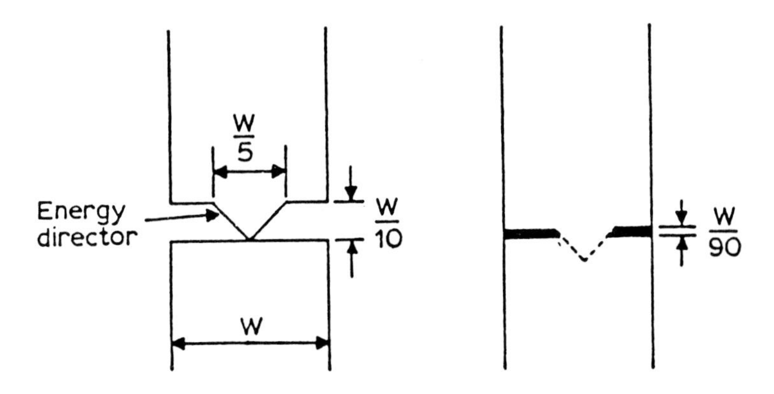

Fig. 10.5 Recommended joint design.

energy. This generally means that plastics with higher elastic moduli are more suitable. These include ABS, PS, SAN, polysuphone and polycarbonate. However, other plastics such as PE can be joined. When high damping capacity plastics are involved the joint must be close to the welding horn to minimise energy loss; this is termed 'near field welding' and means that the horn and weld should be less than ~ 6 mm from each other. 'Far field' welding is restricted to the rigid plastics. Dissimilar plastics can be joined, but the melt temperatures should be within ~ 15°C of each other. However, plastics which are obviously incompatible, such as PC and PS are not weldable.

Tooling is expensive, but once set up this method is ideally suited to mass production and finds wide use in many branches of industry, particularly in the assembly of domestic products such as tape cassettes and vacuum cleaner bodies. No heat is required and it is clean and fast giving joint strengths close to 100% of those of the parent materials. It can be adapted for the joining of metals to plastics by ultrasonic staking, using a metal stake ultrasonically embedded in the plastic. Ultrasonic spot welding and swaging can also be carried out. Metal inserts can also be embedded into plastic parts.

Implant welding

This method uses a resistive or inductive heating element placed between the joint surfaces. This arrangement usually consists of the heating element embedded in the same plastic as that to be joined. The element is then heated to form the joint. The element is usually metallic but can be in the form of a tape which contains iron oxide or similar ferromagnetic particles. Frequencies of 2–30 MHz are used for induction heating, or currents of up to 150 A for

Fig. 10.6 Electrofusion joint between MDPE pipes and electrofusion connector.

resistive heating. This is a straightforward process which can be applied to complex geometries. It has been used to join the two injection moulded halves of the 'Topper' sailing dinghy hull and for joining car bumper sub-components. Welding times of 20 s are attainable and most thermoplastics can be joined. However, the joint strength may be reduced by the presence of the heating element.

A recent application of this process is in the joining of smaller diameter PE gas pipes (up to 500mm diameter). This application uses a specially constructed joint, as shown in Fig. 10.6. The joint contains a coiled resistance heating element embedded near its inside surface. The joint is made of the same type of PE and the plastic near the element melts and fuses with the pipe material. This process has been named 'electrofusion' and is convenient and easier to operate than hot tool welding. However, cleanliness and adequate fusion must be ensured for satisfactory results.

Dielectric welding

This method uses high or radio frequencies to generate heat inside the polymers as a result of dielectric relaxations. It is used for joining thin films which are placed between two plates. A high frequency alternating current is applied to the plates. As the polarity reverses, the molecules try to realign themselves in sympathy. Polar polymers with high dipole moments are most suitable. These are asymmetrical molecules and a good example is PVC which is said to show a high degree of dielectric loss. Dielectric welding is suitable for joining high loss materials which are mechanically 'lossy' and therefore difficult to weld ultrasonically. Another advantage of electric welding is the absence of external heat which could degrade the material. Hence uPVC films are frequently joined by this method. Continuous sealing is possible.

It should be noted that films are also welded by ultrasonic and heat sealing methods. Ultrasonic methods are used particularly where materials which are difficult to join are involved, or where orientations in the films are to be retained, as heating is minimal.

Design and other considerations

In the design and production of effective joints the following points should be considered:

i) the method must be suitable for the plastics involved,
ii) the method must be applicable to the geometries of the components to be joined,
iii) the joint design must avoid stress concentrations in the case of stressed components,
iv) care must be taken to avoid internal defects which may reduce the strengths of stressed components and
v) joints should be carefully prepared and cleaned to avoid air or impurity entrapment.

ADHESIVE BONDING

Adhesives are a very versatile means of joining a wide range of materials, including one type of plastic to another, or to other materials. The adhesives themselves are based on polymers and take several forms. There are many reasons why adhesives are finding rapidly increasing applications. Some are:

i) adhesive bonding may be the only effective means of bonding dissimilar materials,
ii) stress concentrations are avoidable as the materials are joined by adhesive over the whole joint area, unlike in a riveted or screwed joint.
iii) the adhesive bonding technique is inexpensive because the equipment required is fairly simple,
iv) there are no galvanic corrosion problems, as with brazing, soldering or welding when joining some metals,
v) an adhesive joint can be made impermeable to gas and fluid,
vi) the joint can be designed to accommodate differences in expansion coefficients and may act as a source of damping and
vii) adhesives can be used to join heat sensitive materials which are difficult to weld.

Adhesives are liquids which flow over the surface of a solid and make intimate contact with it. Because of the close contact, strong intermolecular forces are set up between the liquid and solid. Liquids have little strength and so must be converted into solids with adequate strengths. Polymers can be readily manipulated between the liquid and solid states and have good solid-state strengths. Consequently they make good adhesives. A phase change from liquid to solid in a polymer can be brought about by one of the following processes:

i) solidification of molten polymer,
ii) loss of solvent or suspension medium from a polymer solution or emulsion, by evaporation or absorption into a porous substrate and
iii) chemical reaction involving polymerisation and/or cross-linking.

Adhesion Science and its practical implications

The main sources of adhesion forces are the chemical bonding forces which may be primary or secondary bonds. The formation of mechanical linkages is of little importance in most cases. Thus the main considerations in selecting, designing and producing an adhesive joint are those steps which ensure that the chemical bonding forces are maximised.

As far as mechanical bonding is concerned, it is often thought that severe abrading of a surface produces a mechanical key for adhesives. However, the main use of abrading is to remove loose surface debris and to thoroughly clean the surfaces to be bonded, except where surface fibres are released and then subsequently embedded in the adhesive. Thorough cleaning enables an adhesive to come into intimate contact with the surfaces. This generally means that the atoms and molecules involved must come within 6×10^{-8}m of one another. The lower the viscosity of the adhesive the more easily this is brought about. Severe surface roughness can trap air bubbles and reduce the bond strength. For most adhesives, degreasing with a suitable solvent is advisable whether or not abrasion is carried out.

Secondary bonds and hydrogen bonds are mainly responsible for adhesion forces. Secondary bonds result from variations in the distribution of electrostatic charges within molecules. Thus molecules are attracted to each other. Hydrogen bonds are similar but particularly strong because of the peculiar properties of hydrogen leading to very strong dipole attractions between hydrogen atoms in one molecule and for example –OH groups in another. The bond energies and their effective distances are shown in Fig. 10.7.

Forces set up in adhesive bonds are not as strong as those predicted from this theory, due presumably to the difficulties of bringing all the atoms and molecules into sufficiently close contact for the full development of secondary

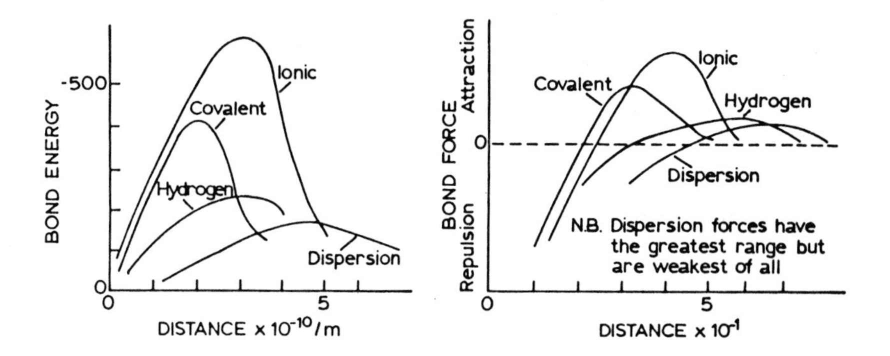

Fig. 10.7 Bond energies and their effective distances (and forces).

bond strength potential. Non-polar polymers such as polyethylene and PTFE cannot generate strong secondary bonds and have to be treated to produce polar sites on their surfaces in order to generate satisfactory adhesive bonds.

In many cases diffusion of the molecules of an adhesive into an adherand is necessary to form a strong bond. Auto-adhesion can even occur spontaneously in this way when one surface is brought into intimate contact with another. If diffusion is required, the polymers involved must be compatible, as indicated by the closeness of their respective solubility parameters and, in addition, the temperature must be above the glass transition temperatures of the polymers. The process of diffusion bonding can be brought about by dissolving a polymer in a solvent. When applying this solution to a material the solvent evaporates and during the drying process adhesive molecules diffuse into the adhered surfaces. This process occurs with some paints as well as certain adhesive systems.

Finally an adhesive must spread over a surface. Thus good wetting is required, i.e. a contact angle close to 0° is desirable (see Fig. 10.8).

Satisfactory wetting is essential to good adhesion. Excess energy is needed to bring the molecules within the bulk of a liquid to its surface. This energy is the surface energy, and is the energy required to create a unit area of fresh surface. The surface energy depends on the molecular nature of the two media in contact. Starting with an adhesive and a material each in contact with air, the adhesive will spread on the material's surface if the resultant surface energy is lower. In other words a surface of high energy is replaced by an interface with lower energy. Surface energy reduction in adhesive systems depends on strong chemical bond formation between the adhesive and adherands.

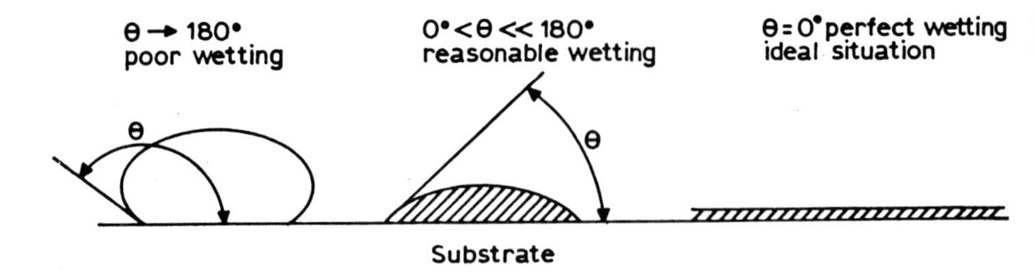

Fig. 10.8 Contact angle.

Categories of adhesives

Adhesives can be divided into five categories each of which enables useful bond strengths to be achieved in different ways.

Hot melt adhesives

These are thermoplastics such as PE, PET, nylons and PE–PVA copolymers which are melted and applied to the adherands using a specially designed gun. The adherands are squeezed together under moderate pressure during the cooling cycle. Good bond strengths are achieved. However, it should be remembered that the adhesive is thermoplastic and thus will weaken at elevated temperatures and may creep appreciably at lower temperatures.

The technique has been adapted in various ways. The hot melt adhesive may be in the form of films, powder, rods or blocks and may be dissolved in a solvent. In the latter case the adhesive is applied in solvent form and then the solvent is allowed to evaporate before heat is applied to form the bond.

This method does not require expensive equipment and can be applied to a wide range of materials to give reasonable bond strengths.

Solvent-based Adhesives

In solvent-based adhesives the viscosity of the polymer adhesive is reduced by dissolving in an appropriate solvent as indicated by consideration of the solubility parameters of the polymer and solvent. Linear or branched (thermoplastic) polymers are required to enable dissolution and it is an advantage if the adherand is dissolved to a limited extent. Some plastics such as PE, PP, nylon and PTFE are virtually insoluble. These are the partially crystalline plastics. Amorphous plastics are more likely to dissolve and these include materials such as PS, ABS, PVC and PC. Solvent adhesives are used widely with the latter materials. Solvent alone will form an adhesive joint as the adherand surfaces are dissolved and interdiffuse. However, polymer solutions have far better gap-filling properties and are far more effective. These adhesives are available in low, medium or high viscosities depending on application.

Many of these adhesives are based on natural or synthetic rubber and are used as 'contact' adhesives. These may be applied to two surfaces as films. When the solvent has evaporated the surfaces are brought together, when the rubbery flexible molecules immediately interdiffuse forming a bond of fairly low strength. Rubber based adhesives form low strength but tough joints unless vulcanised (crosslinked). This is because they operate at temperatures above their glass transition temperatures. Cross-linking is carried out in some cases giving high strength tough bonds. Commercial rubber-based solvent adhesives are based on natural, SBR, nitrile and chloroprene rubbers.

Emulsion adhesives

These are manufactured from polymers made by emulsion polymerisation and are in the form of stabilised suspensions, i.e. emulsions, of minute polymer particles dispersed in water. They have low viscosities, flow easily and are usually white in colour. After application to the adherands the water is either absorbed into the surface if it is porous, or evaporates. As the polymer used is necessarily linear or branched and above its glass transition temperature, the particles coalesce, inter-diffuse and make intimate contact with the adherand surfaces. These adhesives work best on absorbent surfaces and form bonds of low to medium strength. In some cases the strength can be improved by crosslinking the adhesive, using special catalysts added to the emulsion.

Materials such as wood, paper and card are bonded very effectively. Polymers such as PVA and SBR with at least 35% water are widely used. Nitrile and neoprene rubbers are used where enhanced polarities are required to produce adequate bond strengths.

Reaction cured adhesives

This type of adhesive is capable of developing very high bond strengths and a large number of reactive polymers are used. In many cases a highly crosslinked structure is generated in the adhesive and the strengths are correspondingly high. In addition, the adhesives are temperature and solvent resistant. Bond strengths are generated by a combination of polymerisation and crosslinking reactions starting with low viscosity precursors. In the case of an epoxy resin a liquid pre-polymer of low viscosity or in low melting point solid form is crosslinked with a liquid curing agent such as an amine. A high molar mass polymer is formed at a particular temperature and this enables pre-mixed systems to be prepared which are subsequently heat cured. This may be more convenient in some applications.

The main reaction cured adhesive types are:

i) anaerobics
ii) cyanacrylates
iii) epoxies
iv) phenolic

v) resorcinolic
vi) polyurethane

They are also available in toughened form. Toughening is achieved by dispersing rubber particles through the adhesive. These particles act as crack stoppers. Toughening reduces strength but increases toughness. Where un-toughened adhesives are used it is particularly important to avoid stress concentrations and tensile stresses in joints. Toughened acrylic and toughened epoxy adhesives have wide applications in the assembly of load bearing structures in the automotive and aircraft industries.

Anaerobic adhesives cure when air is excluded. They are particularly useful in thread-locking applications where the action is one of jamming rather than adhesion. They are available in various cured strengths so that coaxial or threaded joints of different sizes can be disassembled if required.

Cyanoacrylate adhesives cure when in contact with minute traces of moisture which are always present on joint surfaces. They are strong but brittle. They are available in low viscosity or thickened forms, although gap-filling properties are generally poor. They are widely used for assembly of plastic parts in the electronics industry. An interesting application is in the bonding of rubber, when they form very strong joints.

All of the reaction cured adhesives are highly polar and develop high bond strengths when applied to suitable surfaces. They are generally of low viscosity but can be readily manipulated by control of molar mass or by adding metal and other fillers, to be gap-filling and highly viscous in nature. Metal and carbon black fillers also increase electrical and thermal conductivity.

Pressure sensitive adhesives

These are viscous polymers at room temperature and so are prepared from polymers with low glass transition temperatures. Their most important property is 'tack' which is achieved with amorphous polymers having moderate molar masses and operating above their glass transition temperatures. These adhesives are made to flow into intimate contact with the adherands by application of pressure but the viscosity must be high enough to provide adequate strength. The commonest way of using this type of adhesive is to lay it down on a plastic tape with a removal backing strip. Most pressure sensitive adhesives are based on acrylic polymers, SBR and natural rubber. They are not suitable for structural applications.

Mechanical properties and joint design

Mechanical Properties

This section will deal briefly with structural adhesives which are formulated for strength and as a result are usually hard and brittle. Shear strengths are high but they fail easily in tension and under shock loads. They fail suddenly by rapid crack propagation when a joint experiences a critical tensile load. This problem can be overcome in two ways:

i) toughened adhesives help alleviate this problem, but load bearing capability decreases. The same general problems are still present but are less severe, and
ii) joints should be designed to minimise tensile and cleavage forces. Joint areas should also be as large as possible to reduce stress levels.

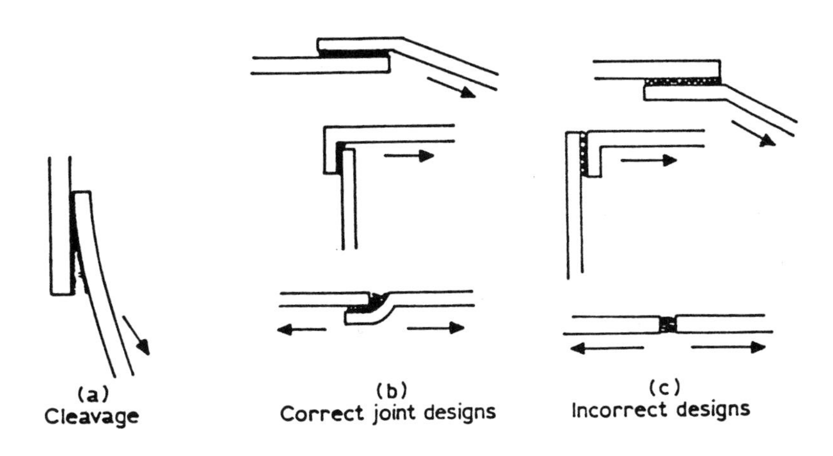

Fig. 10.9 a) Cleavage b) Correct Joint Designs c) Incorrect Designs.

Joint design

Fig. 10.9 shows some correct and incorrect joint designs.

It is interesting to consider the shear stress distribution in a single lap joint, as shown in Fig. 10.10. The stresses are concentrated near the ends of the

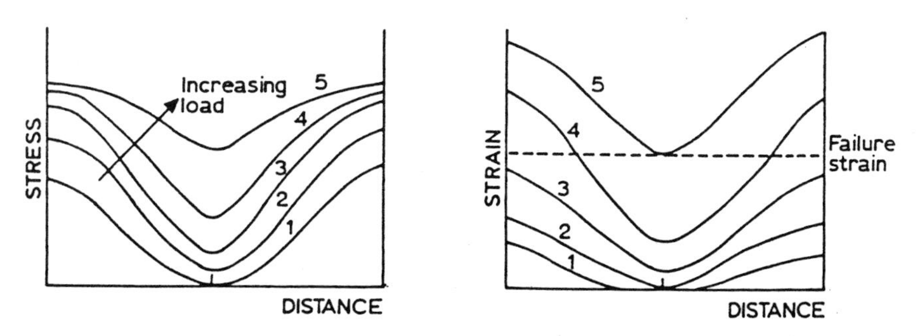

Fig. 10.10 (a) Stress distribution in a single lap joint (b) strain distribution.

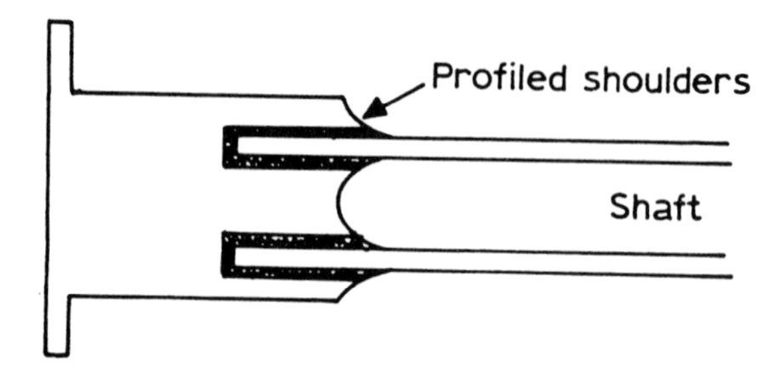

Fig. 10.11 Reduction of stress concentration in bonded component.

joint. The consequence of this is that steps should be taken to reduce stress concentrations at joint extremities. This may be done by judicious profiling of the ends of the adherands or by using toughened adhesives or both. Drive shafts have been constructed from carbon fibre epoxy resin filament wound tubes bonded into metal ends. Profiling of the metal ends is found to reduce stress concentrations and to increase fatigue life considerably, providing entirely satisfactory performance. This principal is show in Fig. 10.11.

Notes: i) for untoughened adhesives the joint will fail if the strain at the edges exceeds the fracture strain. ii) toughened adhesives will not fail as long as the joint exceeds a certain length (17 mm). Below that the stress throughout the joint will be high and creep or fatigue failure will occur.

Surface preparation

As mentioned earlier, the interface between adhesive and adherand is critical in order to maximise bond strengths. Weakness can be caused by poorly bonded oxide layers, entrapped air, contamination with oil or impurities and non-polar adherand materials. Toughened acrylic adhesives are oil tolerant, as the adhesive dissolves oil which is incorporated into the cured adhesive, causing little change in properties. However, careful cleaning is recommended using light abrasion and degreasing with chlorinated solvents in liquid or vapour form. Oxide layers are removed by light abrasion, and air entrapment can be avoided by careful joint design and adhesive application. Some non-polar materials must be given surface pretreatments.

Materials such as polyethylene must be specially prepared. This may involve corona discharge, flame, chemical dip or plasma treatments. The latter is a newer method which activates the surface by bombarding the workpiece with electrically charged gaseous particles in a vacuum. Free radicals are generated which form strong bonds with the adhesive. This is an expensive process but very useful when bonding difficult materials.

MECHANICAL FASTENERS

Self tapping screws

Self tapping screws are the most commonly used form of mechanical fastener. They are available in two forms:

i) *Thread-forming screws*
These are used with thermoplastics and function by displacing and compressing material as the screw is inserted. Elastic relaxation processes then ensure a tight fit and secure attachment. They are not suitable for thermosets which are brittle and therefore crack. Various designs are available for different purposes and plastics.

ii) *Thread-cutting screws*
These are used with thermosets and once again various types are available. Both types of self-tapping screw are inserted into a preformed and preferably tapered hole, chamfered at the entry to guide the screw in and ensure alignment.

Inserts

When repeated assembly and disassembly is required, threaded inserts are more suitable. They can have several other advantages including higher load carrying capacity and electrical conductivity. However, they are expensive to produce as they must be moulded or fixed into the part, involving more expensive procedures. A widely used method of securement in the case of thermoplastics is by ultrasonic insertion. Alternatively, induction or conduction heating can be used for thermoplastics. Various other types of inserts are used. These include self-tapping inserts and wire coils.

Moulded-in threads

Moulded internal and external threads can be used as part of a mechanical fastening system. The choice of thread should take account of the plastic being used. Unfilled thermoplastics are flexible and threads may be 'jumped' by over tightening. Filled plastics are more resistant in this respect but it is advisable to use a buttress or Acme type thread.

Rivets and similar fasteners

The types of rivets which may be used include those with solid and tubular shanks. It is important to allow for sufficient distance between the rivet head and the edges of the plastic being joined. The clinch allowance is also important. The distance from the rivet centre to the edge should be at least three times the rivet diameter and the clinch allowance should be about 60% of the rivet diameter. When using rivets the parts should be designed to resist shear loads rather than tensile loads, otherwise the rivets are likely to pull out of the plastic.

Speed nuts and spring clips can be used where vibration is a problem; when fully tightened the spring clip acts as a lock nut. Speed clips are also available. These clip onto moulded-in bosses on a plastic part with a one-way action.

Problems with mechanical fasteners

There are two problems to be considered in connection with the use of inserts:

i) Excessive stresses can be set up as a result of differential thermal expansion between the metal and plastic parts. This depends on the particular metals and plastics involved. For instance the linear expansivity of polystyrene is nearly seven times that of steel. Polystyrene and some other plastics are brittle and also exhibit environmental stress cracking in contact with organic solvents. The stress levels must be minimised by reducing moulding stresses in the case of moulded-in inserts (by preheating the insets) or by trying to match the expansivities of metal and plastic. For example, filled plastics have lower expansivities and often higher fracture strengths.
ii) Stresses should also be minimised by ensuring that the minimum wall thickness around an insert is at least as large as the insert diameter. This depends on the plastic but is a good general guideline. Inserts with sharp external knurls or threads should also be avoided as the resulting stress concentration can lead to early failure.

Snap fits

The resilience and high elastic strains shown by many semi-crystalline thermoplastics (PP, PE, Nylons, etc.) can be used to design snap fitments. Mouldings can be flexed sufficiently to snap into or over undercuts as shown in Fig. 10.12.

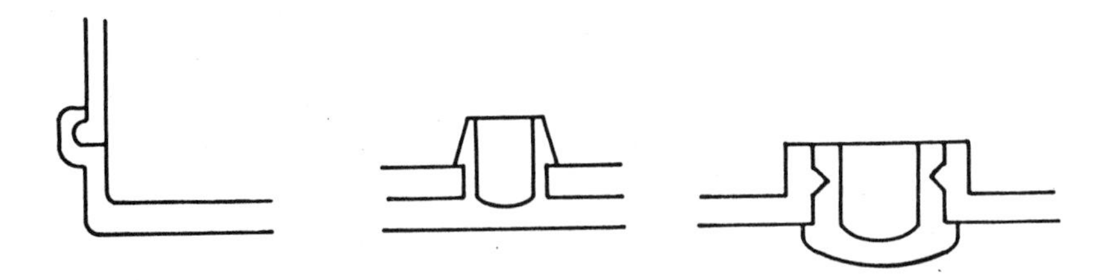

Fig. 10.12 Typical snap fitments in flexible thermoplastics.

REFERENCES

G. DEFRAYNE: *High Performance Adhesive Bonding*, Society of Manufacturing Engineers, 1983.
W. LEES: *Adhesives in Engineering Design*, Design Council, London, 1984.
A.J. KINLOCH: *Structural Adhesives*, Elsevier Applied Science, London, 1986.
J. SHIELDS: *Adhesives Handbook*, 3 edn, Butterworth, London, 1984.
W.C. WAKE: *Synthetic Adhesives and Sealants*, Wiley, New York, 1987.

11 *Designing with Plastics*

INTRODUCTION

The advantages that plastics materials offer over metals are:

a) a high strength/weight ratio,
b) toughness,
c) corrosion and abrasion resistance,
d) low friction,
e) electrical insulation, and
f) ease of processing.

In order to capitalise on these advantages, the designer must be aware of the way in which plastics materials behave differently from traditional materials such as metals, woods and ceramics. As explained in Chapter 1, the forces between the long molecular chains are much weaker than those in metals, so that plastics deform more readily than metals. For this reason, when designing in plastics, the engineer designs for stiffness rather than for strength.

In the past, many designs in metals have been copied because they were successful. Early plastics designers copied these traditional designs too and made poor products because they did not fully appreciate all the differences between the properties of plastics and metals. Eventually, this led to new designs for commonplace items such as milk crates, beer crates, tanks and containers. Once the idea of designing afresh was accepted, designers looked more closely at product requirements and found a gross over-design in many traditional items. Over-design in plastics is rarely done because it is not economically viable.

Designing in plastics materials is more difficult than in metals because of the following phenomena:

a) a non-linear stress-strain curve,
b) temperature-dependent mechanical properties,
c) time-dependent mechanical properties, and
d) anisotropic behaviour, particularly in fibre-reinforced plastics materials.

In order that the design engineer can cope with these phenomena, more data are required for the optimum design for the product.

DATA FOR DESIGN

As mentioned in Chapter 4, there are two ways of illustrating long-term deformation data. The first is to measure the stress at a constant strain as a function of time (stress relaxation), and the second is to measure the strain at a constant stress as a function of time (creep). The creep experiments are regarded as the more representative of real life situations, and moreover can provide data to predict recovery on the removal of stress.

The creep data are obtained by observing the strain as a function of log time, giving the family of curves for a given temperature as shown in Fig. 11.1a. The instantaneous and retarded elastic responses are not separated in this representation, and implicit in the data, it is assumed that the polymer does not change during the long duration of the experiments.

If a horizontal line is drawn through the creep curves, representing a given strain, a graph between stress and log time can be drawn. This gives the isometric curve shown in Fig. 11.1b (stress versus log time at constant strain, ε). This representation is useful in determining the decay of stress in a component under constant strain as time increases. This is like stress relaxation, but it is not a true representation of that phenomenon because the data are derived from creep experiments.

If the value of stress obtained as a function of time is divided by the strain value obtained from a line like *AB* in Fig. 11.1b, the time-dependent creep modulus is obtained. This is shown in Fig. 11.1d in which *CD* is derived from *AB*.

If a vertical line is drawn through the family of creep curves, the stress as a function of strain for different times is obtained. This gives the isochronous stress–strain curves shown in Fig. 11.1c.

All these representations are useful in predicting long-term mechanical properties under steady loads. Often, however, the loading is intermittent. When a load is removed from a plastics component, some of the strain is recovered almost immediately, but the retarded elastic strain requires more time. Thus, like creep, the recovery is time-dependent.

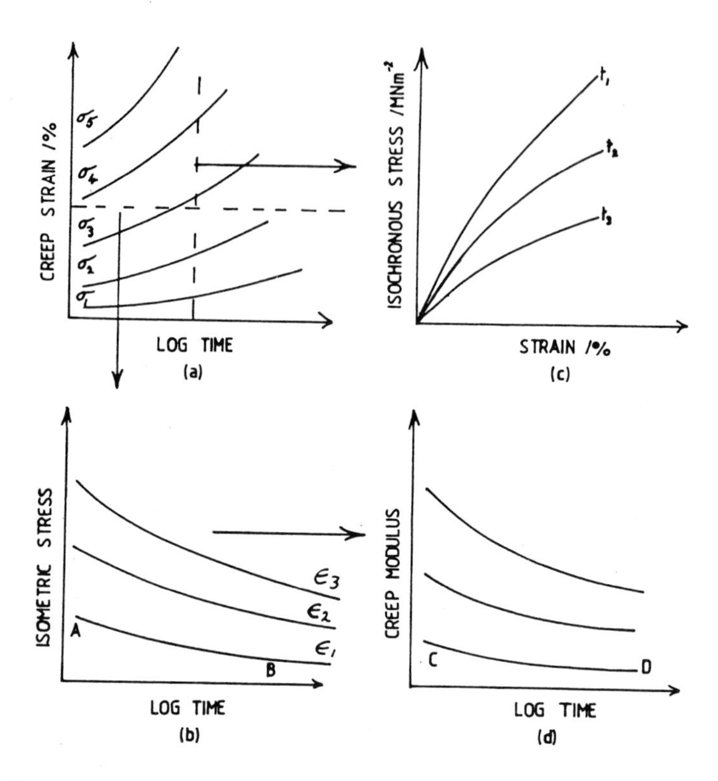

Fig. 11.1 Family of curves showing different representation of creep.

Recovery data are provided in the form of fractional recovered strain versus reduced time, as shown in Fig. 11.2 where

$$\text{fractional recovered strain} = \frac{\text{recovered strain}}{\text{total strain before removal of load}} \qquad (11.1)$$

and

$$\text{reduced time} = \frac{\text{recovery time}}{\text{duration of creep}} \qquad (11.2)$$

The effect of intermittent loading on creep properties is less than that for continuous loading situations. Intermittent loading, however, may introduce fatigue, and the effect of fatigue on mechanical properties varies from material to material.

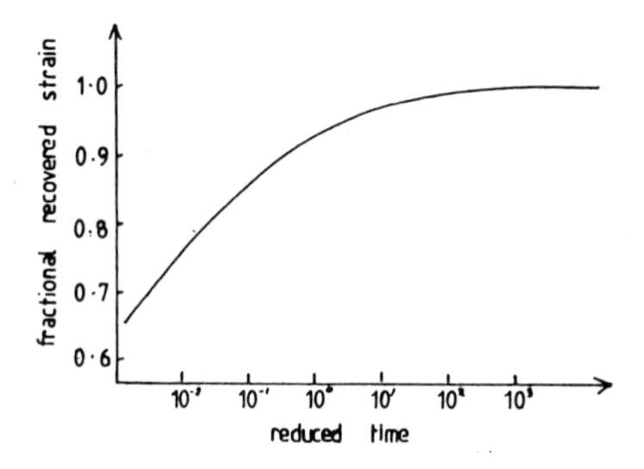

Fig. 11.2 Traditional recoverable strain as a function of reduced time.

USE OF MANUFACTURERS' DATA

Often, in a given situation, the manufacturers' data does not fit the service conditions exactly. For example, the service temperature may lay between temperatures for which data are available, or the density of the material to be used may be different from that in the data, or the required service life may be longer than that for which data are given. The manufacturers usually give guidance on this in their literature.

For 'Propathene' polypropylene, ICI literature gives the following information and advice. For temperatures between 20°C and 60°C, for which data are given, calculate the design stresses for each temperature and linearly interpolate between the two for the service temperature required. For temperatures above 60°C, use a 1.3% decrease in stress or modulus per degree rise in temperature, or use the 100s isochronous stress–strain graph to find the change in modulus with temperature for a 2.0% strain. This cannot be done indefinitely.

For a difference in density, increase the design stress or modulus by 4% for every 0.001 increase in relative density at 20°C.

For extrapolating to predict mechanical behaviour over longer periods than given in the data, care must be taken, and the extrapolation should not be more than one decade in time, particularly for strains in polypropylene of over 2%.

DESIGN PROCEDURES

Designing for Tensile Strength

When working with metals, designs can be made based on the metal's tensile strength, *TS*, and the design stress is calculated by using a factor of safety, *F*, where

$$F = \frac{TS}{\text{design stress}} \tag{11.3}$$

A similar approach can be used with plastics when the *TS* is known at the service temperature of the product, and due allowance is made for the effect of time on the *TS*. Such data are given in long term strength graphs, such as the one in Fig. 11.3.

For a simple bar shape, a factor of safety of 1.5 can be used, whereas for a gas main British Gas specify a factor of safety of about 8. In this case F is given by

$$F = \frac{\text{50 year creep rupture strength}}{\text{hoop stress in service}} \tag{11.4}$$

This kind of approach is often used with brittle polymers, or in cases where a brittle fracture occurs, or is likely to occur, in a ductile polymer.

Example 11.1

A beam made from polypropylene is required to last in service for 100 h under a continuous load at 20°C. From the data given in Fig. 11.3, determine the design stress for these conditions by allowing a factor of safety of 1.5.

Draw a vertical line from the time axis from the point 100 h. From where this line intersects the long-term strength-curve for polypropylene, draw a horizontal line to cut the stress axis. This gives a value of 21.5 MN m^{-2},

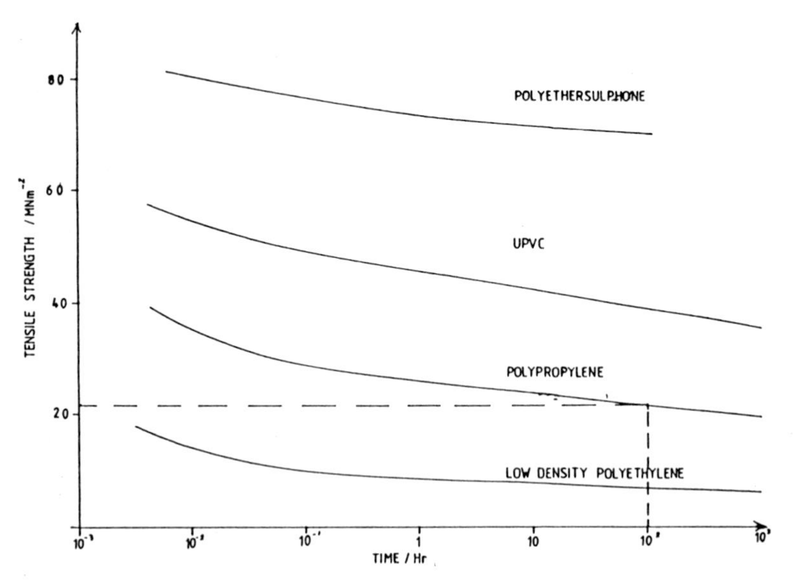

Fig. 11.3 Data for example 11.1.

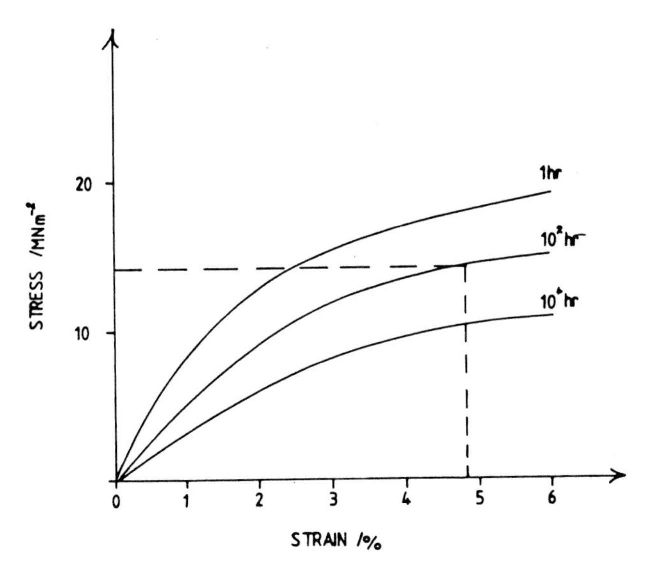

Fig. 11.4 Isochronous stress-strain curve for polypropylene.

which is the long-term tensile strength. Divide this value by 1.5 to obtain the design stress

$$\sigma_D = \frac{\text{long term tensile strength}}{\text{factor of safety}} = \frac{21.5}{1.5}$$

$$= 14.3\ MN\ m^{-2}$$

Fig. 11.4 shows the isochronous stress-strain curve for polypropylene. This is going to be used to analyse this kind of design method for polypropylene, although the design has already been completed using Fig. 11.1.

In Fig. 11.4, draw a horizontal line from the stress axis at 14.3 MN m^{-2}. Note the strain at which this line cuts the stress-strain curve for 100 h. This gives a value of 4.8%.

The only reason for doing this is that it shows that this design method is not always good for ductile materials such as polypropylene. The value of strain of 4.8% is really too high. It would give rise to stress whitening, which would be unsightly and give the appearance of weakness.

Designing for Maximum Strain from Manufacturers' Data

A second approach, which is used for designing with ductile polymers and is becoming increasingly popular in all plastics design, is the use of the maximum strain criterion. It is used particularly for long-term service stresses. Manufacturers and suppliers of polymers often quote values of maximum strain for their materials. Sometimes the conditions applicable to these values are defined, such as for injection moulded parts with little orientation or stress concentrations, products with welded joints, or products with glued joints. The last quoted has the smallest maximum service strain.

For the best mouldings, a maximum design strain of 0.75% is suggested for amorphous polymers in general, because at this strain crazing and voiding is unlikely. This design strain is also recommended for reinforced, partly crystalline plastics. Table 11.1 gives some typical values of maximum design strain for thermoplastics.

Table 11.1 Maximum design strains for common thermoplastics.

Thermoplastic	Design strain %
Glass reinforced nylon	1.0
ABS	1.5
Acetal copolymer	2.0
Polyethersulphone	2.5
Polypropylene	3.0
Polypropylene with welded joints	1.0

Maximum Strain from the 0.85 Secant Modulus

Often it is not known what the maximum design strain is. Under these conditions there are two ways of estimating a suitable value.

The first method involves an examination of the material's stress–strain behaviour for the appropriate service life. Fig. 11.5 gives data for 'Noryl', a modified form of polyphenylene oxide (a graft copolymer of polyphenylene oxide with polystyrene). Using the curve for 10^4 h, a secant modulus (OB) of

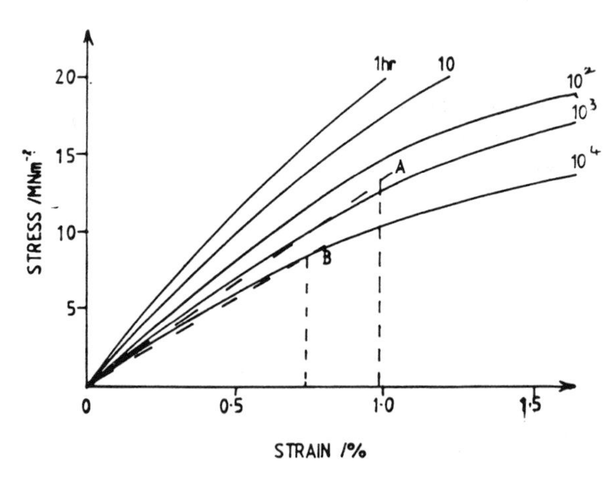

Fig. 11.5 Isochronous stress–strain curves for 'Noryl'.

value 0.85 of the initial modulus (*OA*) is drawn on the stress–strain graph and the value of strain obtained where this line cuts the graph. *OA* is the slope giving the initial modulus. This value of strain is quoted as the maximum for a service life of 10^4 hr. From Fig. 11.5 the initial modulus (from *OA*) is 1.35×10^9 Nm^{-2} and the 0.85 secant modulus is 1.2×10^9 Nm^{-2} giving a maximum design strain of 0.75%.

This is too restrictive for crystalline polymers and for these the limiting strain is decided upon by consultation between the designer and the supplier or manufacturer.

Maximum Strain from the Requirements of the Specific Application

This leads conveniently to the second method of determining the maximum design strain. This is based on the fact that at a certain strain the part may:

a) appear to be distorted and not give confidence to the user,
b) cease to work properly, or
c) impede other components close by.

In the last case the specific application defines the upper strain, whereas the former two are open to opinion.

This is the most popular way of designing in plastics, and a summary of the method is given below:

1) The problem is examined and an estimate is made of the permitted load and the expected lifetime of the component at the maximum service temperature.
2) The maximum design strain is found from manufacturers' data, the 0.85 second modulus method or a consideration of the specific application (a, b, c above).

There are now two possibilities.

3a) The tensile creep modulus versus log time curve at the service temperature is inspected and a value of creep modulus is read off appropriate to the component's lifetime. The design stress σ_D is then the product of the creep modulus and the strain.
3b) The isochronous stress–strain curve is examined at the appropriate values of time and temperature, and the design stress is read directly.
4) In both cases, the secant modulus ($= \frac{\text{design stress}}{\text{design strain}}$) can be compared with the slope of the isochronous curve at the point of interest to see to what degree the part may be over-designed.

This completes the design brief and in it the designer uses values of secant modulus or creep modulus in classical elastic formulae. This must be done with care because these equations were derived for use with metals when

a) the strains are small,
b) the moduli are constant,
c) the strains are a single-valued function of stress,
d) the strains are instantly recoverable,
e) the material is isotropic and homogeneous, and
f) the material behaves identically in compression as in extension.

The above are not true of plastics but the classical equations can be used successfully if due allowance is made for time and temperature effects, mode of deformation, method of fabrication and the effects of environment.

The use of the classical elasticity equations in this case is often referred to as 'pseudo-elastic design', and it is common to all the design methods described previously.

In general $$\sigma(t,T) = E(t,T)\,\varepsilon(t,T)$$

Example 11.2

A thin-walled pipe must withstand an internal pressure of 1.5 MN m^{-2}. It is suggested that the maximum strain over the service life of three years should not exceed 0.5%. Calculate the wall thickness of a 10 mm internal diameter pipe made from glass filled N66, for which data are given in Fig. 11.6.

From the curves, it is evident that when using nylon it is often necessary to allow for the presence of water (hence the use of the 50% relative humidity data), and it must be remembered that the mechanical properties of fibre-

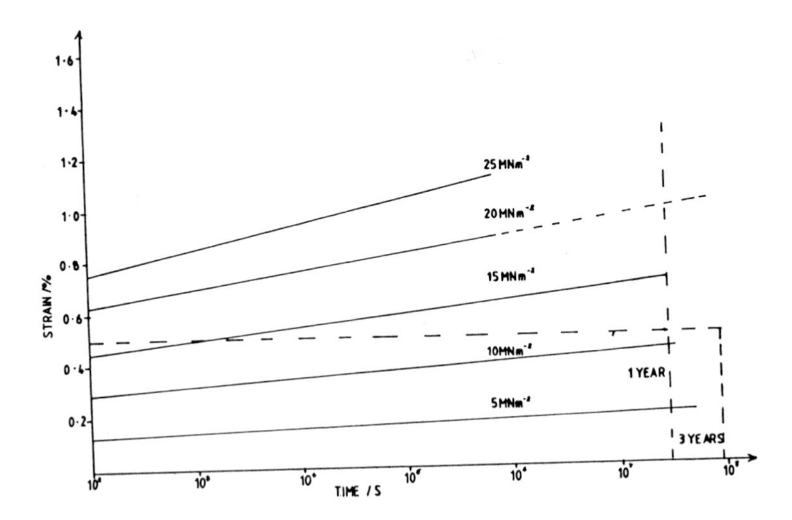

Fig. 11.6 Data for Example 11.2.

reinforced thermoplastics are inferior at right angles to the fibre alignment than along it. In a pipe, the reinforcement will enhance the mechanical properties along its length but unfortunately not in the hoop direction, where the strength is required.

Solution

Service life = $3 \times 365 \times 24 \times 3600 = 9.5 \times 10^7$ s

The hoop stress σ in a thin-walled pipe subjected to an internal pressure P is given by

$$\sigma = \frac{Pd}{2h}$$

where d is the internal diameter and h is the wall thickness.

Therefore $$h = \frac{Pd}{2\sigma}$$

The design stress σ_0 must be found from Fig. 11.6. Draw lines at 0.5% strain and 9.5×10^7 s. The extrapolated 10 MN m^{-2} stress line passes close to the intersection of the lines just drawn.

Therefore $$\sigma_D = 10 \text{ MN m}^{-2}$$

$$h = \frac{1.5 \times 10^6 \times 10 \times 10^{-3}}{2 \times 10 \times 10^6}$$

$$= 0.75 \ mm \ Answer$$

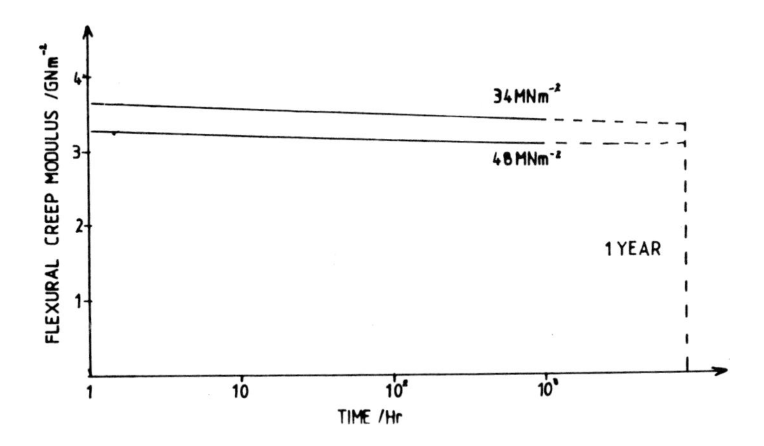

Fig. 11.7 Data for Example 11.3.

Example 11.3

A beam made from polyetherimide of length 3 m, thickness 24 mm and width 150 mm is to be supported at its ends only. When loaded the beam must not deflect more than 7 mm at its centre. Calculate the maximum strain in the beam and the maximum weight that can be supported at its centre for one year. Fig. 11.7 gives data for polyetherimide.

Solution

In bending beam theory it is assumed that the beam is curved into the arc of a circle of radius R (Fig. 11.8).

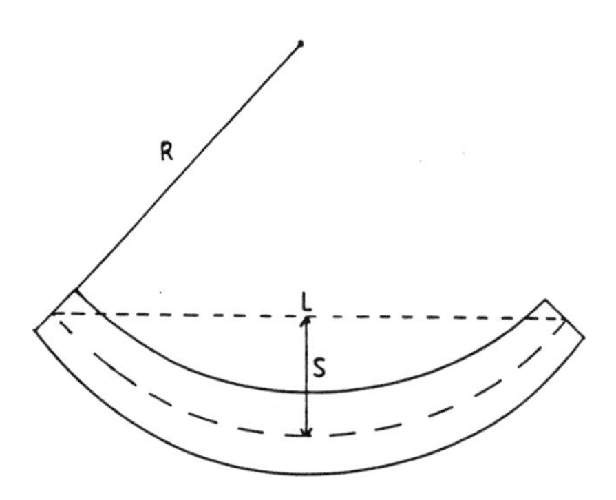

Fig. 11.8 A beam curved into an arc of radius R for a deflection s.

If the deflection is s, then from the rectangle properties of a circle

$$\left(\frac{L}{2}\right)^2 = s(2R - s)$$

$$\simeq 2sR$$

because $R >> s$.

Hence

$$R \simeq \frac{L^2}{8s}$$

Once R is known the maximum strain can be found.

$$\varepsilon = \frac{d}{2R}$$

Where $\frac{d}{2}$ is the half-thickness of the beam.

$$\varepsilon = \frac{8s}{L^2} \times \frac{d}{2} = \frac{4sd}{L^2}$$

$$= \frac{4 \times 7 \times 10^{-3} \times 24 \times 10^{-3} \times 100\%}{3^2}$$

$$= 7.4 \times 10^{-3}\%$$

This is a very small strain indeed. Now the data for polyetherimide must be examined in Fig. 11.7. The flexural creep modulus values are given for constant stresses of 34 and 48 MN m^{-2} only. These are equivalent to strains in the region of 1 and 1.45% (strain = stress/modulus).

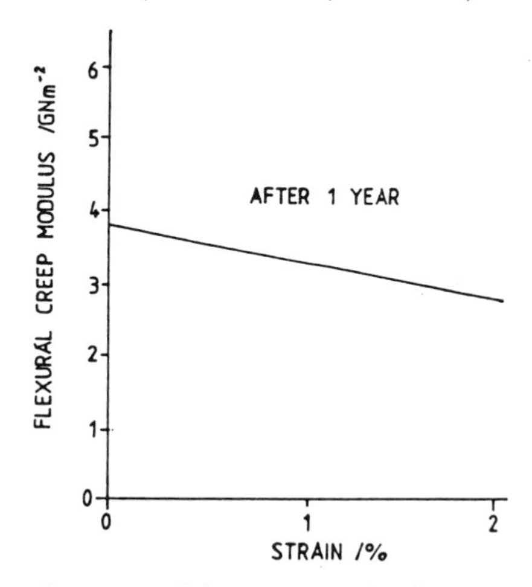

Fig. 11.9 Flexural creep modulus versus strain after one year.

Construct a graph of flexural creep modulus against strain at one year. This gives Fig. 11.9. As the data are sparse, assume the modulus-strain response is linear. Then an extrapolation back to the flexural creep modulus axis gives a value of 3.75 GN m^{-2}, which is taken as the required design value for the strain $7.4 \times 10^{-3}\%$.

This modulus value is substituted in the bending beam equation for central loading to find the load that will cause a deflection of 7 mm.

$$s = \frac{mgL^3}{4Ebd^3}$$

where L is the length of the beam, b its width and d its thickness.

$$m = \frac{4Ebd^3s}{L^3g}$$

$$= \frac{4 \times 3.75 \times 10^9 \times 0.15 \times (24 \times 10^{-3}) \times 7 \times 10^{-3}}{3^3 \times 9.81}$$

$$= \mathit{0.82\ kg}$$

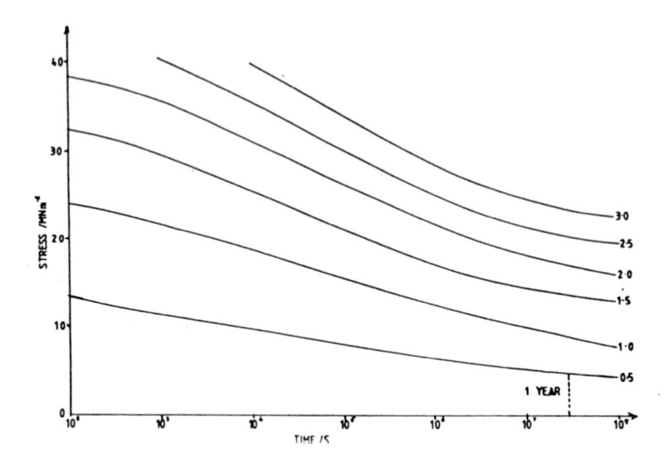

Fig. 11.10 Data for Example 11.4.

Example 10.4

An acetal copolymer seal of thickness 3.0 mm is compressed by clamping screws, which reduce the thickness of the seal by 0.015 mm. Using the data in Fig. 11.10 find the stress in the seal after 100 s and 1 year. Repeat for glass fibre reinforced Nylon 66 using the data in Fig. 11.6.

Solution

Again it will be seen that when using acetal copolymer, some consideration of relative humidity must be taken.

The strain $$E = \frac{0.015}{3} \times 100 = 0.5\ \%$$

Read from the 0.5% strain line in Fig. 11.10. This gives the values.

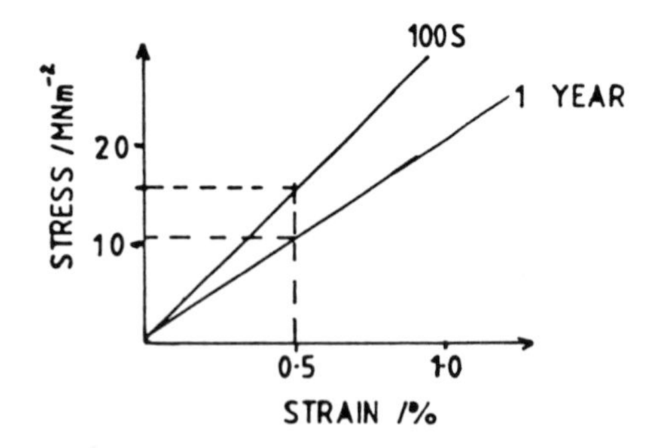

Fig. 11.11 Isochronous stress-strain data for glass reinforced Nylon 66 derived from Fig. 11.6.

13 MN m^{-2} (100 s) and 4.5 MN m^{-2} (1 year)

This is not as easily done from the nylon data in Fig. 11.6. From these data construct isochronous stress–strain data for 100 s and 1 year (Fig. 11.11). In this figure, draw a line at 0.5% strain. This gives

16 MN m^{-2} (100 s) and 10 MN m^{-2} (1 year)

OTHER DESIGN CRITERIA

Designing for stiffness is not the only criterion and often considerations must be made for impact behaviour, fatigue and hardness, but these are beyond the scope of the present work and dealt with by Crawford.

Apart from the obvious choice of material to withstand the chemical environment, there are many cases of materials that are usually resistant to chemicals unless under stress. Examples are polycarbonate, with paints or petrol, and natural rubber with ozone. The phenomenon in which the combination of stress and chemical environment gives rise to failure is called environmental stress cracking (ESC).

Some polymers are more susceptible to ESC than others, and often what proves a fatal combination for one polymer presents no problem for another. For instance, the use of polyacetal in a selector switch that was cleaned

Fig. 11.12 A Fan blade made from 'Zytel' polyamide (by courtesy of Du Pont (UK) Ltd.).

Fig. 11.13 Plastic wheels in 'Zytel' polyamide (by courtesy of Du Pont (UK) Ltd.).

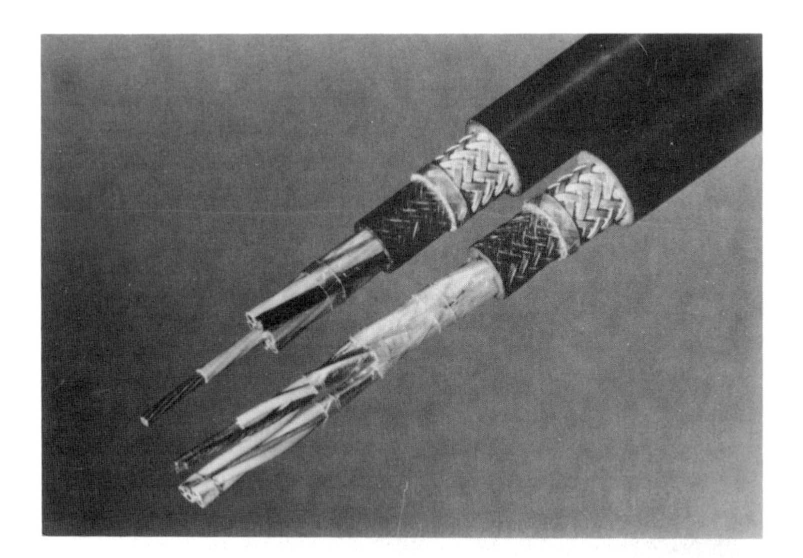

Fig. 11.14 'Victrex' PEEK as primary insulation on 'Flambic' cables produced by BICC General Cables Ltd, and used for power and signal transmission on oil and gas platforms; toughness and flammability properties are important here (by courtesy of ICI plc).

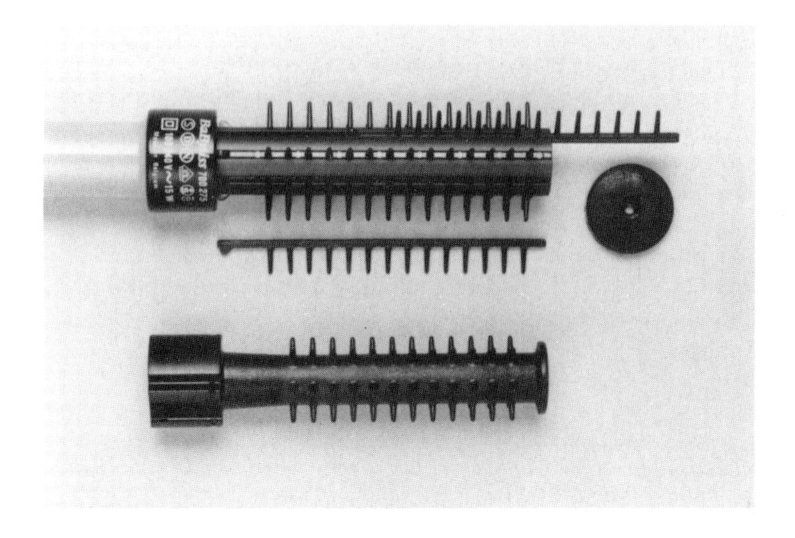

Fig. 11.15 Babyliss use 'Victrex' PES for the toothed elements of electrically heated hair-styling brushes. Rigidity and creep resistance up to 180°C are important properties in this application (by courtesy of ICI plc).

Fig. 11.16 An injection moulded bumper in 'Xency' polycarbonate polybutyleneterephthalate blend. This blend combines very high impact strength heat resistance and dimensional stability (by courtesy of General Electric Plastics).

regularly with a switch cleaning fluid presented no problem, but polycarbonate would have been unsuitable.

Even when the part is not stressed in service, there is no guarantee of success, because residual stresses can be moulded in during processing, and these will give rise to part failure.

Care must always be taken when designing to avoid using a polymer in even the mildest of solvents because of ESC.

Although there are many pitfalls in designing with plastics materials, numerous products once made in metals are now made in engineering thermoplastics. Figs. 11.12 to 11.16 show some typical examples.

REFERENCES

R.L.E. BROWN: *Design and Manufacture of Plastic Parts*, Wiley, New York, 1980.

A.A. COLLYER: *A Practical Guide to the Selection of High Temperature Engineering Thermoplastics*, Elservier Advanced Technology, Oxford, 1990.

R.J. CRAWFORD: *Plastics Engineering*, Pergamon Press, 1983.

P.C. POWELL: *The Selection and Use of Thermoplastics.* Engineering Design Guides, Oxford University Press, Oxford, 1977.

S.S. SCHWARTZ and S.H. GOODMAN: *Plastics Materials and Processing*, Van Nostrand Reinold, New York, 1982.

Index